Mass Spectrometry-Based Metabolomics in Clinical and Herbal Medicines

Mass Spectrometry-Based Metabolomics in Clinical and Herbal Medicines

Strategies, Technologies, and Applications

Edited by
Aihua Zhang
Wanying Wang

WILEY-VCH

Editors

Prof. Aihua Zhang
Heilongjiang Univ. of Chinese Medicine
National Chinmedomics Research Center
Heping Road 24
150040 Harbin
China

Prof. Wanying Wang
Macau University of Science & Technology
State Key Lab of Quality Research in Chinese Medicine
Avenida Wai Long, Taipa
Macau
China

Library of Congress Card No.: applied for

British Library Cataloguing-in-Publication Data A catalogue record for this book is available from the British Library.

Bibliographic information published by the Deutsche Nationalbibliothek
The Deutsche Nationalbibliothek lists this publication in the Deutsche Nationalbibliografie; detailed bibliographic data are available on the Internet at <http://dnb.d-nb.de>.

Print ISBN: 978-3-527-34991-3
ePDF ISBN: 978-3-527-83573-7
ePub ISBN: 978-3-527-83574-4
oBook ISBN: 978-3-527-83575-1

Typesetting Straive, Chennai, India
Printing and binding CPI Group (UK) Ltd, Croydon, CR0 4YY

Printed on acid-free paper

C117515_290721

Contents

Preface

High-definition mass spectrometry (MS) has been undertaken to obtain comprehensive metabolite biomarker, global profiling, and metabolic pathway from large biological data sets. Analysis of the key metabolites *in vivo* has become an important part for understanding the disease mechanisms and has a strong potential to contribute to improve early diagnosis of diseases. With technological advances in MS analytical techniques, metabolomics enables the parallel assessment of the levels of a broad range of metabolites and has a great impact in the investigation of discovering biomarkers, identifying perturbed pathways, diagnosing diseases, and measuring the response to treatment in clinical fields. Abundant evidence shows that herbal medicine could be beneficial for the treatment of various diseases. However, the complexities present in chemical compositions are an obstacle for revealing the bioactive compounds, pharmacodynamic mechanism, metabolism feature, therapy targets, etc. To overcome these issues, MS-based metabolomics is a useful tool for future understanding of herbal medicine from small-molecule metabolism level. In recent years, MS-based metabolomics strategies and technologies have been widely applied in clinical and herbal medicines.

This comprehensive book constitutes a unique and indispensable practical guide for the phytochemistry or related laboratory and provides a hands-on description of new techniques so that students and researchers can become familiar with these invaluable methods. We systematically arranged the current state, challenges, and applications of high-throughput MS-based metabolomics such as metabolites analysis, biomarker discovery, discovery of natural product, mechanism interpretation of action, discovery of active ingredients, and precision medicine application, to enhance their biomedical value in the real world of biomedicine, shedding light on the potential for MS-based metabolomics. Up-to-date information and future developments are described, thereby producing a volume that is a landmark of MS-based metabolomics research and a beneficial guideline to graduate students and researchers in academia, industry, and technology transfer organizations in all biomedical science fields. We sincerely expect our work to promote the technological advances of MS-based metabolomics in clinical and herbal medicine research, especially in the academic progress and international development.

Prof. Aihua Zhang
Laboratory Director
Functional Metabolomics Laboratory
National Chinmedomics Research Center
Heilongjiang University of Chinese Medicine
Heping Road 24, Harbin, China

1

Current State of the Art of High-Throughput Metabolomics

Aihua Zhang, Qiang Yang, Hui Sun, and Xijun Wang

Heilongjiang University of Chinese Medicine, National Chinmedomics Research Center, Functional Metabolomics Laboratory, Department of Pharmaceutical Analysis, Heping Road 24, Harbin 150040, China

1.1 Introduction

At present, a series of systems biology disciplines, such as genomics, proteomics, transcriptomics, metabolomics, and lipidomics have provided new methods for researchers to explore the pathogenesis of diseases. As an important component of systems biology, metabolomics can discover the metabolite biomarkers and various related metabolic pathways in bio-samples [1]. Metabolomics is usually used with nuclear magnetic resonance (NMR), mass spectrometry (MS), and chromatography, and it plays an important role in clinic, disease treatment, drug metabolism, plant research, agricultural, and nutrition research [2]. Due to the important position of metabolomics in the life science, a large number of domestic and foreign scholars have improved the innovation research of techniques and methods of metabolomics [3]. In this chapter, we show the recent advancements and current application as well as future development of metabolomics.

1.2 Metabolomics

In 1999, Dr. Jeremy Nicholson proposed the concept of metabolomics [4]. Metabolomics is a method to quantitatively analyze all the metabolites in bio-sample and to reveal the relationship between metabolites and pathological changes. It mainly focuses on the small-molecule metabolites with relative molecular weight of less than 1000 [5]. Its research process mainly includes sample preparation and analysis, data processing and analysis, and metabolic mechanism analysis. Metabolomics studies used analytical tools including NMR spectroscopy, liquid chromatography–mass spectrometry (LC–MS), and gas-chromatography–mass spectrometry (GC–MS) [6–9]. Due to the very complex and huge original data volume of metabolome, it cannot be analyzed by conventional methods. Instead, it often uses pattern recognition analysis, such as principal component analysis

Mass Spectrometry-Based Metabolomics in Clinical and Herbal Medicines: Strategies, Technologies, and Applications, First Edition. Edited by Aihua Zhang and Wanying Wang.

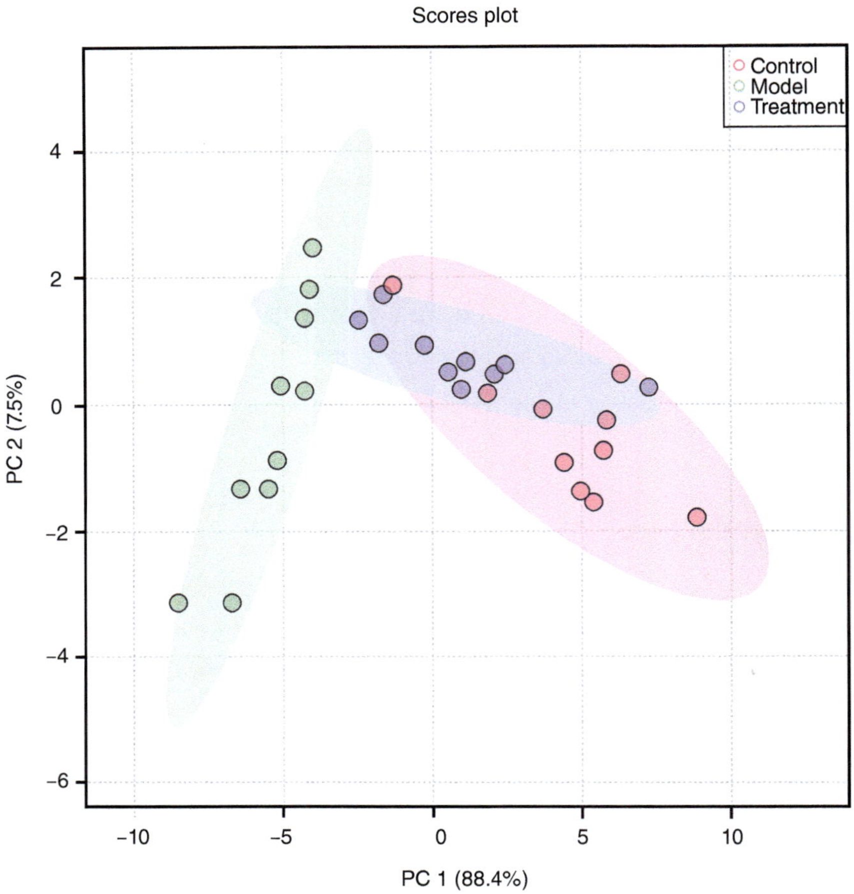

Figure 1.1 Principal component analysis in high-throughput metabolomics.

(Figure 1.1) and loading (molecules) analysis (Figure 1.2), for screening metabolites (Figure 1.3) and for exploring the metabolic pathway (Figures 1.4 and 1.5) and the metabolic network change mechanism (Figure 1.6). Thus metabolomics analysis has formed a series analysis approaches including data extraction, data preprocessing, supervised and unsupervised pattern recognition, and biological information exploration [10–12].

1.3 Mass Spectrometry Technology

MS has become a widely used analytical tool for chemists and is increasingly used for metabolomics analysis. It plays a key role in the identification of structural information for metabolite molecules. With the advance of technology, MS combination with separation technology is gradually improving [13, 14]. Due to the high specificity and sensitivity, low sample consumption, fast analysis, and the advantages

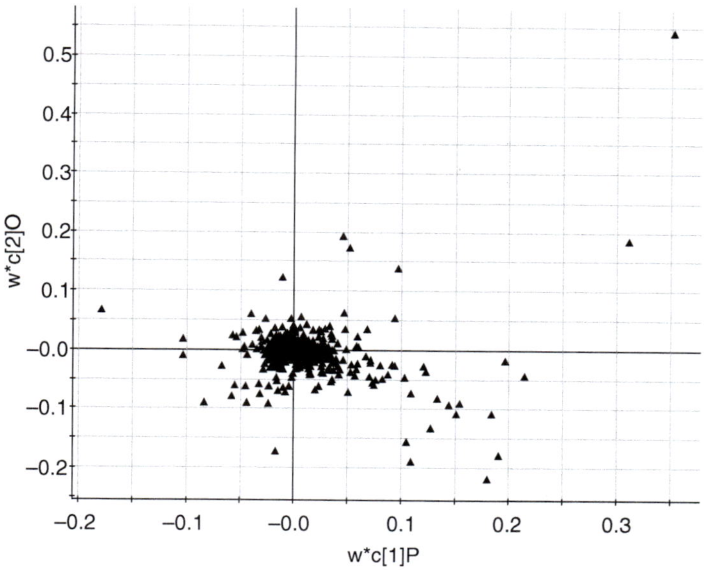

Figure 1.2 The loading ion analysis for metabolomics.

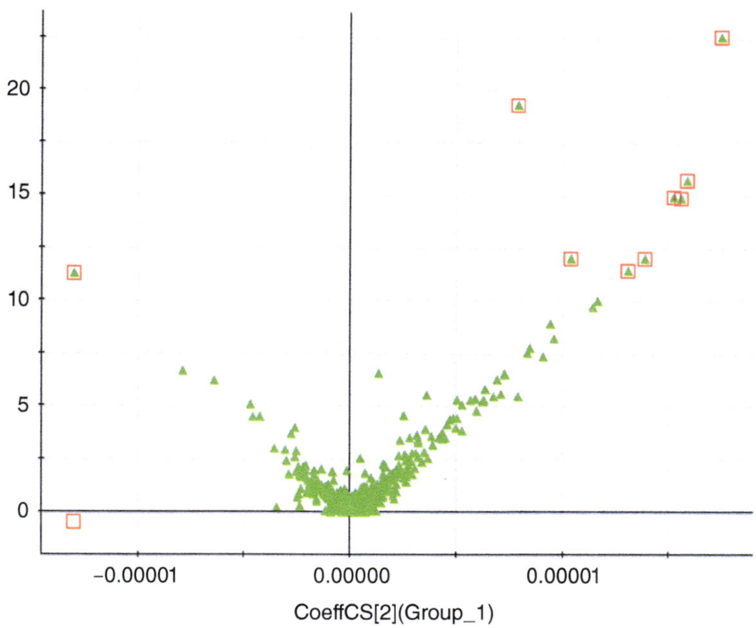

Figure 1.3 The variable importance in projection for screening metabolites from large biological data sets.

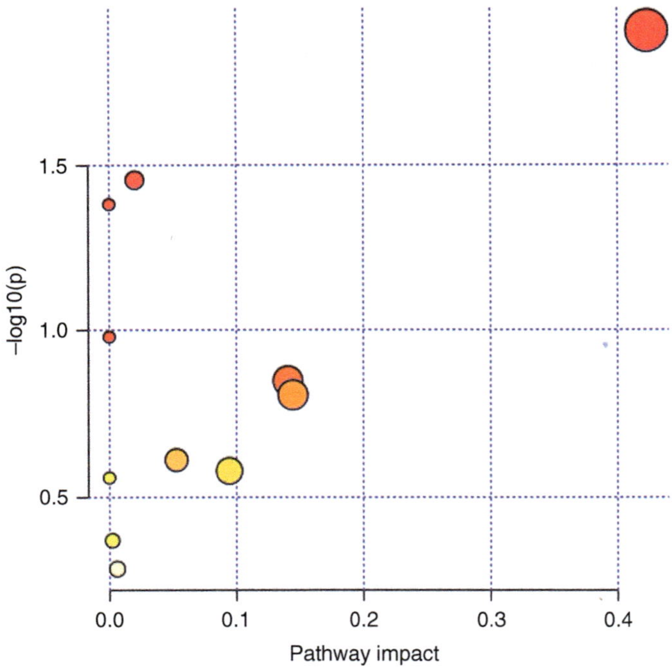

Figure 1.4 Metabolic pathway analysis for the biologically meaningful metabolite sets.

of identification, mass spectrometer is widely used in metabolomics science, and it opens up a new gate for the rapid analysis of complex samples [15–17].

1.4 Recent Advancements and Applications

MS-based metabolomics methods have been used to reveal the disease diagnosis, drug effects, metabolic mechanism, toxins, and various diseases including cardiovascular disease, cancer, natural product discovery, toxicological effects, and nutrition [18–24].

1.4.1 Disease Diagnoses and Metabolism

The disease causes the pathophysiological changes and eventually result in corresponding metabolites and pathway changes. Some scholars have established a diagnostic method for diseases by detection of metabolites. Cholesterol sulfate and phospholipids are considered as novel biomarkers for atherosclerosis [25]. The levels of norvaline, 1,5-anhydroglucitol, and L-aspartic acid are linked with macroalbuminuric diabetic kidney disease [26, 27]. Abdulwahab et al. used MS-based metabolomics and revealed new associations between 37 proteins and T2DM and found the significant up-regulation of immunoglobulin [28]. Yang et al. found that there were 33 distinct metabolites in the urine of femoral head necrosis patients [29].

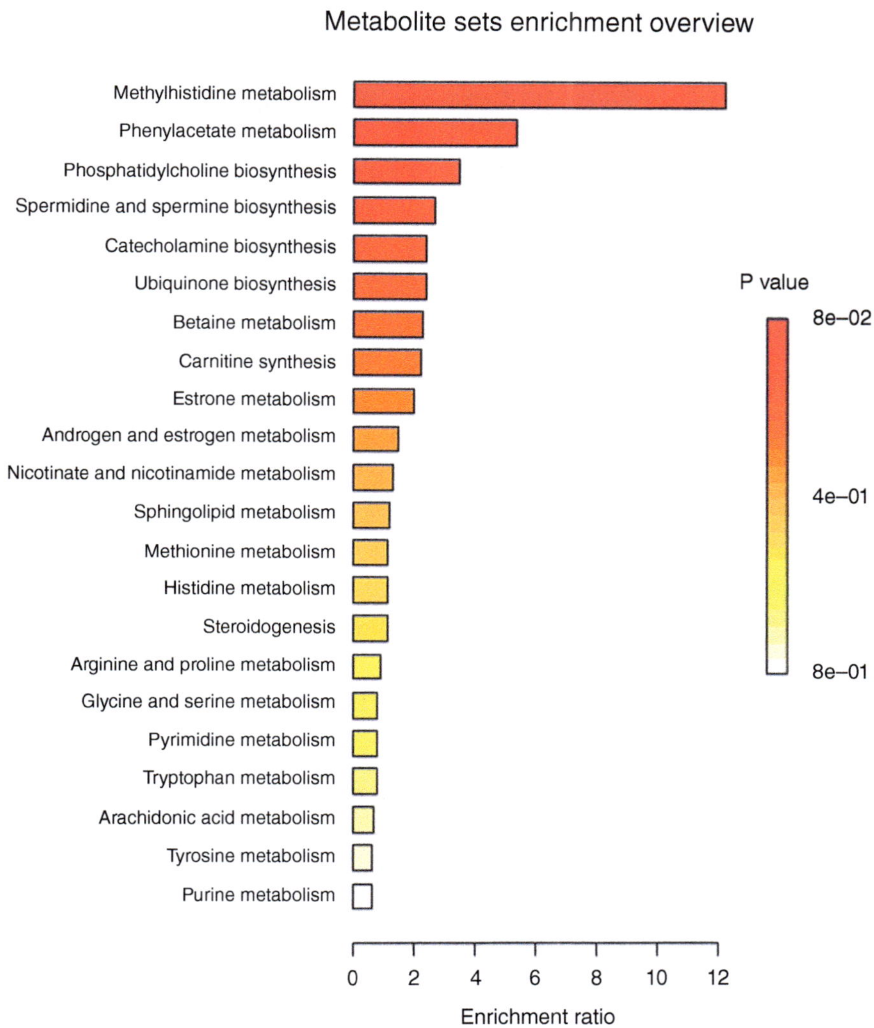

Figure 1.5 Metabolite enrichment analysis for the biologically meaningful metabolite sets based on the libraries.

Lin et al. used MS to detect plasma metabolites and found that the alanine, aspartic acid, and carbamate were significantly different among all groups of respiratory distress patients [30]. Øvrehus et al. had found that early hypertensive nephrosclerosis showed disturbances in dopamine intrarenal biosynthesis [31]. Metabolic pathways such as sphingolipid, vitamin D-related compounds, and steroid precursors were discovered in glaucoma patients [32]. Dong et al. discovered the 31 metabolites between nonalcoholic fatty liver disease and nonalcoholic steatohepatitis [33]. Based on the metabolic method, Wang et al. found that the changes of acute ischemic stroke were mainly related to amino acid-related metabolism [34]. There are 30 different metabolites of gout patients, mainly involving tricarboxylic acid cycle, amino acid metabolism, and lipid metabolism [35]. A total of 12 metabolites,

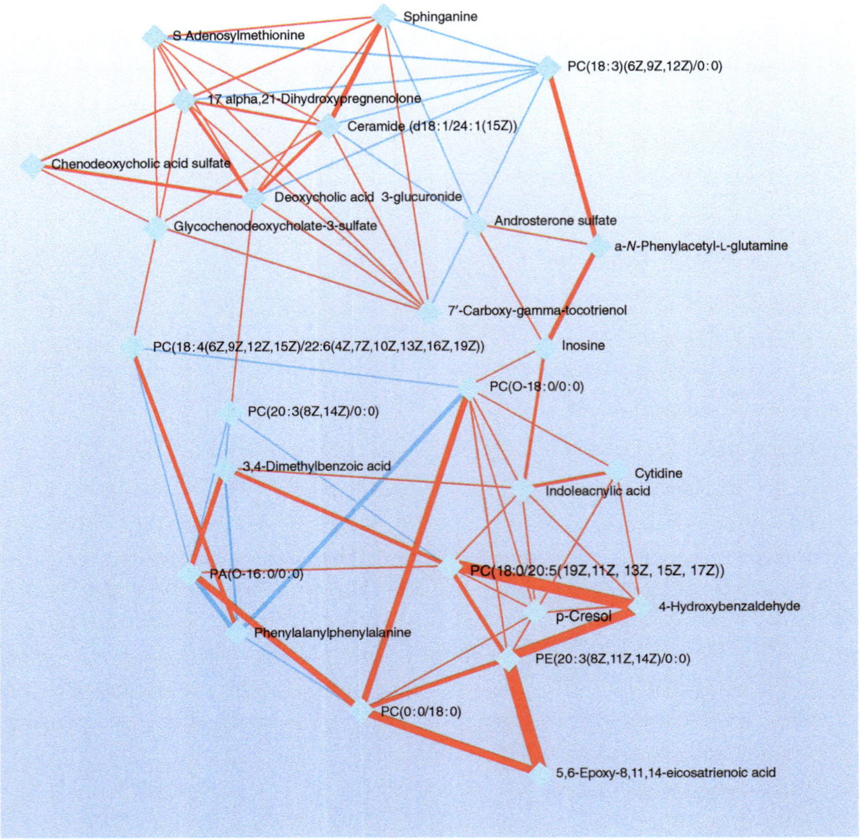

Figure 1.6 Metabolic networks for the biologically meaningful metabolite sets performed with MetaboAnalyst online tool (www.metaboanalyst.ca).

mainly related to fatty acid metabolism, exerted significant changes in ischemic stroke [36]. Metabolomics research is of great significance for the disease diagnosis, to better understand the disease pathogenesis and to provide new evidence for the primary prevention of disease [37–43]. Cancer is a major health problem in the world, so understanding the metabolic causes of cancer and its influencing factors is important for disease treatment. Yang et al. used MS to explore the metabolic changes of ovarian and had identified 18 metabolites closely related to ovarian cancer [44]. Some scholars used LC/MS to study the metabolic changes of pancreatic cancer tissue and found that seven metabolites may be potential biomarkers [45]. Metabolic pathways are closely related to tumor growth, metastasis, and immune escape mechanism [46, 47].

1.4.2 Metabolic Mechanisms

The disorder of metabolic mechanism is one of the crucial factors affecting the disease, for clarifying the metabolic mechanism plays a crucial role in the treatment of the disease. Breviscapine mainly improves the metabolism of phospholipids by regulating the level of serotonin [48]. A total of 19 metabolites were found with the hypoglycemic effect of *Crassostrea gigas* polysaccharide and mainly involved in carbohydrate metabolism, amino acid metabolism, and purine metabolism [49]. Aloe emodin has a therapeutic effect on hyperlipidemia by regulating metabolic disorders [50]. Pang et al. had found that methotrexate could regulate the inflammatory-related metabolic networks [51]. Metabolic mechanism of disease research in metabolomics-based approach provided a new method for disease treatment [52–56].

1.4.3 Drug Efficacy

The efficacy evaluation is an essential requirement for discovering the effective constituents and therapeutic targets. Therefore, based on the metabolomics and other omics techniques, the innovative methods such as chinmedomics, fectomics, and functiomics could be established and used for evaluating the efficacy; discovering the chemicalome and metabolome; screening of active compounds, fectome and functiome; revealing the effective mechanism of drug, herbal medicine, and traditional medicine; etc. and seek to elucidate the therapeutic properties with modern techniques. Some scholars have found that the process of drug treatments could be uncovered by metabolic methods [57]. Bao et al. used UPLC/MS metabolomic method to rapidly identify the anticancer compounds in *Forsythia* and found that betulinic acid was the most effective anticancer compound. The forsythia extract can exert anti-inflammatory effect through acting on different metabolic pathways [58, 59]. A total of 27 potential biomarkers were discovered and related with effective mechanism of Xanthii Fructus for allergic rhinitis, which mainly involved glycerophospholipid and branched-chain amino acid metabolism [60]. Gross saponins of *Tribulus terrestris* fruit could regulate the multiple metabolic pathways to reverse the middle cerebral artery occlusion [61]. Gastrointestinal injury was mainly related to phenylalanine, arachidonic acid, and taurine. Fructus Gardenia affected function state of gastrointestinal via regulating a variety of metabolic pathways [62]. Metabolomics has a wide range of applications in drug efficacy, drug screening, and toxicity and effectiveness evaluations [63–68].

1.4.4 Nutrition

The application of metabolomics in the nutrition field is called nutritional metabolomics, which refers to the systematic study of the interaction between diet and organism metabolism. The changes of cucumber fruit metabolites under nano-Cu stress were examined by NMR and GC–MS and found that nano copper exposure had an effect on the metabolic profile of fruit metabolites [69]. UPLC–MS-based metabolomics strategy was used to explore the nutrient profiles of kale and found that the contents of components in different cabbage varieties were different [70]. MS technology was used to explore the differences of secondary metabolites and heavy metals content in dried ginger from different habitats.

It showed that abiotic stress due to metals may lead to differences in metabolite abundance [71]. GC–MS technology was used to investigate the metabolic changes of strawberry fruit caused by controlled atmosphere treatment and found that several metabolites were mainly related to the biologically active compounds [72]. These studies have provided the scientific basis for the regulation and balance of nutrition and diet [73–80].

1.4.5 Plant Biology

The metabolic characteristics of leaves and roots of soybean seedlings cultured under normal and excessive Mo conditions were analyzed by UPLC–MS/MS. 2-Oxo-arginine, L-nicotine, gluconic acid, D-gluconic acid, and citric acid play an important role in chelation of molybdenum and reducing its toxicity [81]. Yang et al. studied the metabolic responses of magnesium deficiency in soybean leaves and roots and found that there were considerable differences in carbon and nitrogen metabolic responses between leaves and roots under magnesium deficiency [82]. Allwood et al. had investigated the yield, grain quality, and response of grain metabolites to increase nitrogen application and found that nitrogen application had a significant effect on grain yield [83]. Zhang et al. performed metabolomic analysis on two wheat varieties with different induced nitrogen levels and revealed that flavonoids were considered as potential biomarkers under low nitrogen stress [84]. Metabolomics is increasingly utilized in agronomy and plant biology to understand the behavior of plants under different stress conditions [85–88].

1.5 Limitation

Although metabolomics technology is widely used, there are still many problems. The analytical techniques have not been able to achieve comprehensive qualitative and quantitative analysis of all metabolites in living organisms [89–94]. The instruments and equipment required for conducting the metabolomics research are expensive, and the proficiency of the operators is a high-level demand. The experiments have shown that various interference factors, such as age, gender, diet, temperature, and even changes in environment, could affect the metabolic profiles [95–102]. In the metabolomics data analysis, although there are several commonly used databases, the number of metabolites is currently not identified by the standard compound. Therefore, the metabolomics database needs to be further improved. In the process of data analysis, most methods are applied to linear data sets, and how they are used from nonlinear data is worth thinking deeply. Effective extraction of metabolites is still a faced problem [7, 103–110]. It requires mathematics and statistics to be fully borrowed and absorbed in metabolomics field [111–116]. Perfection of metabolomics still requires further efforts.

1.6 Conclusions and Future Perspectives

A better understanding of the disease-related mechanisms on the endogenous metabolism is an urgent need. Fortunately, the MS-based metabolomics has been used in various areas, providing scientific support for the discovery of endogenous metabolites. Advances in this field have enabled the discovery of numerous disease biomarkers and then provided insight into the disease pathogenesis [117–120]. It addresses the challenges of methods and applications for the discovery and identification for metabolite biomarkers, but the analytical instruments, techniques, and data acquisition or analysis require additional improvement. It is still challenging to establish a predictive model for diagnostic rationalization. Other challenges such as sample preparation, sample matrix, and residual effects can also lead to data variability. The challenge is usually accompanied by opportunities, and the current challenges are also huge opportunities for future development. With the improvement of methods, it is believed that metabolomics will be comprehensive, efficient, and even become a powerful tool to monitor disease from beginning to end.

Acknowledgments

This work was supported by grants from the Key Program of Natural Science Foundation of State (Grant No. 81973745, 81302905), Natural Science Foundation of Heilongjiang Province (YQ2019H030), and Heilongjiang Touyan Innovation Team Program.

References

1 Ju, R., Liu, X., Zheng, F. et al. (2019). Removal of false positive features to generate authentic peak table for high-resolution mass spectrometry-based metabolomics study. *Analytica Chimica Acta* 1067: 79–87.
2 Yang, Q., Zhang, A.H., Miao, J.H. et al. (2019). Metabolomics biotechnology, applications, and future trends: a systematic review. *RSC Advances* 9 (64): 37245–37257.
3 Amberg, A., Riefke, B., Schlotterbeck, G. et al. (2017). NMR and MS methods for metabolomics. *Methods in Molecular Biology (Clifton, N.J.)* 1641: 229–258.
4 Nicholson, J.K., Lindon, J.C., and Holmes, E. (1999). 'Metabonomics': understanding the metabolic responses of living systems to pathophysiological stimuli via multivariate statistical analysis of biological NMR spectroscopic data. *Xenobiotica* 29 (11): 1181–1189.
5 Zhang, Y., Liu, P., Li, Y. et al. (2017). Exploration of metabolite signatures using high-throughput mass spectrometry coupled with multivariate data analysis. *RSC Advances* 7 (11): 6780–6787.

6 Zhang, A.H., Ma, Z.M., Sun, H. et al. (2019). High-throughput metabolomics evaluate the efficacy of total lignans from *Acanthophanax senticosus* stem against ovariectomized osteoporosis rat. *Frontiers in Pharmacology* 10: 553.

7 Zhang, A., Sun, H., Wang, P. et al. (2013). Salivary proteomics in biomedical research. *Clinica Chimica Acta* 415: 261–265.

8 Xie, J., Zhang, A.H., Qiu, S. et al. (2019). Identification of the perturbed metabolic pathways associating with prostate cancer cells and anticancer affects of obacunone. *Journal of Proteomics* 206: 103447.

9 Bar, N., Korem, T., Weissbrod, O. et al. (2020). A reference map of potential determinants for the human serum metabolome. *Nature* 588 (7836): 135–140.

10 Zhang, A., Sun, H., Yan, G. et al. (2016). Mass spectrometry-based metabolomics: applications to biomarker and metabolic pathway research. *Biomedical Chromatography* 30 (1): 7–12.

11 Zhang, A., Sun, H., Wang, P. et al. (2012). Modern analytical techniques in metabolomics analysis. *Analyst* 137 (2): 293–300.

12 Wang, X.J., Gao, X., Zhang, A.H. et al. (2019). High-throughput metabolomics for evaluating the efficacy and discovering the metabolic mechanism of Luozhen capsules from the excessive liver-fire syndrome of hypertension. *RSC Advances* 9 (55): 32141–32153.

13 Sun, H., Li, X.N., Zhang, A.H. et al. (2019). Exploring potential biomarkers of coronary heart disease treated by Jing Zhi Guan Xin Pian using high-throughput metabolomics. *RSC Advances* 9 (20): 11420–11432.

14 Sun, H., Wang, M., Zhang, A. et al. (2013). UPLC-Q-TOF-HDMS analysis of constituents in the root of two kinds of *Aconitum* using a metabolomics approach. *Phytochemical Analysis* 24 (3): 263–276.

15 Sovio, U., Goulding, N., McBride, N. et al. (2020). A maternal serum metabolite ratio predicts fetal growth restriction at term. *Nature Medicine* 26 (3): 348–353.

16 Zhang, A., Sun, H., and Wang, X. (2014). Potentiating therapeutic effects by enhancing synergism based on active constituents from traditional medicine. *Phytotherapy Research* 28 (4): 526–533.

17 Wang, X.J., Ren, J.L., Zhang, A.H. et al. (2019). Novel applications of mass spectrometry-based metabolomics in herbal medicines and its active ingredients: current evidence. *Mass Spectrometry Reviews* 38 (4–5): 380–402.

18 Zhang, A.H., Sun, H., Qiu, S. et al. (2013). NMR-based metabolomics coupled with pattern recognition methods in biomarker discovery and disease diagnosis. *Magnetic Resonance in Chemistry* 51 (9): 549–556.

19 Sun, H., Zhang, A.H., Liu, S.B. et al. (2018). Cell metabolomics identify regulatory pathways and targets of magnoline against prostate cancer. *Journal of Chromatography B: Analytical Technologies in the Biomedical and Life Sciences* 1102, 1103: 143–151.

20 Zhang, A.H., Sun, H., Qiu, S. et al. (2013). Metabolomics in noninvasive breast cancer. *Clinica Chimica Acta* 424: 3–7.

21 Zhang, A.H., Qiu, S., Xu, H.Y. et al. (2014). Metabolomics in diabetes. *Clinica Chimica Acta* 429: 106–110.

22 Yan, G.L., Zhang, A.H., Sun, H. et al. (2013). An effective method for determining the ingredients of Shuanghuanglian formula in blood samples using high-resolution LC-MS coupled with background subtraction and a multiple data processing approach. *Journal of Separation Science* 36 (19): 3191–3199.

23 Zhang, A.H., Sun, H., Yan, G.L. et al. (2013). Serum proteomics in biomedical research: a systematic review. *Applied Biochemistry and Biotechnology* 170 (4): 774–786.

24 Sun, H., Zhang, A.H., Zou, D.X. et al. (2014). Metabolomics coupled with pattern recognition and pathway analysis on potential biomarkers in liver injury and hepatoprotective effects of Yinchenhao. *Applied Biochemistry and Biotechnology* 173 (4): 857–869.

25 Johno, H., Yoshimura, K., Mori, Y. et al. (2018). Detection of potential new biomarkers of atherosclerosis by probe electrospray ionization mass spectrometry. *Metabolomics: Official Journal of the Metabolomic Society* 14 (4): 38.

26 Tavares, G., Venturini, G., Padilha, K. et al. (2018). 1,5-Anhydroglucitol predicts CKD progression in macroalbuminuric diabetic kidney disease: results from non-targeted metabolomics. *Metabolomics: Official Journal of the Metabolomic Society* 14 (4): 39.

27 Hao, M., Ji, D., Li, L. et al. (2018). Mechanism of *Curcuma wenyujin* rhizoma on acute blood stasis in rats based on a UPLC-Q/TOF-MS metabolomics and network approach. *Molecules* 24 (1): 82.

28 Abdulwahab, R.A., Alaiya, A., Shinwari, Z. et al. (2019). LC-MS/MS proteomic analysis revealed novel associations of 37 proteins with T2DM and notable upregulation of immunoglobulins. *International Journal of Molecular Medicine* 43 (5): 2118–2132.

29 Yang, G., Zhao, G., Zhang, J. et al. (2019). Global urinary metabolic profiling of the osteonecrosis of the femoral head based on UPLC-QTOF/MS. *Metabolomics: Official Journal of the Metabolomic Society* 15 (3): 26.

30 Lin, S., Yue, X., Wu, H. et al. (2019). Explore potential plasma biomarkers of acute respiratory distress syndrome (ARDS) using GC-MS metabolomics analysis. *Clinical Biochemistry* 66: 49–56.

31 Øvrehus, M.A., Bruheim, P., Ju, W. et al. (2018). Gene expression studies and targeted metabolomics reveal disturbed serine, methionine, and tyrosine metabolism in early hypertensive nephrosclerosis. *Kidney International Reports* 4 (2): 321–333.

32 Barbosa-Breda, J., Himmelreich, U., Ghesquière, B. et al. (2018). Clinical metabolomics and glaucoma. *Ophthalmic Research* 59 (1): 1–6.

33 Dong, S., Zhan, Z.Y., Cao, H.Y. et al. (2017). Urinary metabolomics analysis identifies key biomarkers of different stages of nonalcoholic fatty liver disease. *World Journal of Gastroenterology* 23 (15): 2771–2784.

34 Wang, D., Kong, J., Wu, J. et al. (2017). GC-MS-based metabolomics identifies an amino acid signature of acute ischemic stroke. *Neuroscience Letters* 642: 7–13.

35 Li, Q., Wei, S., Wu, D. et al. (2018). Urinary metabolomics study of patients with gout using gas chromatography-mass spectrometry. *BioMed Research International* 2018: 3461572.

36 Sun, H., Zhao, J., Zhong, D. et al. (2017). Potential serum biomarkers and metabonomic profiling of serum in ischemic stroke patients using UPLC/Q-TOF MS/MS. *PLoS One* 12 (12): e0189009.

37 Lu, H., Ding, J., Liu, W. et al. (2018). UPLC/MS-based metabolomics investigation of the protective effect of hydrogen gas inhalation on mice with calcium oxalate-induced renal injury. *Biological & Pharmaceutical Bulletin* 41 (11): 1652–1658.

38 Zhao, H., Liu, Y., Li, Z. et al. (2018). Identification of essential hypertension biomarkers in human urine by non-targeted metabolomics based on UPLC-Q-TOF/MS. *Clinica Chimica Acta* 486: 192–198.

39 Dittharot, K., Jittorntam, P., Wilairat, P. et al. (2018). Urinary metabolomic profiling in chronic hepatitis B viral infection using gas chromatography/mass spectrometry. *Asian Pacific Journal of Cancer Prevention* 19 (3): 741–748.

40 Zhao, H., Wang, C., Zhao, N. et al. (2018). Potential biomarkers of Parkinson's disease revealed by plasma metabolic profiling. *Journal of Chromatography B: Analytical Technologies in the Biomedical and Life Sciences* 1081–1082: 101–108.

41 Li, X.Z., Yang, X.Y., Wang, Y. et al. (2017). Urine metabolic profiling for the pathogenesis research of erosive oral lichen planus. *Archives of Oral Biology* 73: 206–213.

42 Wang, Y., Sun, W., Zheng, J. et al. (2018). Urinary metabonomic study of patients with acute coronary syndrome using UPLC-QTOF/MS. *Journal of Chromatography B: Analytical Technologies in the Biomedical and Life Sciences* 1100–1101: 122–130.

43 Liang, Q., Liu, H., Xing, H. et al. (2016). Urinary UPLC-MS metabolomics dissecting the underlying mechanisms of Huaxian capsule protects against sepsis. *RSC Advances* 6 (46): 40436–40441.

44 Yang, W., Mu, T., Jiang, J. et al. (2018). Identification of potential biomarkers and metabolic profiling of serum in ovarian cancer patients using UPLC/Q-TOF MS. *Cell Physiol. Biochem.* 51 (3): 1134–1148.

45 Nan, Y., Zhou, X., Liu, Q. et al. (2016). Serum metabolomics strategy for understanding pharmacological effects of Shen Qi pill acting on kidney yang deficiency syndrome. *Journal of Chromatography B: Analytical Technologies in the Biomedical and Life Sciences* 1026: 217–226. https://doi.org/10.1016/j.jchromb.2015.12.004.

46 Liang, Q., Liu, H., Li, X. et al. (2016). High-throughput metabolomics analysis discovers salivary biomarkers for predicting mild cognitive impairment and Alzheimer's disease. *RSC Advances* 6 (79): 75499–75504.

47 Liang, Q., Liu, H., Xing, H. et al. (2016). UPLC-QTOF/MS based metabolomics reveals metabolic alterations associated with severe sepsis. *RSC Advances* 6 (49): 43293–43298.

48 Xia, H., Wu, L., Chu, M. et al. (2017). Effects of breviscapine on amyloid beta 1-42 induced Alzheimer's disease mice: a HPLC-QTOF-MS based plasma

metabonomics study. *Journal of Chromatography B: Analytical Technologies in the Biomedical and Life Sciences* 1057: 92–100.

49 Zhao, G., Hou, X., Li, X. et al. (2018). Metabolomics analysis of alloxan-induced diabetes in mice using UPLC-Q-TOF-MS after *Crassostrea gigas* polysaccharide treatment. *International Journal of Biological Macromolecules* 108: 550–557.

50 Ji, H., Liu, Y., He, F. et al. (2018). LC-MS based urinary metabolomics study of the intervention effect of aloe-emodin on hyperlipidemia rats. *Journal of Pharmaceutical and Biomedical Analysis* 156: 104–115.

51 Pang, Z., Wang, G., Ran, N. et al. (2018). Inhibitory effect of methotrexate on rheumatoid arthritis inflammation and comprehensive metabolomics analysis using ultra-performance liquid chromatography-quadrupole time of flight-mass spectrometry (UPLC-Q/TOF-MS). *International Journal of Molecular Sciences* 19 (10): 2894.

52 Rodrigues, D., Pinto, J., Araújo, A.M. et al. (2018). Volatile metabolomic signature of bladder cancer cell lines based on gas chromatography-mass spectrometry. *Metabolomics: Official Journal of the Metabolomic Society* 14 (5): 62.

53 Liang, Q., Liu, H., Jiang, Y. et al. (2016). Novel liquid chromatography-mass spectrometry for metabolite biomarkers of acute lung injury disease. *Analytical Methods* 8 (31): 6017–6022.

54 Pareek, V., Tian, H., Winograd, N. et al. (2020). Metabolomics and mass spectrometry imaging reveal channeled de novo purine synthesis in cells. *Science* 368 (6488): 283–290.

55 Liang, Q., Liu, H., Xing, H. et al. (2016). High-resolution mass spectrometry for exploring metabolic signatures of sepsis-induced acute kidney injury. *RSC Advances* 6 (36): 29863–29868.

56 Liang, Q., Yu, Q., Wu, H. et al. (2014). Metabolite fingerprint analysis of cervical cancer using LC-QTOF/MS and multivariate data analysis. *Analytical Methods* 6 (12): 3937–3942.

57 Li, C.R., Li, M.N., Yang, H. et al. (2018). Rapid characterization of chemical markers for discrimination of Moutan Cortex and its processed products by direct injection-based mass spectrometry profiling and metabolomic method. *Phytomedicine: International Journal of Phytotherapy and Phytopharmacology* 45: 76–83.

58 Bao, J., Ding, R.B., Jia, X. et al. (2018). Fast identification of anticancer constituents in Forsythiae Fructus based on metabolomics approaches. *Journal of Pharmaceutical and Biomedical Analysis* 154: 312–320.

59 Yuan, A., Gong, L., Luo, L. et al. (2017). Revealing anti-inflammation mechanism of water-extract and oil of forsythiae fructus on carrageenan-induced edema rats by serum metabolomics. *Biomedicine & Pharmacotherapy* 95: 929–937.

60 Zhuang, Y., Qin, K., Yu, B. et al. (2018). A metabolomics research based on UHPLC-ESI-Q-TOF-MS coupled with metabolic pathway analysis: treatment effects of stir-frying Xanthii Fructus on allergic rhinitis in mice model. *Biomedical Chromatography* 32 (12): e4352.

61 Wang, Y., Zhao, H., Liu, Y. et al. (2019). GC-MS-based metabolomics to reveal the protective effect of gross saponins of *Tribulus terrestris* fruit against ischemic stroke in rat. *Molecules* 24 (4): 793.

62 Zhou, J., Yao, N., Wang, S. et al. (2019). Fructus Gardeniae-induced gastrointestinal injury was associated with the inflammatory response mediated by the disturbance of vitamin B6, phenylalanine, arachidonic acid, taurine and hypotaurine metabolism. *Journal of Ethnopharmacology* 235: 47–55.

63 Zhang, A., Sun, H., and Wang, X. (2012). Saliva metabolomics opens door to biomarker discovery, disease diagnosis, and treatment. *Applied Biochemistry and Biotechnology* 168 (6): 1718–1727.

64 Zhang, A.H., Sun, H., Yan, G.L. et al. (2019). Chinmedomics: a powerful approach integrating metabolomics with serum pharmacochemistry to evaluate the efficacy of traditional Chinese medicine. *Engineering* 5 (1): 60–68.

65 Zhang, H.L., Zhang, A.H., Miao, J.H. et al. (2019). Targeting regulation of tryptophan metabolism for colorectal cancer therapy: a systematic review. *RSC Advances* 9 (6): 3072–3080.

66 Li, X., Zhang, A., Sun, H. et al. (2017). Metabolic characterization and pathway analysis of berberine protects against prostate cancer. *Oncotarget* 8 (39): 65022–65041.

67 Fang, H., Zhang, A., Yu, J. et al. (2016). Insight into the metabolic mechanism of scoparone on biomarkers for inhibiting Yanghuang syndrome. *Scientific Reports* 6: 37519.

68 Wang, X., Wang, Q., Zhang, A. et al. (2013). Metabolomics study of intervention effects of Wen-Xin-formula using ultra high-performance liquid chromatography/mass spectrometry coupled with pattern recognition approach. *Journal of Pharmaceutical and Biomedical Analysis* 74: 22–30.

69 Zhao, L., Hu, J., Huang, Y. et al. (2017). ^{1}H NMR and GC-MS based metabolomics reveal nano-Cu altered cucumber (*Cucumis sativus*) fruit nutritional supply. *Plant Physiology and Biochemistry* 110: 138–146.

70 Wang, Y.Q., Hu, L.P., Liu, G.M. et al. (2017). Evaluation of the nutritional quality of Chinese kale (*Brassica alboglabra* Bailey) using UHPLC-quadrupole-orbitrap MS/MS-based metabolomics. *Molecules* 22 (8): 1262.

71 Alolga, R.N., Chavez, M., and Muyaba, M. (2018). Untargeted UPLC-Q/TOF-MS-based metabolomics and inductively coupled plasma optical emission spectroscopic analysis reveal differences in the quality of ginger from two provinces in Zambia. *Journal of Pharmacy and Pharmacology* 70 (9): 1262–1271.

72 Ramírez-Acosta, S., Arias-Borrego, A., Gómez-Ariza, J.L. et al. (2019). Metabolomic study of bioactive compounds in strawberries preserved under controlled atmosphere based on GC-MS and DI-ESI-QqQ-TOF-MS. *Phytochemical Analysis* 30 (2): 198–207.

73 Kim, G.D., Lee, J.Y., and Auh, J.H. (2019). Metabolomic screening of anti-inflammatory compounds from the leaves of *Actinidia arguta* (hardy kiwi). *Foods* 8 (2): 47. https://doi.org/10.3390/foods8020047.

74 Mendonca, C.M., Yoshitake, S., Wei, H. et al. (2020). Hierarchical routing in carbon metabolism favors iron-scavenging strategy in iron-deficient soil *Pseudomonas* species. *Proceedings of the National Academy of Sciences of the United States of America* 117 (51): 32358–32369.

75 Wang, D., Zhang, L., Huang, X. et al. (2018). Identification of nutritional components in black sesame determined by widely targeted metabolomics and traditional Chinese medicines. *Molecules* 23 (5): 1180.

76 Moreira, V., Brasili, E., Fiamoncini, J. et al. (2018). Orange juice affects acylcarnitine metabolism in healthy volunteers as revealed by a mass-spectrometry based metabolomics approach. *Food Research International* 107: 346–352.

77 Saa, D., Nissen, L., and Gianotti, A. (2019). Metabolomic approach to study the impact of flour type and fermentation process on volatile profile of bakery products. *Food Research International* 119: 510–516.

78 Liao, Y., Hu, R., Wang, Z. et al. (2018). Metabolomics profiling of serum and urine in three beef cattle breeds revealed different levels of tolerance to heat stress. *Journal of Agricultural and Food Chemistry* 66 (26): 6926–6935.

79 Shinozaki, Y., Beauvoit, B.P., Takahara, M. et al. (2020). Fruit setting rewires central metabolism via gibberellin cascades. *Proceedings of the National Academy of Sciences of the United States of America* 117 (38): 23970–23981.

80 Defossez, E., Pitteloud, C., Descombes, P. et al. (2021). Spatial and evolutionary predictability of phytochemical diversity. *Proceedings of the National Academy of Sciences of the United States of America* 118 (3): e2013344118.

81 Xu, S., Hu, C., Hussain, S. et al. (2018). Metabolomics analysis reveals potential mechanisms of tolerance to excess molybdenum in soybean seedlings. *Ecotoxicology and Environmental Safety* 164: 589–596.

82 Yang, N., Jiang, J., Xie, H. et al. (2017). Metabolomics reveals distinct carbon and nitrogen metabolic responses to magnesium deficiency in leaves and roots of soybean [*Glycine max* (Linn.) Merr.]. *Frontiers in Plant Science* 8: 2091.

83 Allwood, J.W., Xu, Y., Martinez-Martin, P. et al. (2019). Rapid UHPLC-MS metabolite profiling and phenotypic assays reveal genotypic impacts of nitrogen supplementation in oats. *Metabolomics: Official Journal of the Metabolomic Society* 15 (3): 42.

84 Zhang, Y., Ma, X.M., Wang, X.C. et al. (2017). UPLC-QTOF analysis reveals metabolomic changes in the flag leaf of wheat (*Triticum aestivum* L.) under low-nitrogen stress. *Plant Physiology and Biochemistry* 111: 30–38.

85 Du, P., Wu, X., Xu, J. et al. (2017). Determination and dissipation of mesotrione and its metabolites in rice using UPLC and triple-quadrupole tandem mass spectrometry. *Food Chemistry* 229: 260–267.

86 Pei, G., Niu, X., Zhou, Y. et al. (2017). Crosstalk of two-component signal transduction systems in regulating central carbohydrate and energy metabolism during autotrophic and photomixotrophic growth of *Synechocystis* sp. PCC 6803. *Integrative Biology: Quantitative Biosciences from Nano to Macro* 9 (5): 485–496.

87 Mibei, E.K., Owino, W.O., Ambuko, J. et al. (2018). Metabolomic analyses to evaluate the effect of drought stress on selected African eggplant accessions. *Journal of the Science of Food and Agriculture* 98 (1): 205–216.

88 Qian, Y., Zhang, S., Yao, S. et al. (2018). Effects of vitro sucrose on quality components of tea plants (*Camellia sinensis*) based on transcriptomic and metabolic analysis. *BMC Plant Biology* 18 (1): 121.

89 Koelmel, J.P., Ulmer, C.Z., Jones, C.M. et al. (2017). Common cases of improper lipid annotation using high-resolution tandem mass spectrometry data and corresponding limitations in biological interpretation. *Biochimica et Biophysica Acta, Molecular and Cell Biology of Lipids* 1862 (8): 766–770.

90 Vaysse, P.M., Heeren, R., Porta, T. et al. (2017). Mass spectrometry imaging for clinical research – latest developments, applications, and current limitations. *Analyst* 142 (15): 2690–2712.

91 Li, Y.F., Qiu, S., Gao, L.J. et al. (2018). Metabolomic estimation of the diagnosis of hepatocellular carcinoma based on ultrahigh performance liquid chromatography coupled with time-of-flight mass spectrometry. *RSC Advances* 8 (17): 9375–9382.

92 Wang, H., Yan, G., Zhang, A. et al. (2013). Rapid discovery and global characterization of chemical constituents and rats metabolites of *Phellodendri amurensis* cortex by ultra-performance liquid chromatography-electrospray ionization/quadrupole-time-of-flight mass spectrometry coupled with pattern recognition approach. *Analyst* 138 (11): 3303–3312.

93 Wang, X., Li, J., and Zhang, A.H. (2016). Urine metabolic phenotypes analysis of extrahepatic cholangiocarcinoma disease using ultra-high performance liquid chromatography-mass spectrometry. *RSC Advances* 6 (67): 63049–63057.

94 Zhang, A.H., Sun, H., Yan, G.L. et al. (2014). Metabolomics study of type 2 diabetes using ultra-performance LC-ESI/quadrupole-TOF high-definition MS coupled with pattern recognition methods. *Journal of Physiology and Biochemistry* 70 (1): 117–128.

95 Zhang, A., Sun, H., Qiu, S. et al. (2014). Metabolomics insights into pathophysiological mechanisms of nephrology. *International Urology and Nephrology* 46 (5): 1025–1030.

96 Sun, H., Zhang, A.H., Song, Q. et al. (2018). Functional metabolomics discover pentose and glucuronate interconversion pathways as promising targets for Yang Huang syndrome treatment with Yinchenhao Tang. *RSC Advances* 8 (64): 36831–36839.

97 Zhang, A.H., Yu, J.B., Sun, H. et al. (2018). Identifying quality-markers from Shengmai San protects against transgenic mouse model of Alzheimer's disease using chinmedomics approach. *Phytomedicine: International Journal of Phytotherapy and Phytopharmacology* 45: 84–92.

98 Zhang, A., Sun, H., Xu, H. et al. (2013). Cell metabolomics. *Omics: A Journal of Integrative Biology* 17 (10): 495–501.

99 Wang, X., Zhang, A., Zhou, X. et al. (2016). An integrated chinmedomics strategy for discovery of effective constituents from traditional herbal medicine. *Scientific Reports* 6: 18997.

100 Fang, H., Zhang, A.H., Sun, H. et al. (2019). High-throughput metabolomics screen coupled with multivariate statistical analysis identifies therapeutic targets in alcoholic liver disease rats using liquid chromatography-mass spectrometry. *Journal of Chromatography B: Analytical Technologies in the Biomedical and Life Sciences* 1109: 112–120.

101 Cao, H., Zhang, A., Zhang, H. et al. (2015). The application of metabolomics in traditional Chinese medicine opens up a dialogue between Chinese and Western medicine. *Phytotherapy Research* 29 (2): 159–166.

102 Liu, X.Y., Zhang, A.H., Fang, H. et al. (2018). Serum metabolomics strategy for understanding the therapeutic effects of Yin-Chen-Hao-Tang against Yanghuang syndrome. *RSC Advances* 8 (14): 7403–7413.

103 Zhao, Y., Lv, H., Qiu, S. et al. (2017). Plasma metabolic profiling and novel metabolite biomarkers for diagnosing prostate cancer. *RSC Advances* 7 (48): 30060–30069.

104 Li, X.N., Zhang, A., Wang, M. et al. (2017). Screening the active compounds of *Phellodendri Amurensis* cortex for treating prostate cancer by high-throughput chinmedomics. *Scientific Reports* 7: 46234.

105 Zhang, A., Sun, H., Han, Y. et al. (2013). Urinary metabolic biomarker and pathway study of hepatitis B virus infected patients based on UPLC-MS system. *PLoS One* 8 (5): e64381.

106 Sun, H., Zhang, A., and Wang, X. (2012). Potential role of metabolomic approaches for Chinese medicine syndromes and herbal medicine. *Phytotherapy Research* 26 (10): 1466–1471.

107 Zhang, A., Sun, H., Yan, G. et al. (2014). Metabolomics in diagnosis and biomarker discovery of colorectal cancer. *Cancer Letters* 345 (1): 17–20.

108 Zhang, A., Sun, H., Han, Y. et al. (2012). Exploratory urinary metabolic biomarkers and pathways using UPLC-Q-TOF-HDMS coupled with pattern recognition approach. *Analyst* 137 (18): 4200–4208.

109 Zhang, A.H., Wang, P., Sun, H. et al. (2013). High-throughput ultra-performance liquid chromatography-mass spectrometry characterization of metabolites guided by a bioinformatics program. *Molecular BioSystems* 9 (9): 2259–2265.

110 Sun, H., Zhang, A., Yan, G. et al. (2013). Metabolomic analysis of key regulatory metabolites in hepatitis C virus-infected tree shrews. *Molecular & Cellular Proteomics* 12 (3): 710–719.

111 Ge, N., Kong, L., Zhang, A.H. et al. (2021). Identification of key lipid metabolites during metabolic dysregulation in the diabetic retinopathy disease mouse model and efficacy of Keluoxin capsule using an UHPLC-MS-based non-targeted lipidomics approach. *RSC Advances* 11 (10): 5491–5505.

112 Zhang, A., Sun, H., Wang, P. et al. (2012). Future perspectives of personalized medicine in traditional Chinese medicine: a systems biology approach. *Complementary Therapies in Medicine* 20 (1–2): 93–99.

113 Chu, H., Zhang, A., Han, Y. et al. (2016). Metabolomics approach to explore the effects of Kai-Xin-San on Alzheimer's disease using UPLC/ESI-Q-TOF mass spectrometry. *Journal of Chromatography B: Analytical Technologies in the Biomedical and Life Sciences* 1015–1016: 50–61.

114 Zhang, A., Sun, H., Yan, G. et al. (2015). Metabolomics for biomarker discovery: moving to the clinic. *BioMed Research International* 2015: 354671.

115 Zhang, A., Sun, H., and Wang, X. (2017). Emerging role and recent applications of metabolomics biomarkers in obesity disease research. *RSC Advances* 7 (25): 14966–14973.

116 Sun, H., Wang, H., Zhang, A. et al. (2015). Berberine ameliorates nonbacterial prostatitis via multi-target metabolic network regulation. *Omics: A Journal of Integrative Biology* 19 (3): 186–195.

117 Xiong, H., Zhang, A.H., Zhao, Q.Q. et al. (2020). Discovery of quality-marker ingredients of *Panax quinquefolius* driven by high-throughput chinmedomics approach. *Phytomedicine: International Journal of Phytotherapy and Phytopharmacology* 74: 152928.

118 Zhang, A., Wang, H., Sun, H. et al. (2015). Metabolomics strategy reveals therapeutical assessment of limonin on nonbacterial prostatitis. *Food and Function* 6 (11): 3540–3549.

119 Zhang, H.L., Zhang, A.H., Zhou, X.H. et al. (2018). High-throughput lipidomics reveal mirabilite regulating lipid metabolism as anticancer therapeutics. *RSC Advances* 8 (62): 35600–35610.

120 Sun, H., Zhang, A.H., Yang, L. et al. (2019). High-throughput chinmedomics strategy for discovering the quality-markers and potential targets for Yinchenhao decoction. *Phytomedicine: International Journal of Phytotherapy and Phytopharmacology* 54: 328–338.

2

Innovations in Analytical Techniques of Metabolomics

Ling Kong, Aihua Zhang, Shi Qiu, and Xijun Wang

Heilongjiang University of Chinese Medicine, National Chinmedomics Research Center, Functional Metabolomics Laboratory, Heping Road 24, Harbin 150040, China

2.1 Introduction

Metabolomics, as an important part of systems biology, analyzes the broad-spectrum metabolites in organisms, discovers the changes of metabolites in different states, and elucidates the relationship between the metabolites and the corresponding pathological states [1–3]. Metabolomics is mainly used to explore small-molecular substances with small relative molecular weight ($\leq$1 kDa), which are mostly intermediates or end products of carbohydrate, lipid, and protein metabolism [4–7]. The small-molecule metabolites have a key role in biological sample, which is to represent potential candidates to understand the disease. The recent rapid development of a range of analytical platforms can enable separation, characterization, and quantification of these metabolites [8–11].

Recently, metabolomics has been rapidly developed by life scientists, providing a powerful technical platform for exploring disease mechanisms, interpreting life phenomena, discovering biomarkers, developing drugs, and quality control of herbal medicine [12–16]. Due to the complexity of metabolites, no single analytic platform could be used to detect all the metabolites in a sample. The precise and high-resolution spectroscopic techniques are essential for metabolomics research. The spectroscopic techniques includes mass spectrometry (MS) combined with chromatographic (liquid chromatography (LC), gas chromatography (GC), supercritical fluid chromatography (SFC) and capillary electrophoresis (CE), and nuclear magnetic resonance (NMR) spectroscopy [17–20]. These techniques can be simultaneously used in the metabolomics analysis. The combined use of analytical instruments and tools has revealed the beneficial outcomes in metabolomics, providing reliable detection of thousands of metabolites in a sample. Herein, the strengths and application of each technique as well as its limitations are discussed; furthermore, the key role of the integrated tools for metabolomics studies is also highlighted.

Mass Spectrometry-Based Metabolomics in Clinical and Herbal Medicines: Strategies, Technologies, and Applications, First Edition. Edited by Aihua Zhang and Wanying Wang.

2.2 Current Analytical Technologies of Metabolomics

Metabolomics is close to the biological phenotype of the bio-systems, which provides a number of benefits compared to other omics, and the most advantageous is the rapid observation of the metabolic perturbations in the pathological states. It highlights the interactions of complex components and focuses on the global system rather than the individual parts. Application of the analytical technologies to the measurement of changes in metabolite concentrations could help reveal the effects of perturbations under defined conditions [21–25].

Metabolomics is devoted to the non-targeted and targeted analysis of endogenous small-molecule metabolites [26–29]. Metabolomics is usually divided into untargeted and targeted metabolomics. Targeted metabolomics is the analysis of a group of pre-defined metabolites, which is usually linked to a specific metabolic network, whereas untargeted metabolomics research could exhibit the whole framework of metabolic network, not only specific pre-defined metabolites. The untargeted metabolomics research usually includes metabolic fingerprint and metabolite analysis. Metabolic fingerprint analysis uses the pattern of metabolite composition to distinguish samples, while metabolite analysis simultaneously measures metabolites to reflect the metabolic activities in sample. Untargeted metabolomics can be used as a discovery and hypothesis-generating tool, and its primary goal is the differential analysis of metabolites.

Identification of the small-molecule metabolites of interest became possible by the use of technological advances. Recent advances in analytical technologies enabled the metabolite profiling to understand the complex process [30–33]. There are numerous innovational technologies, which mainly include the NMR, high-resolution MS, CE, and high performance liquid chromatography (HPLC), that could detect the metabolites within a few minutes for metabolic applications. Every technique has advantages and drawbacks. Thus, a combination of different analytical technologies can obtain a broad metabolome. For NMR technology, the benefits are as follows: noninvasive, comprehensive, fast, and simple sample processing, but it is not sensitive enough. Due to the complexity of metabolites, the advanced separation techniques have been coupled with high-resolution MS. Advances in MS including mass analyzers, ionization techniques, and detection approach have provided beneficial tools to assess the small-molecular changes from large-scale metabolome [16, 34–38]. Up to now, GC–MS technology has been identified as the first choice for the identification of volatile compounds. Due to the high throughput, sensitivity, and wide dynamic range of LC–MS, its application has exceeded NMR and GC–MS. LC–MS is appropriate for thermally unstable, nonvolatile substances and could avoid the sample pretreatment in GC–MS. The combination of ultra-performance liquid chromatography (UPLC) and MS technology has showed the improving precision of mass measurements and resolution of metabolite species. The UPLC in combination with MS improved the resolution and increased the sensitivity, resulting in higher peak capacity, compared with conventional HPLC columns; therefore it is more suitable for metabolomics analysis [39–42]. Favorable outcomes have been gained with these techniques. In recent years, the hyphenated techniques have received a lot of attention in this field.

2.3 Chromatography Platform

The essence of the chromatographic process is the distribution and equilibrium between the stationary phase and the mobile phase. The distribution of different substances between the two phases will be different, which makes them move at different speeds with the mobile phase. With the movement of the mobile phase, the different components in the mixture are separated from each other on the stationary phase. The interactions between metabolites and chromatographic system are based on physicochemical properties of metabolites as well as on the type of the stationary phase, which determine the order of elution in the chromatography. In general, volatile organic compounds, lipids, and derivable molecules are analyzed by GC, while the majority of semi-polar metabolites are analyzed by LC. SFC has higher separation efficiency than LC, and it can be used for the analysis of thermally unstable and high molecular weight compounds not suitable for GC. Samples can be prepared into a solution or suspension solution that can be separated by CE, ranging from inorganic ions to biological macro-molecules.

2.3.1 Gas Chromatography Technique

GC is one of the earliest analytical techniques for metabolomics. It mainly uses the differences of boiling point, polarity, and adsorption properties to achieve the separation of mixtures. The main prerequisite for GC analysis is that the compound should be volatile and thermally stable. As most of the metabolites are polar and nonvolatile in fluids, they cannot be directly analyzed without sample preparation. Therefore, metabolic profiling via GC–MS usually requires modification of the polar molecules by chemical derivatization to reduce the polarity and increase the thermal stability and volatility of the analytes. After vaporization in the vaporizer, the sample is carried into the chromatographic column by the inert gas. Due to the fast transfer speed of the sample in the gas phase, the components could reach instantaneous equilibrium between the mobile and stationary phase. Recently, the use of high selectivity detector exhibits the advantages of high sensitivity and wide application for GC. The oven temperature program is extremely essential for samples containing molecules in a wide range of boiling points. With the increase of temperature, the gas solubility in the liquid decreases, and the retention time of analytes will be shortened. Multidimensional approaches, such as two-dimensional GC, have recently attracted substantial interest and could provide greater resolving power and enhance peak capacity and sensitivity.

2.3.2 Liquid Chromatography Technique

Chromatographic separation directly affects the accuracy and effectiveness of the analysis. With the advancement of HPLC, UPLC covers new technologies such as small particle packing, very low system volume, and rapid detection means, which increase the analytical throughput, sensitivity, and peak capacity [43–46]. UPLC is more sensitive than HPLC and allows smaller injection volume. It is not limited by the volatility of the sample and has a wide range of applications, especially for the substances with low volatility and poor thermal stability. It is a rapid separation method with high accuracy and wide separation range. It has little damage to the structure of compounds and is suitable for the separation of organic molecules and biomolecules. Reversed-phase chromatography is widely used in metabolomics because it is reliable, stable, and repeatable and covers a wide range of molecular structures [47–50]. Nonpolar chromatographic columns are more suitable for metabolomics analysis because they can separate semi-polar compounds, such as phenolic acids, alkaloids, glycosylated steroids, and other glycosylated species.

The characteristics of mobile phase are also important for separation in terms of peak capacity, resolution, retention time, peak shape, and sensitivity. The mobile phase used for metabolomics analysis in LC is usually composed of water solution and organic reagent. They are selected according to the polarity of the analytes. The pH of the mobile phase is also another factor that affects retention behavior of the metabolites, especially for acidic and basic metabolites. The pH also affects the ionization degree of metabolites in MS system. To improve the ionization efficiency of metabolites, acidic buffers and basic buffers were used in the positive ionization mode and negative ionization mode, respectively [51–54]. Gradient elution is performed in metabolomics research to prevent any ion inhibition in MS and improve the resolution between metabolites.

2.3.3 Supercritical Fluid Chromatography Technique

SFC is a supplement to GC and LC. It can effectively separate thermally unstable compounds at a lower temperature than the gas phase, due to the lower column temperature improving separation selectivity. SFC is an essential form of a normal phase as it usually utilizes a nonpolar mobile phase and polar stationary phase. The mobile phase consists of a mixture of an organic modifier and a supercritical fluid. When the temperature and pressure increase to their respective critical points, the substance enters the so-called supercritical state. In this case, the density is close to that of liquid, and the viscosity and diffusivity are close to that of gas, respectively. Due to low critical parameters and favorable properties, CO_2 is the most commonly used solvent in SFC. In order to maintain the pressure in the column and keep the CO_2-based mobile phase in a supercritical state, SFC instruments are equipped with a backpressure regulator, which plays a key role in the experimental reproducibility. SFC can choose the pressure program, temperature program, and different mobile phases or modifiers, so the selection range of operating conditions is wider than that of GC. By using a high proportion of CO_2 in the mobile phase, SFC adopts the principles of green analytical chemistry more than LC.

2.3.4 Capillary Electrophoresis Chromatography Technique

CE is driven by a high voltage electric field and separated by a capillary tube according to the difference of electrophoresis degree or distribution behavior of components in a sample. It has considerable advantages for the separation of charged substances: small sample quantity, high column efficiency, fast analysis speed, and environmental protection. Capillary electrophoresis chromatography

(CEC) is a new micro-column separation technology that combines the advantages of HPLC and CE. It is a highly efficient micro-separation technology, in which the conventional chromatographic packing is filled into the capillary or the stationary phase is bonded and coated on the inner surface of the capillary, and the electroosmosis flow is used as the driving force of the mobile phase. According to the difference of partition coefficient between the stationary phase and the mobile phase and the different migration rate, the separation of each component in the sample should be realized.

2.4 Mass Spectrometry Platform

A mass spectrometer is usually composed of three major parts: ion source, mass analyzer, and detector. Ion sources are often used for electrospray ionization (ESI), atmospheric pressure chemical ionization (APCI), etc. The samples need to be analyzed under the scan range of positive ionization mode and negative ionization mode to maximize metabolite coverage. Mass analyzers usually includes quadrupole, time-of-flight (TOF), Fourier transform ion cyclotron resonance (FTICR), ion-mobility (IM), orbitrap, etc. [29, 55–59]. For metabolomics analysis, combination of chromatography and MS can exhibit the advantage of high-resolution chromatographic system and high sensitivity of MS detector. In the past few decades, MS has been extensively developed in the metabolomics field, and MS-metabolomics approach has created the potential to detect the levels of hundreds of metabolites range from the metabolites' characterization to biomarker discovery [60–63].

2.4.1 GC–MS Technique

GC–MS is a gold standard for the analysis of volatiles, and it is relatively a simpler technique than LC–MS, using the reference libraries. It is one of the most important tools for the volatile compound analysis from various matrices such as urine, feces, breath, saliva, and blood. But most compounds need a chemical derivatization at the elevated temperatures to produce the volatility and thermal stability. The most widely used ionization technique in GC–MS is electron impact (EI) ionization, which could produce highly reproducible and characteristic fragmentation spectra. Moreover, EI ionization can provide the structural information about the breaking down of molecules and composing molecular fragments for metabolite identification. In GC–MS metabolomics studies, quadrupole, TOF, and FTMS are the most common mass analyzer. Quadrupole is the most commonly used mass analyzer for untargeted metabolomics, but the TOF sensitivity and selectivity is higher. Quadrupole mass analyzers provide fast scanning and are easy to use and suitable for the determination of compound masses and fragment ions. High-mass-resolution TOF technology exhibits the most advantageous for resolving peak deconvolution, the narrow and sharp peaks. The softer ionization technique such as chemical ionization is another tool to keep the molecular ions intact and is beneficial for identification of unknown metabolite.

2.4.2 LC–MS Technique

Both HPLC and MS techniques are often combined to characterize the unknown endogenous or exogenous metabolites in the complex biological samples [64–67]. Furthermore, LC–MS could provide a list of retention time, m/z values, and relative abundances of metabolites. LC–MS technique is the mostly used tool to achieve precise and accurate identification of the metabolic phenotype-related metabolites. In LC–MS metabolomics, ESI is by far the most preferred choice as its soft ionization capability generates a large number of ions and usually forms the intact molecular ions that help in the identification. APCI ionization approach is the complimentary to ESI in analyzing thermally stable and nonpolar compounds. TOF is mostly used in LC–MS-based metabolomics, owing to its high resolution and fast detection of metabolites. The combination of UPLC with MS detection could cover a number of polar metabolites and enlarge the number of detected metabolites. This technique represents a promising micro-separation platform in metabolomics, and recently, this technology is accepted by the analytical community. Large-scale UPLC–MS metabolomics studies are gaining attention for their use in the various fields, due to their sensitivity and quantitative reproducibility [68–70].

2.4.3 SFC–MS Technique

This detector commonly used in LC is suitable for SFC detection, which can be placed before or after a backpressure regulator. Similar to other chromatographic techniques, this detection mode is usually not sufficiently sensitive and selective for bioanalytical applications, where MS becomes essential. SFC improves the coverage and analysis throughput of metabolite. The earliest application of SFC is tailored toward the analysis of nonpolar compounds, and it will now be used for the analysis of the polar compounds. In fact, highly organic mobile phases and additives has extended the applications of SFC–MS to analyze a wide range of diverse metabolites with different polarities.

2.4.4 CE–MS Technique

CE–MS is considered a complementary tool for the analysis of complex biological samples to detect polar and ionic compounds in metabolomics studies. ESI is the most common soft ionization technique employed for CE coupling with MS. However, it requires a conductive liquid at the end of the capillary to obtain compatible potentials. When analyzing a large number of samples, ESI showed better reproducibility and stability, although it dilutes the sample and results in a loss in sensitivity. In recent years, due to the emergence of fast scan of new technologies, it is enough to meet the characteristics of CE narrow peak shape, which makes CE–MS develop rapidly. CE–MS has provided numerous advantages over other separation techniques. The significant advantage has a very small sample requirement to injection volume ranging from 1 to 20 nl. Thus, CE–MS has been used for both non-targeted and targeted analysis of metabolites such as amino acids, organic acids, nucleotides, thiols, carbohydrates, vitamins, and peptides.

2.5 Nuclear Magnetic Resonance Platform

In general, NMR compared with mass spectroscopy has several advantages including high repeatability, low sample, quantitation without the standards, easy metabolite identification, ability to identify both polar and nonpolar compounds, and a non-destructive analysis. The main disadvantage of NMR compared with MS is low sensitivity and resolution, making it inappropriate for the analysis of low-abundant metabolites. NMR could simultaneously quantify a wide range of organic compounds in the micro-molar range, provides a holistic view of the metabolites in certain conditions, and is well suited for metabolomics studies. High-resolution NMR has been extensively used for metabolic profiling, fingerprinting, and metabolic flux analysis within biofluids or tissue extracts.

2.6 Comprehensive and Multidimensional Analytical Platform

A large number of molecules from metabolome require analytical platforms to reach its extensive coverage. Although NMR, LC–MS, and GC–MS are often used for large-scale analysis, metabolomics study is not limited to these techniques. The integrated analytical techniques allow researchers to analyze the boarder range of metabolites as compared with the single analytical method and have accelerated the research in metabolomics. Hundreds of metabolites can be measured simultaneously by analytical platforms such as GC × GC and GC × GC–TOF–MS, matrix assisted laser desorption ionization (MALDI)–FTICR–MS, to obtain their concentration levels in a multidimensional way [71–73]. Comprehensive GC × GC–MS is a powerful technique that has gained increasing attention, could provide much chemical selectivity and sensitivity and separation capacity for complex sample, and brings more accurate information on the retention time and compound mass spectra. Comprehensive and multidimensional analytical platform techniques can partially overcome the shortcomings of single-analyzed technique and substantially enhance the coverage level of the metabolites.

2.7 Conclusions

This chapter highlights the importance and benefits of innovations in analytical techniques of metabolomics research. The success of metabolomics analysis is determined by the number of metabolites covered. The more metabolites analyzed, the better the chance of defining the phenotype. Monitoring more metabolites requires high-throughput techniques that enable screening compounds within a wide dynamic range. Therefore, innovations in analytical techniques of metabolomics allow us to better understand the changes in metabolic phenotype. The recent development of the multidimensional analytical platform is an ideal approach and will provide the reproducible and sensitive detection of thousands of metabolites in metabolomics as needed for particular experiments.

Acknowledgments

This work was supported by grants from the National Key Research and Development Program of China (2018YFC1706103), Key Program of Natural Science Foundation of State (Grant No. 81830110, 81861168037, 81973745), Natural Science Foundation of Heilongjiang Province (LH2020H095, YQ2019H030), and Heilongjiang Touyan Innovation Team Program.

References

1 Bar, N., Korem, T., Weissbrod, O. et al. (2020). A reference map of potential determinants for the human serum metabolome. *Nature* 588 (7836): 135–140.

2 Zhang, A.H., Sun, H., Han, Y. et al. (2013). Ultraperformance liquid chromatography–mass spectrometry based comprehensive metabolomics combined with pattern recognition and network analysis methods for characterization of metabolites and metabolic pathways from biological data sets. *Analytical Chemistry* 85 (15): 7606–7612.

3 Pareek, V., Tian, H., Winograd, N. et al. (2020). Metabolomics and mass spectrometry imaging reveal channeled de novo purine synthesis in cells. *Science* 368 (6488): 283–290.

4 Zhang, A., Sun, H., and Wang, X. (2013). Power of metabolomics in biomarker discovery and mining mechanisms of obesity. *Obesity Reviews* 14 (4): 344–349.

5 Liang, Q., Yu, Q., Wu, H. et al. (2014). Metabolite fingerprint analysis of cervical cancer using LC-QTOF/MS and multivariate datsa analysis. *Analytical Methods* 6 (12): 3937–3942.

6 Zhang, A., Sun, H., Yan, G. et al. (2014). Metabolomics analysis of marker metabolites for patients with pancreatic cancer. *Medicinal Chemistry* 4 (6): 506–510.

7 Guo, H., Chou, W.C., Lai, Y. et al. (2020). Multi-omics analyses of radiation survivors identify radioprotective microbes and metabolites. *Science* 370 (6516): 9097.

8 Zhang, A., Sun, H., Qiu, S. et al. (2014). Metabolomics insights into pathophysiological mechanisms of nephrology. *International Urology and Nephrology* 46 (5): 1025–1030.

9 Zhang, A., Sun, H., Yan, G. et al. (2014). Metabolomics in diagnosis and biomarker discovery of colorectal cancer. *Cancer Letters* 345 (1): 17–20.

10 Zhang, H.L., Zhang, A.H., Miao, J.H. et al. (2019). Targeting regulation of tryptophan metabolism for colorectal cancer therapy: a systematic review. *RSC Advances* 9 (6): 3072–3080.

11 Liang, Q., Liu, H., Xing, H. et al. (2016). High-resolution mass spectrometry for exploring metabolic signatures of sepsis-induced acute kidney injury. *RSC Advances* 6 (36): 29863–29868.

12 Zhang, A., Yan, G., Han, Y. et al. (2014). Metabolomics approaches and applications in prostate cancer research. *Applied Biochemistry and Biotechnology* 174 (1): 6–12.

13 Wang, X., Zhang, A., and Sun, H. (2013). Power of metabolomics in diagnosis and biomarker discovery of hepatocellular carcinoma. *Hepatology* 57 (5): 2072–2077.

14 Zhang, A., Sun, H., Wang, P. et al. (2012). Recent and potential developments of biofluid analyses in metabolomics. *Journal of Proteomics* 75 (4): 1079–1088.

15 Nan, Y., Zhou, X., Liu, Q. et al. (2016). Serum metabolomics strategy for understanding pharmacological effects of ShenQi pill acting on kidney yang deficiency syndrome. *Journal of Chromatography B* 1026: 217–226.

16 Zhang, A., Liu, Q., Zhao, H. et al. (2016). Phenotypic characterization of nanshi oral liquid alters metabolic signatures during disease prevention. *Scientific Reports* 6: 19333.

17 Andres, D.A., Young, L., Veeranki, S. et al. (2020). Improved workflow for mass spectrometry-based metabolomics analysis of the heart. *The Journal of Biological Chemistry* 295 (9): 2676–2686.

18 Zhang, A., Sun, H., Wang, P. et al. (2012). Modern analytical techniques in metabolomics analysis. *Analyst* 137 (2): 293–300.

19 Garcia-Perez, I., Posma, J.M., Serrano-Contreras, J.I. et al. (2020). Identifying unknown metabolites using NMR-based metabolic profiling techniques. *Nature Protocols* 15 (8): 2538–2567.

20 Wang, S., Blair, I.A., and Mesaros, C. (2019). Analytical methods for mass spectrometry-based metabolomics studies. *Advances in Experimental Medicine and Biology* 1140: 635–647.

21 Harder, J.M., Guymer, C., Wood, J. et al. (2020). Disturbed glucose and pyruvate metabolism in glaucoma with neuroprotection by pyruvate or rapamycin. *Proceedings of the National Academy of Sciences of the United States of America* 117 (52): 33619–33627.

22 Liang, Q., Liu, H., Zhang, T. et al. (2016). Discovery of serum metabolites for diagnosis of mild cognitive impairment to Alzheimer's disease progression using an optimized metabolomics method. *RSC Advances* 6: 3586–3591.

23 Zhang, A.H., Sun, H., Qiu, S. et al. (2013). Metabolomics in noninvasive breast cancer. *Clinica Chimica Acta* 424: 3–7.

24 Wang, X., Li, J., and Zhang, A.H. (2016). Urine metabolic phenotypes analysis of extrahepatic cholangiocarcinoma disease using ultra-high performance liquid chromatography–mass spectrometry. *RSC Advances* 6 (67): 63049–63057.

25 Zhang, A., Sun, H., Han, Y. et al. (2013). Urinary metabolic biomarker and pathway study of hepatitis B virus infected patients based on UPLC-MS system. *PLoS One* 8 (5): e64381.

26 Wang, X.J., Zhang, A.H., Kong, L. et al. (2019). Rapid discovery of quality-markers from Kaixin san using chinmedomics analysis approach. *Phytomedicine* 54: 371–381.

27 Liang, Q., Wang, C., Li, B. et al. (2015). Metabolomics of alcoholic liver disease: a clinical discovery study. *RSC Advances* 5 (98): 80381–80387.

28 Ren, J.L., Zhang, A.H., Kong, L. et al. (2018). Advances in mass spectrometry-based metabolomics for investigation of metabolites. *RSC Advances* 8 (40): 22335–22350.

29 Zhang, Y., Liu, P., Li, Y. et al. (2017). Exploration of metabolite signatures using high-throughput mass spectrometry coupled with multivariate data analysis. *RSC Advances* 7 (11): 6780–6787.

30 Zhang, A.H., Sun, H., Sun, W.J. et al. (2013). Trajectory analysis of metabolomics profiling in liver injured rats using ultra-performance liquid chromatography coupled with mass spectrometry. *Analytical Methods* 5 (19): 5294–5301.

31 Vascellari, S., Melis, M., Palmas, V. et al. (2021). Clinical phenotypes of Parkinson's disease associate with distinct gut microbiota and metabolome enterotypes. *Biomolecules* 11 (2): 144.

32 Zhang, A., Fang, H., Wang, Y. et al. (2017). Discovery and verification of the potential targets from bioactive molecules by network pharmacology-based target prediction combined with high-throughput metabolomics. *RSC Advances* 7 (81): 51069–51078.

33 Liang, Q., Liu, H., Xing, H. et al. (2016). UPLC–QTOF/MS based metabolomics reveals metabolic alterations associated with severe sepsis. *RSC Advances* 6 (49): 43293–43298.

34 Zhang, A., Sun, H., and Wang, X. (2012). Saliva metabolomics opens door to biomarker discovery, disease diagnosis, and treatment. *Applied Biochemistry and Biotechnology* 168 (6): 1718–1727.

35 Singh, K.D., Tancev, G., Decrue, F. et al. (2019). Standardization procedures for real-time breath analysis by secondary electrospray ionization high-resolution mass spectrometry. *Analytical and Bioanalytical Chemistry* 411 (19): 4883–4898.

36 Zhang, A., Sun, H., Xu, H. et al. (2013). Cell metabolomics. *Omics: A Journal of Integrative Biology* 17 (10): 495–501.

37 Vial, T., Tan, W.L., Deharo, E. et al. (2020). Mosquito metabolomics reveal that dengue virus replication requires phospholipid reconfiguration via the remodeling cycle. *Proceedings of the National Academy of Sciences of the United States of America* 117 (44): 27627–27636.

38 Liang, Q., Wang, C., Li, B. et al. (2015). Metabolic fingerprinting to understand therapeutic effects and mechanisms of silybin on acute liver damage in rat. *Pharmacognosy Magazine* 11 (43): 586.

39 Zhang, A., Sun, H., Yan, G. et al. (2015). Metabolomics for biomarker discovery: moving to the clinic. *BioMed Research International.*

40 Sun, H., Zhang, A., Yan, G. et al. (2013). Metabolomic analysis of key regulatory metabolites in hepatitis C virus-infected tree shrews. *Molecular & Cellular Proteomics* 12 (3): 710–719.

41 Zhang, A.-H., Qiu, S., Xu, H.-Y. et al. (2014). Metabolomics in diabetes. *Clinica Chimica Acta* 429: 106–110.

42 Qiu, S., Zhang, A.H., Guan, Y. et al. (2020). Functional metabolomics using UPLC-Q/TOF-MS combined with ingenuity pathway analysis as a promising strategy for evaluating the efficacy and discovering amino acid metabolism as a potential therapeutic mechanism-related target for geniposide against alcoholic liver disease. *RSC Advances* 10 (5): 2677–2690.

43 Fang, H., Zhang, A.H., Sun, H. et al. (2019). High-throughput metabolomics screen coupled with multivariate statistical analysis identifies therapeutic targets

in alcoholic liver disease rats using liquid chromatography-mass spectrometry. *Journal of Chromatography B* 1109: 112–120.

44 Zhang, A.H., Ma, Z.M., Sun, H. et al. (2019). High-throughput metabolomics evaluate the efficacy of total lignans from Acanthophanax senticosus stem against ovariectomized osteoporosis rat. *Frontiers in Pharmacology* 10: 553.

45 Anyfanti, P., Nikolaidou, B., and Gkaliagkousi, E. (2020). Urine metabolomic phenotyping for detection of adrenocortical carcinoma: still a long way to go. *Lancet Diabetes and Endocrinology* 8 (11): 876–877.

46 Zhang, A., Wang, H., Sun, H. et al. (2015). Metabolomics strategy reveals therapeutical assessment of limonin on nonbacterial prostatitis. *Food & Function* 6 (11): 3540–3549.

47 Qiu, S., Zhang, A., Zhang, T. et al. (2017). Dissect new mechanistic insights for geniposide efficacy on the hepatoprotection using multiomics approach. *Oncotarget* 8 (65): 108760.

48 Matsuda, F. (2016). Technical challenges in mass spectrometry-based metabolomics. *Mass Spectrometry* 5 (2): S0052.

49 Liang, Q., Liu, H., Jiang, Y. et al. (2016). Novel liquid chromatography-mass spectrometry for metabolite biomarkers of acute lung injury disease. *Analytical Methods* 8 (31): 6017–6022.

50 Walter, T.H., Blaze, M.T.M., and Boissel, C. (2021). Electrospray ionization mass spectrometry ion suppression/enhancement caused by column bleed for three mixed-mode reversed-phase/anion-exchange high performance liquid chromatography columns. *Rapid Communications in Mass Spectrometry*: e9098. https://doi.org/10.1002/rcm.9098.

51 Zhang, A.H., Sun, H., Yan, G.L. et al. (2014). Metabolomics study of type 2 diabetes using ultra-performance LC-ESI/quadrupole-TOF high-definition MS coupled with pattern recognition methods. *Journal of Physiology and Biochemistry* 70 (1): 117–128.

52 Yachida, S., Mizutani, S., Shiroma, H. et al. (2019). Metagenomic and metabolomic analyses reveal distinct stage-specific phenotypes of the gut microbiota in colorectal cancer. *Nature Medicine* 25 (6): 968–976.

53 Zhang, A., Sun, H., Yan, G. et al. (2016). Mass spectrometry-based metabolomics: applications to biomarker and metabolic pathway research. *Biomedical Chromatography* 30 (1): 7–12.

54 Zhang, T., Zhang, A., Qiu, S. et al. (2016). Current trends and innovations in bioanalytical techniques of metabolomics. *Critical Reviews in Analytical Chemistry* 46 (4): 342–351.

55 Burton, C. and Ma, Y. (2019). Current trends in cancer biomarker discovery using urinary metabolomics: achievements and new challenges. *Current Medicinal Chemistry* 26 (1): 5–28.

56 Song, Q., Zhang, A.H., Yan, G.L. et al. (2017). Technological advances in current metabolomics and its application in tradition Chinese medicine. *RSC Advances* 7 (84): 53516–53524.

57 Chu, H., Zhang, A.H., Han, Y. et al. (2015). Metabolomics and its potential in drug discovery and development from TCM. *World Journal of Traditional Chinese Medicine* 1 (4): 26.

58 Li, Y.F., Qiu, S., Gao, L.J. et al. (2018). Metabolomic estimation of the diagnosis of hepatocellular carcinoma based on ultrahigh performance liquid chromatography coupled with time-of-flight mass spectrometry. *RSC Advances* 8 (17): 9375–9382.

59 Wang, X., Wang, Q., Zhang, A. et al. (2013). Metabolomics study of intervention effects of Wen-Xin-formula using ultra high-performance liquid chromatography/mass spectrometry coupled with pattern recognition approach. *Journal of Pharmaceutical and Biomedical Analysis* 74: 22–30.

60 Li, X., Zhang, A., Sun, H. et al. (2017). Metabolic characterization and pathway analysis of berberine protects against prostate cancer. *Oncotarget* 8 (39): 65022.

61 Zhang, A., Sun, H., and Wang, X. (2012). Serum metabolomics as a novel diagnostic approach for disease: a systematic review. *Analytical and Bioanalytical Chemistry* 404 (4): 1239–1245.

62 Fang, H., Zhang, A., Zhou, X. et al. (2020). High-throughput metabolomics reveals the perturbed metabolic pathways and biomarkers of Yang Huang syndrome as potential targets for evaluating the therapeutic effects and mechanism of geniposide. *Frontiers in Medicine* 14 (5): 651–663.

63 Sun, H., Zhang, A.H., Song, Q. et al. (2018). Functional metabolomics discover pentose and glucuronate interconversion pathways as promising targets for Yang Huang syndrome treatment with Yinchenhao Tang. *RSC Advances* 8 (64): 36831–36839.

64 Wang, X., Sun, H., Zhang, A. et al. (2011). Ultra-performance liquid chromatography coupled to mass spectrometry as a sensitive and powerful technology for metabolomic studies. *Journal of Separation Science* 34 (24): 3451–3459.

65 Zhang, A., Sun, H., Dou, S. et al. (2013). Metabolomics study on the hepatoprotective effect of scoparone using ultra-performance liquid chromatography/electrospray ionization quadruple time-of-flight mass spectrometry. *Analyst* 138: 353–361.

66 Zhang, A., Sun, H., Yan, G. et al. (2013). Urinary metabolic profiling identifies a key role for glycocholic acid in human liver cancer by ultra-performance liquid-chromatography coupled with high-definition mass spectrometry. *Clinica Chimica Acta* 418: 86–90.

67 Yang, Q., Zhang, A.H., Miao, J.H. et al. (2019). Metabolomics biotechnology, applications, and future trends: a systematic review. *RSC Advances* 9 (64): 37245–37257.

68 Liu, R., Hong, J., Xu, X. et al. (2017). Gut microbiome and serum metabolome alterations in obesity and after weight-loss intervention. *Nature Medicine* 23 (7): 859–868.

69 Zhang, A., Sun, H., Han, Y. et al. (2012). Exploratory urinary metabolic biomarkers and pathways using UPLC-Q-TOF-HDMS coupled with pattern recognition approach. *Analyst* 137: 4200–4208.

70 Zhang, A.H., Ma, Z.M., Kong, L. et al. (2020). High-throughput lipidomics analysis to discover lipid biomarkers and profiles as potential targets for evaluating efficacy of Kai-Xin-San against APP/PS1 transgenic mice based on UPLC–Q/TOF–MS. *Biomedical Chromatography* 34 (2): e4724.

71 Zhang, P., Carlin, S., Lotti, C. et al. (2020). On sample preparation methods for fermented beverage VOCs profiling by GC × GC-TOFMS. *Metabolomics: Official journal of the Metabolomic Society* 16 (10): 102.

72 Yang, B., Tsui, T., Caprioli, R.M. et al. (2020). Sample preparation and analysis of single cells using high performance MALDI FTICR mass spectrometry. *Methods in Molecular Biology* 2064: 125–134.

73 Tsuchida, S., Umemura, H., and Nakayama, T. (2020). Current status of matrix-assisted laser desorption/ionization-time-of-flight mass spectrometry (MALDI-TOF MS) in clinical diagnostic microbiology. *Molecules* 25 (20): 4775.

3

Sample Preparation Method for Mass Spectrometry-Based Metabolomics

Shi Qiu, Aihua Zhang, Hui Sun, and Xijun Wang

Heilongjiang University of Chinese Medicine, National Chinmedomics Research Center, Functional Metabolomics Laboratory, Department of Pharmaceutical Analysis, Heping Road 24, Harbin 150040, China

3.1 Introduction

Metabolomics is the global assessment of endogenous metabolites in the biological system [1]. Metabolome is a collection of small molecules, mainly endogenous metabolites with less than 1000 molecules that participate in the biological metabolism [2–4]. The process of the metabolite changes was evaluated for all the metabolites in the living body, cells, tissue, and organs to understand or to reflect the changing pattern of the physiological or pathological progression of diseases [5–8]. Analysis of these key metabolites from the plasma, serum, urine, cerebrospinal fluid (CSF), saliva, synovial fluid, semen, tissue homogenates, etc. has played an important role to monitor the state of biological organisms [9–12]. The advantages of using a range of biological samples have ensured the widespread use of metabolites as diagnostic tools. Metabolite changes as indicators are an important part in the clinical practice [13]. Metabolomics has a great impact on discovering metabolite biomarkers, identifying the perturbed pathways, diagnosing diseases, or treatment [14–16]. The small-molecule metabolites including amino acids, lipids, peptides, nucleic acids, organic acids, vitamins, thiols, and carbohydrates represent attractive candidates to understand the disease phenotypes [17–20].

The hyphenated mass spectrometry (MS) is becoming a useful tool in the study of body fluids. It represents a promising micro-separation platform in metabolomics and has a strong potential to contribute to understanding the disease mechanisms [21]. This MS platform has been widely used to provide the sensitive and reproducible detection of thousands of metabolites in biofluid samples [22]. Thus, this chapter gives a brief description of the development and applications of sample preparation method for MS-based metabolomics analysis (Figure 3.1). More importantly, it focuses on the important roles of the endogenous metabolites, and the potential and limitations as well as some new trends in the development of sample preparation are also discussed.

Mass Spectrometry-Based Metabolomics in Clinical and Herbal Medicines: Strategies, Technologies, and Applications, First Edition. Edited by Aihua Zhang and Wanying Wang.

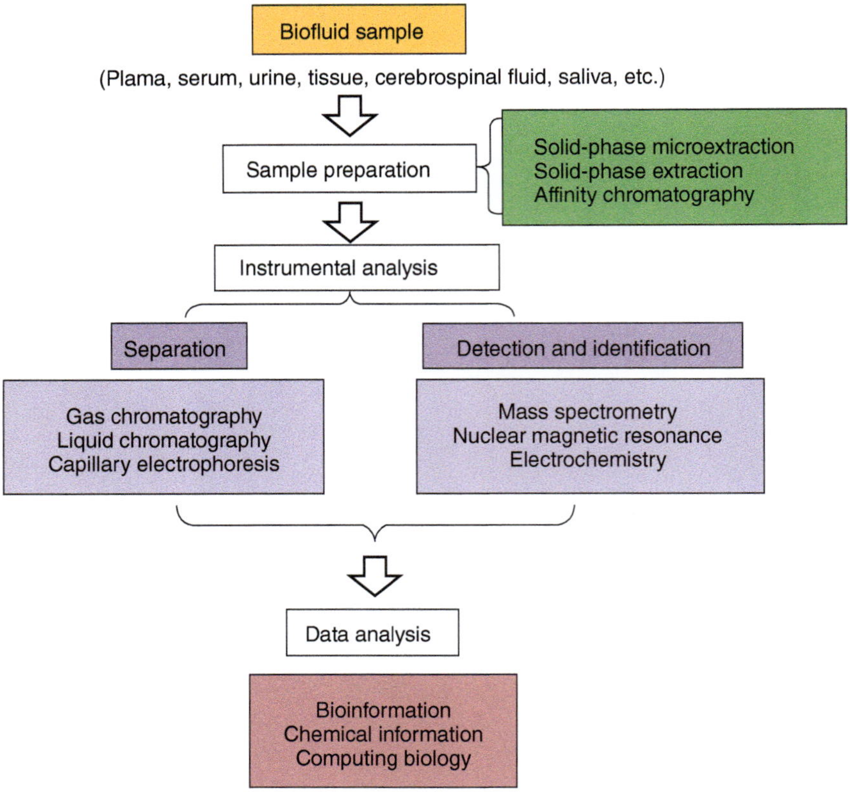

Figure 3.1 The entire process of sample analyses in metabolomics.

3.2 Sample Collection and Storage

3.2.1 Urine

Monitoring certain metabolite levels in urine sample, which is the commonly used biofluids in metabolomics, has become an important way to detect early stages of the disease. Due to its relatively less complex sample pretreatment, lower protein content, and sample complexity, urine as an analytical tool could obtain some advantages over other biofluids. Urine has simple and non-invasive collection techniques that make it a particularly suitable biofluids for metabolomics analysis to obtain the meaningful information. Urinary samples have vast chemical diversity in the concentration range of the metabolome and the integrity of the metabolites under some collection and storage conditions. They are usually collected into a lone container before loaded into the sampling tube marked with a time point. It is suggested to avoid more than two cycles of freezing and thawing of the stored sample for six months [23]. Laparre et al. had explored two short-term storage conditions that impact on the urine metabolome [24]. It was found that 78% of metabolites were stable at the storage conditions and, particularly, the amino acid was sensitive to change after prolonged storage. Shipping or storing samples on cool packs or at room temperature for more than eight hours should be avoided, and multiple numbers of freeze and thaw cycles should also be prevented [25]. MS-based urinary metabolomics has been capable of identifying the altered endogenous metabolites, for earlier screening of potential biomarkers of disease diagnosis.

3.2.2 Plasma and Serum

Plasma or serum contains a wide range of macro-molecules that can overlap with peaks from small-molecule metabolites, under specific physiological or pathologic states. Large-scale metabolites of plasma or serum are gaining more and more attention for their use in the diagnosis of disease. Pre-analytical factors impact on the metabolomics analysis of plasma and serum samples. López-Bascón et al. had found that serum and plasma in conventional tubes can affect the critical pathways. They concluded that at the serum metabolite level, there are no critical differences that can be observed between collected with polymeric gel and EDTA tubes [26]. Kamlage et al. had investigated metabolic differences in plasma and serum samples treated by different anticoagulants and found that the heparin plasma had the best detection [27]. Along with quality assurance, methods to assess sample quality regarding storage confounders are needed. Hebels et al. had investigated the effects on metabolic profiles in samples processed more than eight hours after collection [28]. Effects on metabolic profiles were noted in samples processed ≥8 hours after collection, but no effects were due to storage temperature.

3.2.3 Cerebrospinal Fluids

CSF analysis is employed in biomarker discovery studies for various neurodegenerative disorders. Potential sensitive CSF-derived biomarkers are available but invasive for sample collection, which limits their use in routine diagnostics. Metabolomics provides key information on the biomarkers to understand the neurological pathology. CSF samples can be accurately screened by metabolomics but requires reliable collection and storage methods. Gray et al. had collected CSF in polypropylene tubes by lumbar puncture, and then CSF samples were centrifuged, frozen, and stored at −80 °C [29]. Herman et al. also collected CSF in polypropylene tubes by lumbar puncture, and then after the samples were centrifuged and supernatant transferred, they were gently mixed, aliquoted, and stored at −80 °C [30]. Rosenling et al. recommended that the CSF samples should be centrifuged immediately after collection, and then freeze the supernatant in liquid nitrogen and avoid freeze and thaw cycles [31].

3.2.4 Tissue Homogenates

The measurement of metabolites in tissues is of great importance in the biomedical sciences, providing more relevant information than is available from the biofluids. The tissue separated according to anatomy methods is also rapidly immersed in 0.9%

NaCl to wash away residues and blood from the surface, snap-frozen, and then stored for further analysis [32, 33]. Broekaert and Fendt had collected the organs as fast as possible, washed the tissue in ice, removed the saline, and then put the tissue with the precooled bio-squeezer and store at 80 °C [34]. The brain, heart, liver, kidney, muscle, and tissue have been used to balance metabolome reproducibility [35]. Tissue samples from inguinal subcutaneous and visceral adipose tissue depot could be isolated, rinsed with 0.9% NaCl, frozen in liquid nitrogen, and stored at −80 °C for further analysis [36].

3.2.5 Saliva

Saliva is a readily accessible and informative biofluid, making it ideal for the early detection of a wide range of diseases including cardiovascular, renal, and autoimmune diseases, viral and bacterial infections, and, importantly, cancers. Saliva biomarkers are the perfect diagnostic medium due to its noninvasive collection, handling, and storage of samples. Figueira et al. collected pure parotid saliva using saliva-stimulating tablets and sterile cups, and then all the saliva samples were collected into sterile ice-chilled vials and were stored in aliquots at −80 °C until analysis [37]. Saliva samples were centrifuged at 3500 g for 10 minutes to remove cellular components and stored in a freezer at −80 °C [38].

3.3 Sample Preparation and Processing

The sample preparation is an important step in the analysis of the extraction analytes from biofluids to achieve the effect of enrichment and analysis of detection limits [39]. Liquid–liquid extraction (LLE), supercritical fluid extraction, solid-phase extraction (SPE), plasma protein precipitation [40], and other methods are commonly used in the sample preparation. LLE is one of the classical extraction methods, which is based on the different distribution coefficient of the test components and interference impurity components in the two incompatible solvents to achieve the sample purification and thereby to improve the selectivity of the analytical method. In the extraction process, the pH of the aqueous phase is a very important parameter and is related to the variety of the extracted solvent and the volume ratio of the solvent and the aqueous phase. SPE is a separation and purification method based on the separation mechanism of liquid chromatography, which is a more suitable technology for pretreatment of biological samples. Compared with LLE, SPE has less emulsification phenomenon, higher recovery rate, less dosage, and is faster [41].

Plasma, serum, biological tissue, and other biological samples contain a lot of protein that may form turbidity or precipitation and interference determination in the determination of metabolites. Commonly used methods for removing proteins are solvent treatment, enzymatic hydrolysis, ultrafiltration, and hot water treatment. The addition of a water-miscible organic solvent allows the protein to be agglomerated by transforming the intra-molecular and intermolecular hydrogen bonds to release the protein-bound components. Commonly used organic solvents are acetonitrile, methanol, ethanol, propanol, acetone, and tetrahydrofuran. Methanol and acetonitrile are commonly utilized as protein precipitants as the mobile phase in the reverse-phase HPLC, and the acetonitrile precipitation efficiency is higher than methanol. In addition, in the determination of some acid instability and protein binding solid ingredients, available enzymatic hydrolysis method is used to avoid some of the test components in the acid or high temperature degradation [42].

Before analysis, the urine samples were always thawed and vortexed at room temperature. The urine samples are often diluted with water or other mixture [43] such as a buffer solution with NaH_2PO_4–Na_2HPO_4 and TSP D_2O [44]. The mixed solution vortexes for a while, and then all the samples should be centrifuged, and an aliquot of urinary supernatant should be transferred to the text tube. Plasma samples were thawed at 4 °C or on ice before protein precipitation [45]. They were prepared by protein precipitation with acetonitrile, cold methanol, or other mixture [46]. After protein precipitation procedure, the sample can be analyzed or transferred to SPE cartridge. The SPE column was first eluted by the addition of formic acid in acetonitrile or others, and the eluates were collected in a vial and evaporated to completely dry. Methanol was also added to serum sample to precipitate the proteins [47]. Lindahl et al. adapted 150 ml MeOH to 50 ml serum per sample and centrifuged it for 15 minutes. The supernatant was then transferred, dried down in a speed vacuum concentrator, and re-dissolved in 50 ml 1 : 1 MeOH:H_2O [48]. After thoroughly mixing and centrifugation, the clear supernatant was transferred into autosampler vials [49]. For analysis of B6 vitamers in CSF, IS working solution was added to CSF in an Eppendorf tube by UPLC–MS analysis [50]. Tissues were immersed in an ice-cold mixture of water, methanol, or other. After centrifugation, the supernatant was collected and processed in a similar manner as described in the serum preparation section. To precipitate cells and remove any solid debris, saliva samples were centrifuged after thawing [51]. The sample was mixed with methanol and vortexes before centrifugation [52]. Proteins were precipitated by adding ice-cold acetonitrile and centrifugation [53].

3.4 Application

Metabolomics brings new insights into disease pathophysiology, drug mechanism, and treatment at the small-molecular level [11, 54, 55]. Biofluids are primarily researched to identify biomarkers for disease diagnosis and prediction, precision therapy, drug research, and discovery [56–59]. Urine is widely used for easy collection and is rich in metabolites, and especially it can reflect imbalances in all biochemical pathways in the body [60]. Metabolic profile of urine samples will be used to the clinical diagnosis of depression and investigate the mechanisms of depression onset and its regulation. Wang et al. has determined the correlations of metals with amino acid metabolic intermediates in the urine of healthy pregnant women during normal pregnancy [61]. Biomarkers' routine application can assess the detection of windows stability of gamma-hydroxybutyrate as a medication against narcolepsy [62]. A urine fingerprint is related to pathophysiological changes in energy generation and oxidative stress and is associated with acute kidney injury [63].

Serum or plasma metabolic-based biomarkers can also be used for early diagnosis. The identified biomarkers could advance diagnosis and triage of pediatric sepsis [64]. Branched chain amino acids are monitored as potential indicators of diabetes with plasma metabolomics approach [65]. Dong et al. had investigated that NOTCH2 may be a blood biomarker for the clinical diagnosis of depression. It revealed that the differential metabolites in plasma are closely associated with blood stasis syndrome [66]. The metabolic research of tissue sample can also contribute to understanding disease mechanisms [67]. CSF could often be used as a source of prognostic biomarkers and showed an association to mitochondrial dysfunction in metabolic analysis in Parkinson's disease. Saliva is a promising biofluid for both unbiased and targeted biomarker detection. Romano et al. had discriminated chronic periodontitis patients from healthy subjects by identifying specific molecular signatures by saliva metabolomic analysis [68].

3.5 Future Development

Metabolomics is the link between genotypes and phenotypes and is considered as an emerging but powerful tool for precision medicine. It could provide potential advantages for discovering the clinical biomarkers from the diseases. Based on metabolomics, the physiological and pathological conditions of organism are obtained by analyzing the overall changes for small-molecular metabolites within a biologic system after external stimuli. It can provide a global quantitative analysis on a wide range of metabolites in biological samples, especially biofluid samples, for example, urine, plasma, serum, CSF, saliva, synovial fluid, tissue homogenates, etc. Analysis of these endogenous small-molecule metabolites in body fluids can be used to monitor the states for diseases. The novel biomarkers that were detected earlier than the traditional clinical methods could improve the development of disease prevention, prediction diagnosis, and personalized treatment, thus benefiting for public health.

Acknowledgments

This work was supported by grants from the Key Program of Natural Science Foundation of State (Grant No. 81973745, 81302905), Natural Science Foundation of Heilongjiang Province (YQ2019H030), the University Nursing Program for Young Scholars with Creative Talents in Heilongjiang Province (UNPYSCT-2015118), Longjiang Scholar Program of Education Department of Heilongjiang Province (Q201916), and Heilongjiang Touyan Innovation Team Program.

References

1 Nicholson, J.K. and Wilson, I.D. (2003). Opinion: understanding 'global' systems biology: metabonomics and the continuum of metabolism. *Nature Reviews. Drug Discovery* 2 (8): 668–676.

2 Qiu, S., Zhang, A.H., Guan, Y. et al. (2020). Functional metabolomics using UPLC-Q/TOF-MS combined with ingenuity pathway analysis as a promising strategy for evaluating the efficacy and discovering amino acid metabolism as a potential therapeutic mechanism-related target for geniposide against alcoholic liver disease. *RSC Advances* 10 (5): 2677–2690.

3 Zhang, A., Sun, H., Yan, G. et al. (2016). Mass spectrometry-based metabolomics: applications to biomarker and metabolic pathway research. *Biomedical Chromatography* 30 (1): 7–12.

4 Yang, Q., Zhang, A.H., Miao, J.H. et al. (2019). Metabolomics biotechnology, applications, and future trends: a systematic review. *RSC Advances* 9 (64): 37245–37257.

5 Zhang, A., Sun, H., Wang, P. et al. (2012). Recent and potential developments of biofluid analyses in metabolomics. *Journal of Proteomics* 75 (4): 1079–1088.

6 Zhang, A., Sun, H., and Wang, X. (2017). Emerging role and recent applications of metabolomics biomarkers in obesity disease research. *RSC Advances* 7 (25): 14966–14973.

7 Zhang, A., Zhou, X., Zhao, H. et al. (2017). Metabolomics and proteomics technologies to explore the herbal preparation affecting metabolic disorders using high resolution mass spectrometry. *Molecular BioSystems* 13 (2): 320–329.

8 Wu, X.H., Sun, X.L., Zhao, C. et al. (2019). Exploring the pharmacological effects and potential targets of paeoniflorin on the endometriosis of cold coagulation and blood stasis model rats by ultra-performance liquid chromatography tandem mass spectrometry with a pattern recognition approach. *RSC Advances* 9 (36): 20796–20805.

9 Liang, Q., Liu, H., Xie, L.X. et al. (2017). High-throughput metabolomics enables biomarker discovery in prostate cancer. *RSC Advances* 7 (5): 2587–2593.

10 Liang, Q., Liu, H., Xing, H. et al. (2016). High-resolution mass spectrometry for exploring metabolic signatures of sepsis-induced acute kidney injury. *RSC Advances* 6 (36): 29863–29868.

11 Zhang, A., Liu, Q., Zhao, H. et al. (2016). Phenotypic characterization of nanshi oral liquid alters metabolic signatures during disease prevention. *Scientific Reports* 6: 19333.

12 Zhang, A.H., Ma, Z.M., Sun, H. et al. (2019). High-throughput metabolomics evaluate the efficacy of total lignans from *Acanthophanax senticosus* stem against ovariectomized osteoporosis rat. *Frontiers in Pharmacology* 10: 553.

13 Liang, Q., Liu, H., Zhang, T. et al. (2016). Discovery of serum metabolites for diagnosis of mild cognitive impairment to Alzheimer's disease progression using an optimized metabolomics method. *RSC Advances* 6: 3586–3591.

14 Zhang, A., Fang, H., Wang, Y. et al. (2017). Discovery and verification of the potential targets from bioactive molecules by network pharmacology-based target prediction combined with high-throughput metabolomics. *RSC Advances* 7 (81): 51069–51078.

15 Liang, Q., Zhu, Y., Liu, H. et al. (2017). High-throughput lipidomics enables discovery of the mode of action of huaxian capsule impacting the metabolism of sepsis. *RSC Advances* 7 (71): 44990–44996.

16 Zhang, A., Wang, H., Sun, H. et al. (2015). Metabolomics strategy reveals therapeutical assessment of limonin on nonbacterial prostatitis. *Food & Function* 6 (11): 3540–3549.

17 Fang, H., Zhang, A.H., Sun, H. et al. (2019). High-throughput metabolomics screen coupled with multivariate statistical analysis identifies therapeutic targets in alcoholic liver disease rats using liquid chromatography-mass spectrometry. *Journal of Chromatography B* 1109: 112–120.

18 Zhang, Y., Liu, P., Li, Y. et al. (2017). Exploration of metabolite signatures using high-throughput mass spectrometry coupled with multivariate data analysis. *RSC Advances* 7 (11): 6780–6787.

19 Qiu, S., Zhang, A., Zhang, T. et al. (2017). Dissect new mechanistic insights for geniposide efficacy on the hepatoprotection using multiomics approach. *Oncotarget* 8 (65): 108760.

20 Liang, Q., Liu, H., Xing, H. et al. (2016). UPLC-QTOF/MS based metabolomics reveals metabolic alterations associated with severe sepsis. *RSC Advances* 6 (49): 43293–43298.

21 Zhang, A.-H., Qiu, S., Xu, H.-Y. et al. (2014). Metabolomics in diabetes. *Clinica Chimica Acta* 429: 106–110.

22 Liang, Q., Liu, H., Jiang, Y. et al. (2016). Novel liquid chromatography-mass spectrometry for metabolite biomarkers of acute lung injury disease. *Analytical Methods* 8 (31): 6017–6022.

23 ŽivkovićSemren, T., BrčićKaračonji, I., Safner, T. et al. (2018). Gas chromatographic-mass spectrometric analysis of urinary volatile organic metabolites: optimization of the HS-SPME procedure and sample storage conditions. *Talanta* 176: 537–543.

24 Laparre, J., Kaabia, Z., Mooney, M. et al. (2017). Impact of storage conditions on the urinary metabolomics fingerprint. *Analytica Chimica Acta* 951: 99–107.

25 Rotter, M., Brandmaier, S., Prehn, C. et al. (2017). Stability of targeted metabolite profiles of urine samples under different storage conditions. *Metabolomics* 13 (1): 4.

26 Zhou, Z., Chen, Y., He, J. et al. (2017). Systematic evaluation of serum and plasma collection on the endogenous metabolome. *Bioanalysis* 9 (3): 239–250.

27 Kamlage, B., Neuber, S., Bethan, B. et al. (2018). Impact of prolonged blood incubation and extended serum storage at room temperature on the human serum metabolome. *Metabolites* 8 (1): 6.

28 Hebels, D.G., Georgiadis, P., Keun, H.C. et al. (2013). Performance in omics analyses of blood samples in long-term storage: opportunities for the exploitation of existing biobanks in environmental health research. *Environmental Health Perspectives* 121 (4): 480–487.

29 Gray, E., Larkin, J.R., Claridge, T.D. et al. (2015). The longitudinal cerebrospinal fluid metabolomic profile of amyotrophic lateral sclerosis. *Amyotrophic Lateral Sclerosis & Frontotemporal Degeneration* 16 (7–8): 456–463.

30 Herman, S., Khoonsari, P.E., Tolf, A. et al. (2018). Integration of magnetic resonance imaging and protein and metabolite CSF measurements to enable early diagnosis of secondary progressive multiple sclerosis. *Theranostics* 8 (16): 4477–4490.

31 Rosenling, T., Slim, C.L., Christin, C. et al. (2009). The effect of preanalytical factors on stability of the proteome and selected metabolites in cerebrospinal fluid (CSF). *Journal of Proteome Research* 8 (12): 5511–5522.

32 Zheng, H., Ni, Z., Cai, A. et al. (2018). Balancing metabolome coverage and reproducibility for untargeted NMR-based metabolic profiling in tissue samples through mixture design methods. *Analytical and Bioanalytical Chemistry* 410 (29): 7783–7792.

33 Müller, O.J., Heckmann, M.B., Ding, L. et al. (2019). Comprehensive plasma and tissue profiling reveals systemic metabolic alterations in cardiac hypertrophy and failure. *Cardiovascular Research* 115 (8): 1296–1305.

34 Broekaert, D. and Fendt, S.M. (2019). Measuring in vivo tissue metabolism using ^{13}C glucose infusions in mice. *Methods in Molecular Biology* 1862: 67–82.

35 Choi, J., Shoaib, M., Yin, T. et al. (2019). Tissue-specific metabolic profiles after prolonged cardiac arrest reveal brain metabolome dysfunction predominantly after resuscitation. *Journal of the American Heart Association* 8 (17): e012809.

36 Airaksinen, K., Jokkala, J., Ahonen, I. et al. (2018). High-fat diet, betaine, and polydextrose induce changes in adipose tissue inflammation and metabolism in C57BL/6J mice. *Molecular Nutrition & Food Research* 62 (23): e1800455.

37 Figueira, J., Gouveia-Figueira, S., Öhman, C. et al. (2017). Metabolite quantification by NMR and LC-MS/MS reveals differences between unstimulated, stimulated, and pure parotid saliva. *Journal of Pharmaceutical and Biomedical Analysis* 140: 295–300.

38 D'alessandro, A., Nemkov, T., Moore, H.B. et al. (2016). Metabolomics of trauma-associated death: shared and fluid-specific features of human plasma vs lymph. *Blood Transfusion* 14 (2): 185–194.

39 Paczia, N., Nilgen, A., Lehmann, T. et al. (2012). Extensive exometabolome analysis reveals extended overflow metabolism in various microorganisms. *Microbial Cell Factories* 11: 122.

40 Calderón-Santiago, M., Priego-Capote, F., and de Castro, M. (2015). Enhancing detection coverage in untargeted metabolomics analysis by solid-phase extraction on-line coupled to LC-MS/MS. *Electrophoresis* 36 (18): 2179–2187.

41 Barri, T., Holmer-Jensen, J., Hermansen, K., and Dragsted, L.O. (2012). Metabolic fingerprinting of high-fat plasma samples processed by centrifugation- and filtration-based protein precipitation delineates significant differences in metabolite information coverage. *Analytica Chimica Acta* 718: 47–57.

42 Ge, D. and Lee, H.K. (2012). Sonication-assisted emulsification microextraction combined with vortex-assisted porous membrane-protected micro-solid-phase extraction using mixed zeoliticimidazolate frameworks 8 as sorbent. *Journal of Chromatography A* 1263: 1–6.

43 Deng, K., Han, P., Song, W. et al. (2018). Plasma metabolomic profiling distinguishes right-sided from left-sided colon cancer. *Clinica Chimica Acta* 487: 357–362.

44 Xiong, Z., Lang, L., Gao, X. et al. (2019). An integrative urinary metabolomic study of the therapeutic effect of Guizhi Fuling capsule on primary dysmenorrheal rats based ^{1}H NMR and UPLC-MS. *Journal of Pharmaceutical and Biomedical Analysis* 164: 750–758.

45 Liu, J.H., Sun, H., Zhang, A.H. et al. (2014). Serum pharmacochemistry combined with multiple data processing approach to screen the bioactive components and their metabolites in Mutan Cortex by ultra-performance liquid chromatography tandem mass spectrometry. *Biomedical Chromatography* 28 (4): 500–510.

46 Kitagawa, H., Ohbuchi, K., Munekage, M. et al. (2019). Phenotyping analysis of the Japanese Kampo medicine maoto in healthy human subjects using wide-targeted plasma metabolomics. *Journal of Pharmaceutical and Biomedical Analysis* 164: 119–127.

47 Seo, S.H., Kim, E.J., Park, S.E. et al. (2018). GC/MS-based metabolomics reveals biomarkers in asthma murine model modulated by *Opuntia humifusa*. *Evidence-Based Complementary and Alternative Medicine* 2018: 1202860.

48 Lindahl, A., Forshed, J., and Nordström, A. (2016). Overlap in serum metabolic profiles between non-related diseases: implications for LC-MS metabolomics biomarker discovery. *Biochemical and Biophysical Research Communications* 478 (3): 1472–1477.

49 Reinicke, M., Schröter, J., Müller-Klieser, D. et al. (2018). Free oxysterols and bile acids including conjugates – simultaneous quantification in human plasma and cerebrospinal fluid by liquid chromatography-tandem mass spectrometry. *Analytica Chimica Acta* 1037: 245–255.

50 van der Ham, M., Albersen, M., de Koning, T.J. et al. (2012). Quantification of vitamin B6 vitamers in human cerebrospinal fluid by ultra performance liquid chromatography-tandem mass spectrometry. *Analytica Chimica Acta* 712: 108–114.

51 Kawanishi, N., Hoshi, N., Masahiro, S. et al. (2019). Effects of inter-day and intra-day variation on salivary metabolomic profiles. *Clinica Chimica Acta* 489: 41–48.

52 Li, Y., Wang, D., Zeng, C. et al. (2018). Salivary metabolomics profile of patients with recurrent aphthous ulcer as revealed by liquid chromatography-tandem mass spectrometry. *Journal of International Medical Research* 46 (3): 1052–1062.

53 Kaplan, K., Dwivedi, P., Davidson, S. et al. (2009). Monitoring dynamic changes in lymph metabolome of fasting and fed rats by electrospray ionization-ion mobility mass spectrometry (ESI-IMMS). *Analytical Chemistry* 81 (19): 7944–7953.

54 Zhang, A.H., Sun, H., Yan, G.L. et al. (2019). Chinmedomics: a powerful approach integrating metabolomics with serum pharmacochemistry to evaluate the efficacy of traditional Chinese medicine. *Engineering* 5 (1): 60–68.

55 Nan, Y., Zhou, X., Liu, Q. et al. (2016, 2016). Serum metabolomics strategy for understanding pharmacological effects of ShenQi pill acting on kidney yang deficiency syndrome. *Journal of Chromatography B: Analytical Technologies in the Biomedical and Life Sciences* 1026: 217–226.

56 Zhang, A., Sun, H., and Wang, X. (2014). Urinary metabolic profiling of rat models revealed protective function of scoparone against alcohol induced hepatotoxicity. *Scientific Reports* 4: 6768.

57 Wang, X., Wang, Q., Zhang, A. et al. (2013). Metabolomics study of intervention effects of Wen-Xin-formula using ultra high-performance liquid chromatography/mass spectrometry coupled with pattern recognition approach. *Journal of Pharmaceutical and Biomedical Analysis* 74: 22–30.
58 Zhang, A.H., Sun, H., Yan, G.L. et al. (2014). Metabolomics study of type 2 diabetes using ultra-performance LC-ESI/quadrupole-TOF high-definition MS coupled with pattern recognition methods. *Journal of Physiology and Biochemistry* 70 (1): 117–128.
59 Zhang, A.H., Sun, H., Han, Y. et al. (2013). Ultraperformance liquid chromatography-mass spectrometry based comprehensive metabolomics combined with pattern recognition and network analysis methods for characterization of metabolites and metabolic pathways from biological data sets. *Analytical Chemistry* 85 (15): 7606–7612.
60 Zhao, M., Lau, K.K., Zhou, X. et al. (2017). Urinary metabolic signatures and early triage of acute radiation exposure in rat model. *Molecular BioSystems* 13 (4): 756–766.
61 Wang, M., Xia, W., Li, H. et al. (2018). Normal pregnancy induced glucose metabolic stress in a longitudinal cohort of healthy women: novel insights generated from a urine metabolomics study. *Medicine* 97 (40): e12417.
62 Wang, M., Xia, W., Liu, H. et al. (2018). Urinary metabolomics reveals novel interactions between metal exposure and amino acid metabolic stress during pregnancy. *Toxicology Research* 7 (6): 1164–1172.
63 Chetwynd, A.J., Dunn, W.B., and Rodriguez-Blanco, G. (2017). Collection and preparation of clinical samples for metabolomics. *Advances in Experimental Medicine and Biology* 965: 19–44.
64 Mickiewicz, B., Thompson, G.C., Blackwood, J. et al. (2018). Biomarker phenotype for early diagnosis and triage of sepsis to the pediatric intensive care unit. *Scientific Reports* 8 (1): 16606.
65 Le, T.T., Shafaei, A., Genoni, A. et al. (2019). Development and validation of a simple LC-MS/MS method for the simultaneous quantitative determination of trimethylamine-*N*-oxide and branched chain amino acids in human serum. *Analytical and Bioanalytical Chemistry* 411 (5): 1019–1028.
66 Luo, H., Gu, C., Liu, C. et al. (2018). Plasma metabolic profiling analysis of *Strychnos nux-vomica* Linn. and *Tripterygium wilfordii* Hook F-induced renal toxicity using metabolomics coupled with UPLC/Q-TOF-MS. *Toxicology Research* 7 (6): 1153–1163.
67 Warren, J.S., Oka, S.I., Zablocki, D. et al. (2017). Metabolic reprogramming via PPARα signaling in cardiac hypertrophy and failure: from metabolomics to epigenetics. *American Journal of Physiology – Heart and Circulatory Physiology* 313 (3): H584–H596.
68 Romano, F., Meoni, G., Manavella, V. et al. (2019). Effect of non-surgical periodontal therapy on salivary metabolic fingerprint of generalized chronic periodontitis using nuclear magnetic resonance spectroscopy. *Archives of Oral Biology* 97: 208–214.

4

Multivariate Data Analysis Approach for Mass Spectrometry-Based Metabolomics

Junling Ren, Aihua Zhang, Ling Kong, and Xijun Wang

Heilongjiang University of Chinese Medicine, National Chinmedomics Research Center, Functional Metabolomics Laboratory, Department of Pharmaceutical Analysis, Heping Road 24, Harbin 150040, China

4.1 Introduction

Metabolomics is a booming omics field following genomics, transcriptomics, and proteomics, which is a vital section of systems biology. Metabolites are the end products of complex cellular regulation networks that can influence or even alter regulation via feedback loops [1]. Through overall analysis of metabolites, the insightful knowledge of physiological, pathological, and biochemical status can be directly understood, and combined with chemical and informatics methods, the endogenous small-molecule metabolic changes were also determined, and the corresponding biomarkers were obtained, which characterize or reveal the functional states at a particular time or environment [2]. Thus, metabolomics might provide the "fingerprint" of metabolic systems, which best reveals the clinical phenotype [3]. The studies of metabolomics were conducive to enhance the understanding of disease mechanisms and drug effects, as well as to improve the ability to predict personal disease progression or variation in drug response phenotypes [4].

The physical and chemical properties of metabolites are diverse and often distribute over a wide range of concentrations [5]. Consequently, how to comprehensively detect metabolites has brought great challenges to the field of analytical chemistry. The common analytical techniques used in metabolomics are nuclear magnetic resonance spectroscopy and mass spectrometry (MS) [6]. Due to the obvious superiorities of speed, sensitivity, and wide dynamic range, applications of MS are rapidly broadening, and molecular and cellular biology were also encompassed. MS-based metabolomics technologies aid the analysis of *in vivo* metabolite changes, understanding the possible mechanism of diseases and identifying potential biomarkers that may improve diagnosis and treatment of diseases [7].

With the continuous breakthroughs of analytical technology, high-throughput data including variables and responses will be obtained after sample detection, and these data are at least two-dimensional. How to process these multi-variable and multidimensional data is one of the most critical problem faced by metabolomics.

Mass Spectrometry-Based Metabolomics in Clinical and Herbal Medicines: Strategies, Technologies, and Applications, First Edition. Edited by Aihua Zhang and Wanying Wang.

Therefore, this chapter not only introduced the current commonly used MS technology in metabolomics but also briefly described the commonly used multivariate statistical methods in metabolomics.

4.2 Separation-Based MS Techniques

4.2.1 Gas Chromatography–Mass Spectrometry

Gas chromatography–mass spectrometry (GC–MS) is the preferred instrument for volatile metabolites analysis, which expresses high sensitivity and resolution, prominent reproducibility, and highly repeatable fragmentation, as well as the availability of the spectral library, making the identification of chemical structures more convenient and helpful for the elucidation of alteration mechanisms of biology or pathology [8]. Although it has the advantages of low cost and easy operation, the tedious sample processing and the derivatization usually needed when the nonvolatile constituents were analyzed. To increase the volatility and thermal stability of analytes, various derivatizations, such as alkylation and acylation, can be employed to "protect" functional groups [9]. Among these derivatization methods, methoximation and trimethylsilylation are commonly used in the large-scale metabolomics studies with GC–MS [10]. On account of the difference in derivatization efficiency of the metabolites, the reproducibility of the overall analysis might be affected [11]. The formation of by-products and structural transformation during the process of derivatization should also be noted, which might lead to mis-interpretation of the generated data.

The peak capacity and sensitivity were greatly increased with the conducted comprehensive GC, which aids to the detection of metabolites [12, 13]. In GC × GC, the compounds eluted from the primary column were imported into the secondary column to perform a second independent separation [14]. Most of the reported applications of GC × GC were executed using the nonpolar stationary phase in the first dimension of separation and the polar stationary phase in the second dimension. With this choice usually made due to the elution behavior of the compound class that is predictable, compounds were subjected to the first dimension separation according to the different boiling points. As the similar boiling point failed to be isolated in the first dimension separation, compounds were separated by polarity in the second dimension separation; thus, separation power was significantly enhanced [15]. Miyazaki et al. examined the temporal changes of serum metabolites in neonatal calves after first ingestion of colostrum [16]. Approximately 1400 serum metabolites were detected, which number was significantly larger than that of conventional GC–MS that detected only about 180 metabolites in the same serum sample, and expand the cognition of serum metabolites.

4.2.2 Liquid Chromatography–Mass Spectrometry

Liquid chromatography–mass spectrometry (LC–MS) has become the dominating analytical technique in the field of global metabolites profiling. Compared with

GC–MS, the mobile phase is changed from gas to liquid, eliminating the need for metabolites to be volatile. Thus, no sample derivatization is required, which means that a broader range of analytes can be detected. In 2004, the introduction of commercialized ultra-high performance liquid chromatography (UPLC) with sub-2 μm dramatically increases the throughput of regular LC method [17]. UPLC usually employs sub-2 μm or 2.6–2.8 μm stationary phase particles at the chromatography column and can withstand high solvent flow rate and high pressure at the range of 6000–19 000 psi, which enables to reduce peak width, shorten analytical run times, increase peak capacity, better ionization, and reduce mass spectral overlap, leading to better structure determination and confirmation [18]. It is reported that the separation of human serum metabolites with UPLC–MS brings about 20% more detected metabolites compared with that of high-performance liquid chromatography (HPLC)–MS [19].

4.2.2.1 Reversed-Phase Liquid Chromatography–Mass Spectrometry

At present, the most LC–MS-based metabolomics studies were performed using reversed-phase liquid chromatography (RPLC) column with C18 or C8 materials [20]. RPLC provides an extensive range of selectivity for the analysis of diverse metabolites and is highly compatible with the analysis of organism [21]. For RPLC separation, gradient elution is often employed, which starts with a high aqueous content and up to a high organic content [22]. For the sake of improving the separation efficiency, various buffer modifiers (e.g. acetic acid, formic acid, ammonium acetate, etc.) can be added into the mobile phase to adjust the separation selectivity or detection sensitivity [23]. However, highly polar and ionic compounds are difficult to retain in RPLC, and the analysis of polar compounds were also practicable using RPLC by adding ion-pairing agents into mobile phase by means of ion-pairing liquid chromatography (IPLC) [24]. The ion-pair reagent is a compound that has an opposite charge compared with the target metabolites in aqueous solution [25]. As for cation analysis, ion-pair reagent usually includes perfluorocarboxylic acids, hydrochloric acid, perchloric acid, pentane sulfonic acids, heptane sulfonic acids, hexane sulfonic acids, octane sulfonic acids, etc., while for anion analysis, it usually includes tetramethylammonium, tetraethylammonium, tetrapropylammonium, tetrabutylammonium, tributylamine, hexylamine, etc. [26]. Guo et al. exploited an ion-pairing reversed-phase ultra-high-performance liquid chromatography- mass spectrometry (IP-RP-UHPLC–MS) method to separate and analyze biochemical-related metabolites, which contain phosphate and carboxylic acid in cellular metabolism [27]. Though the usage of ion-pair reagent is helpful, the remaining unpaired reagent may contaminate the ion source, which may influence the sensitivity or reproducibility of LC–MS [25, 28].

4.2.2.2 Hydrophilic Interaction Chromatography–Mass Spectrometry

Hydrophilic interaction chromatography (HILIC) utilizes a polar stationary phase and a high proportion of organic mobile phase, which is usually added with at least 3% of water and followed by increasing the percentage of water, as the eluent [5, 29]. Contrary to RPLC, the more lipophilic compounds are barely attracted to the stationary phase and eluted rapidly in HILIC, whereas the more polar compounds are retained and only eluted when the aqueous phase is increased. HILIC–MS is a complementation to RPLC–MS since the early eluting of analytes in RPLC–MS are usually well retained in HILIC–MS. However, HILIC usually generates a wider peak than RPLC, leading to lower peak capacity and higher dependence on the peak resolution of the mass spectrometer [25]. Overall, HILIC–MS analysis provides a broader view for metabolites analysis. Cífková et al. systematically optimized the separation of acidic lipid classes and other lipid classes by HILIC–MS [30]. It is rather a tough task to separate acidic lipid classes due to broadening or tailing chromatographic peaks by them. The established method obtained a higher number of lipids, providing a basis for the future detail research of lipids.

4.2.2.3 Supercritical Fluid Chromatography–Mass Spectrometry

A substance with temperature and pressure that are above its critical point is termed as super critical fluid [31]. Supercritical fluid chromatography (SFC) uses supercritical fluid, usually CO_2, as the primary mobile phase, which possesses a modest supercritical point. Moreover, the polarity of supercritical CO_2 is similar to that of hexane that can be adjusted by polar solvents. Since it is nontoxic and easy to handle, CO_2 is often used as the mobile phase [32, 33]. The properties of low viscosity and high diffusivity of the supercritical fluids permits higher flow rates and lower pressure drops than that in traditional HPLC [34]. At present, SFC is recognized as a suitable means for both nonpolar and polar compound separation [34, 35].

Lísa et al. developed a novel analytical strategy for high-throughput and comprehensive analysis of lipids by SFC–MS, which enables a quick separation of 30 classes of nonpolar and polar lipid within six minutes [36]. Taguchi et al. established a rapid method to targeted profile bile acids in rat serum with SFC–MS [37]. Twenty-four bile acids were simultaneously quantified without any solid-phase extraction and complex sample preparation within 13 minutes. SFC can be considered as a mature separation technology, while the knowledge of separation process is far behind than that of GC and LC [31]. The preceding applications of SFC–MS had already shown its potential for metabolomics, which can be an alternative to the existing analytical methods for rapid metabolomics research.

4.2.3 Capillary Electrophoresis–Mass Spectrometry

Electrophoresis is defined as the migration of ions under the influence of a spatially uniform electric field [38]. Capillary electrophoresis (CE) applies high voltage to produce electrophoretic flow of different ionic species in a narrow-bore capillary [39]. CE–MS is a robust technique for polar and charged metabolites analysis, which is a complementary tool to chromatographic separation technique [40]. In capillary zone electrophoresis, when combined with MS, metabolites were separated first on the basis of charge-to-size ratio and then on account of mass-to-charge ratio, which provides an alternative selectivity to LC [41]. CE is considered to be a new arrival technique among metabolomics and has not yet routinely applied in metabolites profiling due to some drawbacks, such as weakness of system stability where the narrow capillaries are easily blocked by salt. While it is a valuable analytical tool for those who have mastered these skills and knowledge [42], it has some distinct advantages over other instrumental techniques employed for metabolomics: low separation volume, which is especially suitable for small experimental animals' biological fluids study, homogeneous separation environment, easy sample preparation process, and capable of concentrating analytes [43]. CE–MS displayed a better separation efficiency than GC–MS and LC–MS with respect to shorter analysis time. Moreover, it was reported that CE–MS was capable of representing a reproducible outcomes of peptide mapping among 13 different laboratories using the same sample, methods, and capillaries but different set of instruments [44].

Significant sensitivity enhancement of modern MS and optimized method transformed CE–MS into a potential tool for metabolomics analysis, and the application of CE–MS in metabolomics has dramatically increased in the past decade [45]. Onjiko et al. developed "microprobe single-cell CE–ESI–MS" for *in situ* analysis of metabolites in single cells in the freely developing vertebrate embryo [46]. Only <0.02% of the single-cell content can allow the detection of ~230 different molecules in positive ion mode, which expands the biological toolbox of system cell biology and opens a new direction to help understand the normal and impaired development of cell biology [46].

4.2.4 Ion Mobility–Mass Spectrometry

Ion mobility–mass spectrometry (IM–MS), a gas-phase electrophoretic technology, enables differentiation of ions in gas phase and electric field on the basis of their size, shape, charge, and mass, which can provide important supplementary information for molecular chromatographic separation and ion mass spectrometry separation [47]. The advantages of IM–MS include distinguishing isobars and isomers, increasing signal-to-noise ratio, and determining the conformational state caused by the charge location [48].

At present, more and more metabolomics studies by IM–MS have been published. Zhang et al. used electrospray ionization-ion mobility spectrometry (ESIIM)–MS to study the striatal metabolomics in Parkinson's disease using rat model, and nine metabolites were selected as the potential biomarkers [49]. A major discovery of this research was the finding of a dopamine isomer, which has not been reported previously and was separated using IM–MS. It implicated that the range ability of dopamine in Parkinson's disease may be masked by the presence of its isomer. With the maturity of metabolomics, it is found that the function of isomers is not exactly the same, making isomer recognition become particularly important. Bowman et al. explored field asymmetric ion mobility spectrometry (FAIMS)–MS to separate lipid isomers, and success rate of approximately 75% was found in the four main types of glycerolipids and phospholipids [50].

4.3 Separation-Free MS Techniques

4.3.1 Direct Infusion–Mass Spectrometry

Direct infusion–mass spectrometry (DI–MS) is an analysis platform based on direct injection or infusion of samples into MS, especially the ultra-high resolution and accuracy MS, without prior chromatographic or electrophoretic separation, which shows obvious advantage in high-throughput screening [51]. This feature greatly shortened total analysis time, thus minimized the drift of the sequence analysis and improved the repeatability and accuracy among inter-samples [52].

Although DI–MS can achieve rapid sample analysis, ion source contamination of sample residue is a problem, and the emergence of chip-based nano-ESI (electrospray ionization) solves this problem excellently [53]. The nozzle of chip-based nano-ESI is engraved on the silicon board, and analyzed liquid is passed through the nozzle and sprayed under a relatively gentle pressure [54]. Since the requirement of sample volume is very small and each nozzle can be used only once, thus the sample residue effect was eliminated. It has been reported that this technique can reduce or even eliminate the differences in MS response among analytes and adduct ions [55]. Zhang et al. employed the chip-based direct-infusion nanoelectrospray Fourier transform ion cyclotron resonance MS to simultaneously quantify six free fatty acids, several types of cancer patients, and corresponding benign diseases, and the results suggest this platform could provide high sensitivity and high-throughput analysis of free fatty acids, and serum unsaturated free fatty acids might have important clinical significance for early detection of cancers [56–58].

4.3.2 Matrix-Assisted Laser Desorption Ionization–Mass Spectrometry

Matrix-assisted laser desorption ionization (MALDI), another powerful soft ionization source, has become an indispensable analytical tool for complex sample analysis [59]. This technique provides distinguishing features such as rapid analysis ability, low sample consumption, and relatively high tolerance toward salts [60]. The sample analysis time of MALDI–MS is mainly dependent on the amount of laser shots required for the experiment, which is suited for the high-throughput analysis. In addition, 0.1–1 μl of sample or sample and matrix mixture is deposited on the target plate for MALDI–MS analysis, and just a small fraction of the spotted sample was consumed during the analysis. Due to the poor reproducibility of the signal intensities of the analytes, the quantitative analysis is the weakness of MALDI–MS, as well as the reproducibility of the signal intensities largely determined by the type of matrices used and sample preparation [61]. The most commonly used MALDI matrices are acyano-4-hydroxycinnamic acid and 2,5-dihydroxybenzoic acid, but these matrices might affect the detection of small molecules. Nanomaterials, such as carbon dots [62], graphene [63], etc., have been extensively developed as effective matrices for MALDI profiling of small molecules, which is conducive for the application for MALDI–MS in metabolomics.

4.3.3 Mass Spectrometry Imaging

Mass spectrometry imaging (MSI)is a powerful tool for determining metabolites *in situ*, which can simultaneously map multiple molecules, from low molecular weight metabolites and drugs to large proteins, while maintaining the morphological integrity of the analyzed tissue slides [64]. Classically, spatial information of biological processes in tissue was obtained from histology staining or immunohistochemistry. However, histology staining is non-molecular, and immunohistochemistry requires the knowledge of the target analyte prior to investigation and limited to simultaneous investigation of only a small number of analytes at one time [65]. Currently, the MSI is an emerging technology to raise the understanding of the *in situ* distribution of metabolites or biomarkers directly in tissue sections or cells [66]. The general workflow of MSI can be briefly summarized as follows: the tissue is cut into thin slices (usually 10–20 μm) on the indium tin oxide-coated slide, then according to the analysis requirement, select whether to wash the tissue or not, and spray appropriate matrix or not on the basis of the selected ion source, and then analysis can be performed [67]. Tissue section is detected according to a pre-defined raster, which generates mass spectrum at each measurement point, and then the measured information performed image reconstruction to obtain the distribution of molecules in tissue [68]. Currently, a variety ionization methods have been employed for MSI, such as MALDI, secondary ion MS, desorption electrospray ionization, laser ablation electrospray ionization, etc.

4.3.4 Direct Analysis in Real Time–Mass Spectrometry

Direct analysis in real time (DRAT) is another versatile, new ambient ionization technology demonstrated by Cody in 2005, which enables rapid analyzing of solids, liquids, and gases at atmospheric pressure without sample preparation [69]. Helium is conducted via an axial tube and supports a corona discharge engenders ions, electrons, and excited atoms. Then, the helium passes through the other two chambers, where the electrons, cations, and anions are removed and get into the atmospheric reaction zone including only electronically excited neutral substances [70]. These released atoms in atmosphere will cause the environmental gas to undergo gas-phase reaction ionization cascade. And these ions serve as chemical ionizing reagents near the surface of the analyzed sample, resulting in analyte ions and finally transferred to the mass analyzer [70]. Penning ionization is the important step in DART ionization, and the flow rate of carrier gas and temperature are the two major factors that affect DART ionization performance [71]. From the very beginning, DART was utilized in plant component analysis, vegetable pesticide residue monitoring, forensic, and now gradually used in metabolomics research. Zhou et al. reported a metabolomics fingerprinting research of human serum by DART-MS [72], and the volatility metabolites were derivatized after protein precipitate and followed by DART-MS analysis.

4.4 Tools for MS Result Analysis

There are various open access and commercial software/websites available for MS data process and analysis (Table 4.1). These software/websites involves peak alignment, peak extraction, metabolites identification, and metabolic pathway analysis by searching metabolomics databases. The commonly used metabolomics databases include Kyoto Encyclopedia of Genes and Genomes (KEGG), human metabolome

Table 4.1 Multivariate data analysis approach and software/websites for mass spectrometry-based metabolomics.

Software/website	Accepted data forms	Statistics	Pathway analysis	Data visualization	Source
MarkerLynx	.raw	√		√	Waters
MarkerView	.d	√		√	AB Sciex
MassHunter	all	√		√	Aglient
Mass Profiler	All	√	√	√	Aglient
MetQuest	.raw	√		√	Thermo
Progenesis QI	All	√		√	Waters
SIEVE	.raw	√		√	Thermo
IDEOM	All	√		√	http://mzmatch.sourceforge.net/ideom.html
MathDAMP	All	√		√	http://mathdamp.iab.keio.ac.jp
MAVEN	All	√	√	√	http://maven.princeton.edu
MetAlign	All	√		√	http://www.metalign.nl
MetboAnalyst	All	√	√	√	http://www.metaboanalyst.ca
MET-COFEA	All	√		√	http://bioinfo.noble.org/manuscript-support/metcofea
metaX	mzXML	√	√	√	http://metax.genomics.cn
MET-XAlign	All	√		√	http://bioinfo.noble.org/manuscript-support/metxalign
MS-DIAL	All	√		√	http://prime.psc.riken.jp/Metbolomics_Software/MS-DIAL
MsXelerator	All	√		√	http://www.msmetrix.com
MZmine	All	√		√	http://mzmine.sourceforge.net
msCompare	All	√			https://trac.nbic.nl/mscompare
OpenMS	All	√		√	https://www.openms.de
SMART	All	√		√	http://www.stat.sinica.edu.tw/hsinchou metabolomics/SMART.htm
WebSpecmine	mzXML, netCDF or mzData	√	√	√	https://gitlab.bio.di.uminho.pt/WebSpecmine/desktop_docker
XCMS	All	√	√		https://xcmsonline.scripps.edu

database (HMDB), METLIN, MassBank, and others [73, 74]. However, there is no standardized software/website for MS data processing and analysis currently. The data processing and analysis software/websites may have influence on the data processing result. Data processing is a vital part of metabolomics research and each software/website has its own characteristics and advantage. Researchers should choose the software/website to process and analyze data according to their own needs. It is believed that with the development of metabolomics methodologies and databases, data processing and analysis software/website will continue to be enriched and become perfect.

4.5 Multivariate Statistical Analysis Approach for Data Analysis

The aims of metabolomics is to identify and to quantify small-molecular metabolites in organisms through high-throughput technology, which usually generates large data sets with high dimensions. Therefore, specialized approaches are needed for data analysis [75]. At present, multivariate statistical analysis is often used to process the MS-based data, among these multivariate statistical analysis methods, the most commonly used are principal component analysis (PCA) (Figure 4.1), partial least squares discriminant analysis (PLS-DA) (Figure 4.2), heatmap

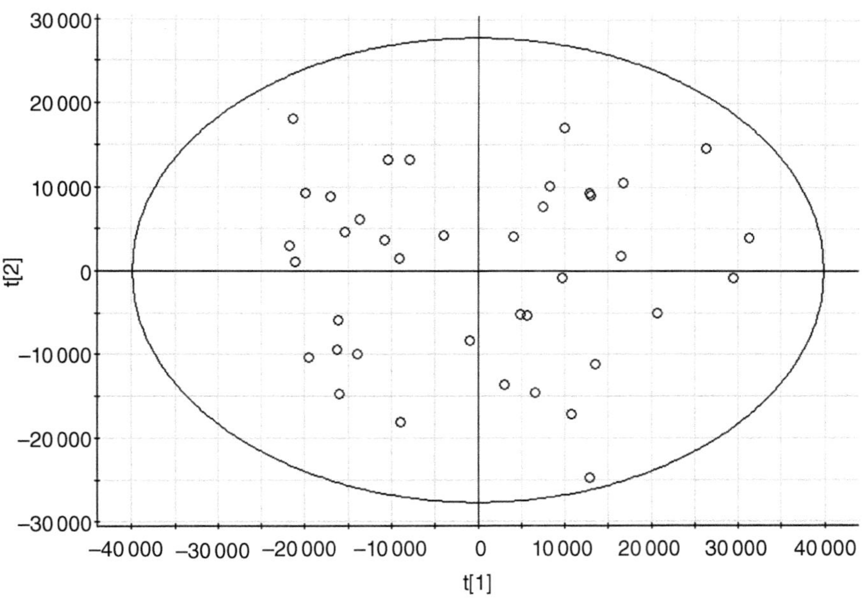

Figure 4.1 The principal component analysis for metabolomics research.

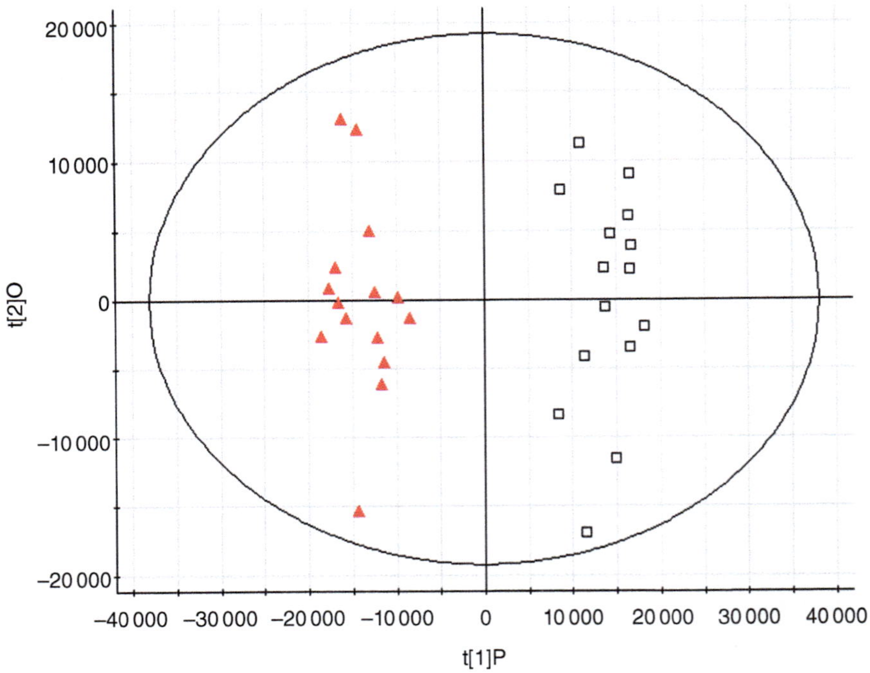

Figure 4.2 The partial least squares discriminant analysis for metabolomics research.

visualization (Figure 4.3), K-means clustering analysis (Figure 4.4), tree clustering analysis (Figure 4.5), correlation analysis for the biologically meaningful metabolite sets (Figure 4.6), etc., of which can make the data set visualized and interpreted in a simple way through dimension reduction.

PCA can extract the main sources of variation in the data set without any prior knowledge, which is suitable for identifying outlying samples and studying samples or groups in the data set, while maintaining the features of the data set that has the largest contribution to the variance [76]. PCA can also take each principal component as a new independent variable for regression analysis and screen regression variables. PLS is a supervised learning method, which has the same principle as PCA analysis; it uses partial least squares method to project data structure [77]. When the sample background information is available, supervised methods can be used to identify metabolites in a targeted way [75]. The variable importance in projection (VIP) is performed to reflect the influence of the variables in the results of metabolic analysis data in contributing to the separation process of PLS-DA model (Figure 4.7). The PLS-DA model can be evaluated by the goodness-fit parameter (R^2Y) and predictive ability parameter (Q^2Y); the closer the value to 1, the better the model. However, since it is a supervised learning method, over-fitting may occur due to excessive principal components and complex grouping. Thus, the validity and reliability of the model should be verified in data analysis process. On the basis of the same model fitting and prediction ability as

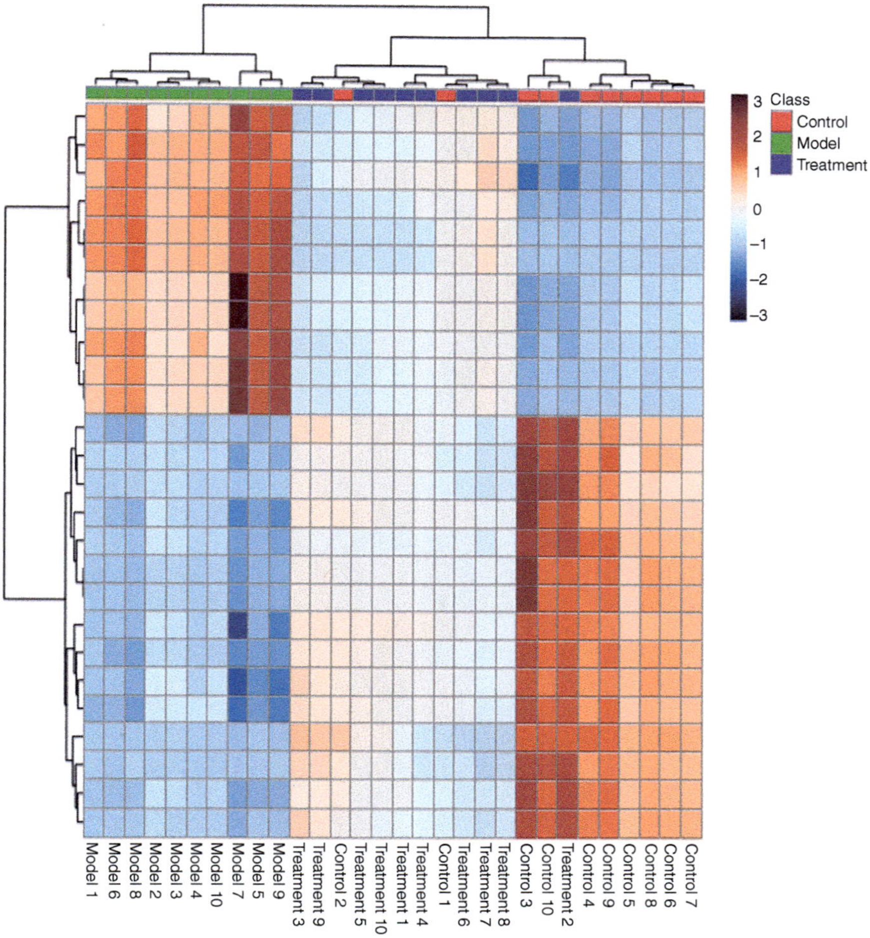

Figure 4.3 The heatmap visualization analysis for metabolomics research.

PLS, orthogonal partial least squares discriminant analysis (OPLS) can distinguish the predictive component from the orthogonal component to make the model more interpretable, which separates Y-predictive from Y-orthogonal variation, thus facilitating the grouping of the unrestrained data [78]. Like the PLS-DA, false-positive results were often yielded and require cross-validation to guarantee reliability [79].

Clustering analysis can objectively reflect the relationship or similarity between things and classify them by statistics and classification criteria, allowing users to visualized the similarity of samples or groups [80, 81], which belongs to the unsupervised classification method. In metabolomics research, cluster analysis can be used to observe the relationship between samples and groups, which is convenient for data mining [82].

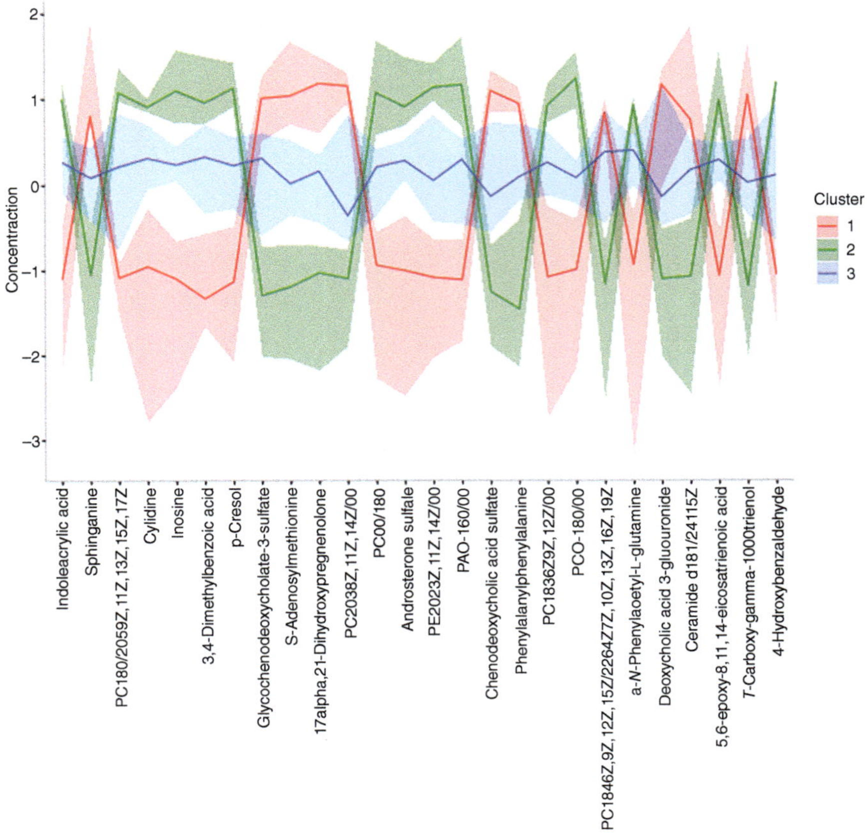

Figure 4.4 The K-means clustering analysis for metabolomics research.

In addition, multiple regression analysis is also an important tool for metabolomics research. Through the establishment of regression model, the association between the multivariate data of metabolomics and function, quality, and grade can be investigated, so as to achieve the purpose of screening important variables, establishing evaluation criteria and predicting results. Random forest is a combined classifier based on decision tree proposed by Breiman; each tree depends on the values of a random vector sampled independently and with the same distribution for all trees in the forest [83]. It is an excellent machine learning method with the following advantages: fast speed, stable and insensitive to noise, and little or no over-fitting [84]. The regression analysis is particularly advantageous in the classification analysis of clinical diagnostic and selection of biomarkers; Chen et al. compared the classification results of PLS, support vector machine, linear discriminant analysis, and RF analysis when analyzing the metabolomics data of clinical diagnosis samples with diverse demographic, genetic backgrounds, various pathological conditions, or treatments, and RF outperformed the other three classifiers in the given clinical data sets [85]. At present, the commonly used multivariate statistical methods were as mentioned earlier; combined with a variety of software, researchers can more easily show the results of metabolomics research.

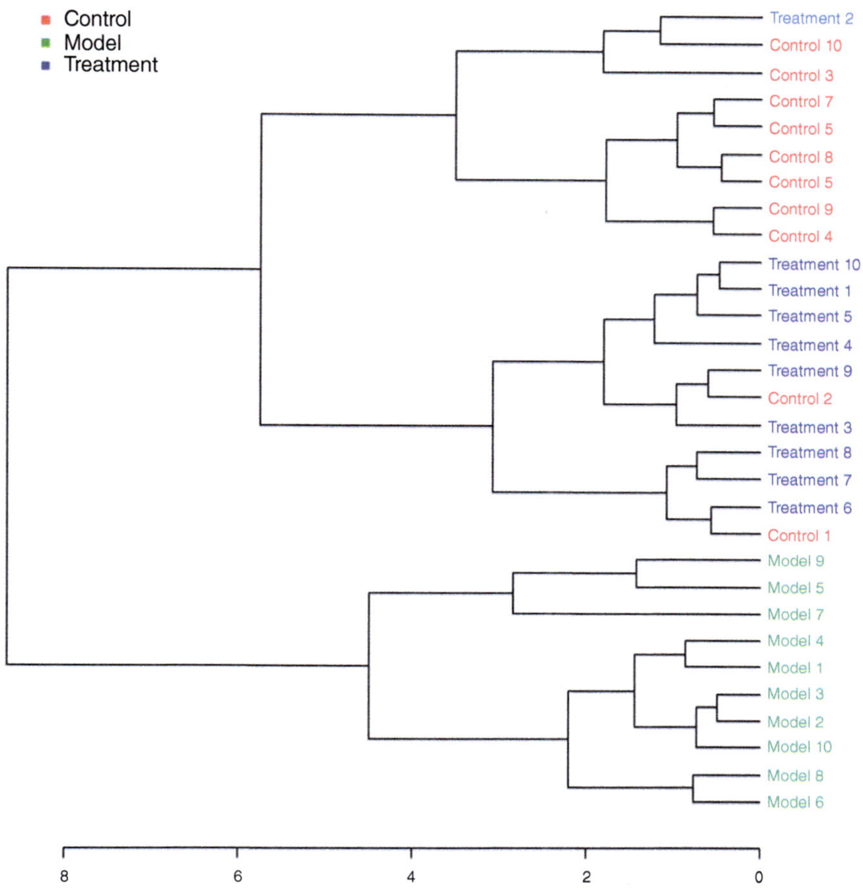

Figure 4.5 The tree clustering analysis for metabolomics research.

4.6 Limitations and Challenges

MS-based metabolomics technology plays an important role in metabolites profiling due to its high sensitivity, high speed, and high throughput. However, some problems are constantly highlighted in the continuous exploration of this field. Firstly, since the concentration of metabolites in organisms varies greatly, and some metabolites with lower concentrations may have important regulatory effects, the detection of trace level metabolites presents a challenge for MS-based metabolomics. Therefore, the combined use of multiple separation, extraction methods, and MS detectors are imperative. Secondly, with the continuous improvement of sample processing methods, chromatographic separation capabilities, and MS sensitivity, massive amounts of data will be obtained. Thus, data processing will become another challenge for metabolomics. Thirdly, on account of the different forms of adduct ions, the large amount of data after peak alignment and the imperfect databases and standards, it has brought great challenges to the identification of metabolites.

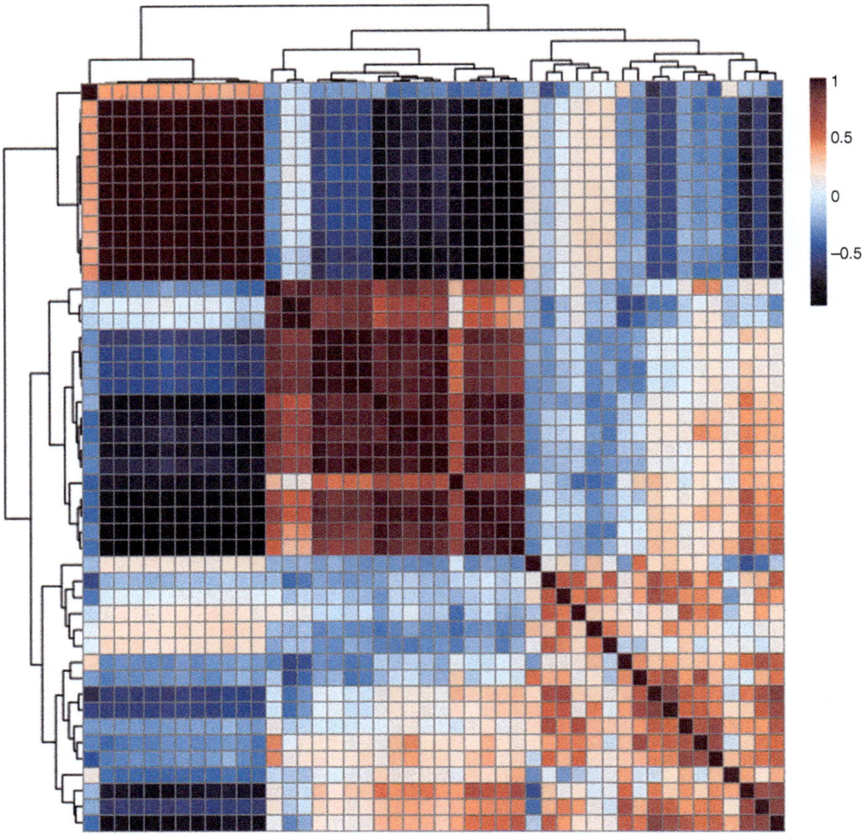

Figure 4.6 Correlation analysis heatmap for the biologically meaningful metabolite sets.

Fourthly, accurate concentration data are more relevant to experimental conditions and data collection methods, which is considered to be comparable among different groups. At present, most MS-based metabolomic studies are limited to the relative quantification of metabolites. The constant enrichment of isotopic affinity tags and standards will be conducive to the quantification of metabolites. Proper choice of multivariate statistical analysis method and the reliability of the method needs to be further improved. Lastly, with the upgrading of the MS instrument, the construction and popularization of the standard approach should be accelerated to achieve data sharing, effective usage, and verification of experimental results [86, 87].

4.7 Conclusions and Future Perspectives

The latest advances of novel MS techniques have greatly improved the progression of metabolomics over the past decade (Figure 4.8). The high selectivity, flexibility, and lower detection limits of MS, along with its compatibility with various separation techniques and the capacity to provide qualitative and quantitative data,

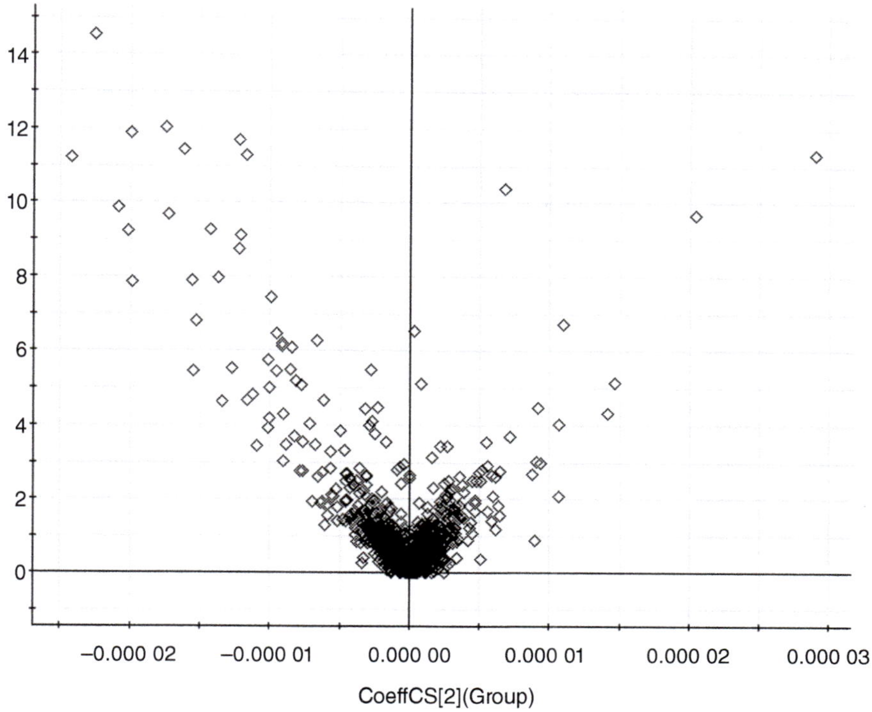

Figure 4.7 The variable importance in projection (VIP) performed by partial least squares discriminant analysis for metabolomics research.

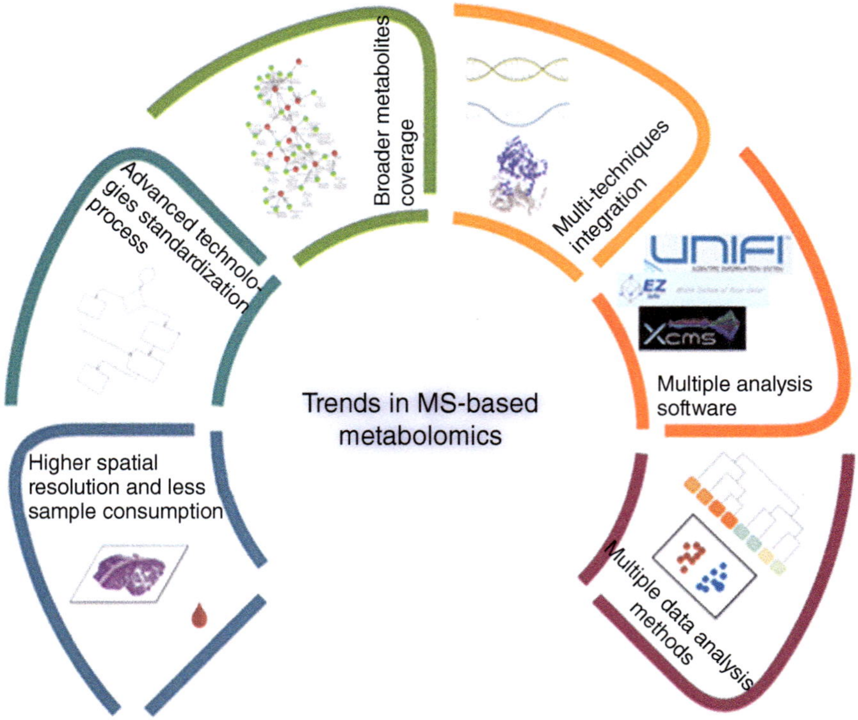

Figure 4.8 Trends and development of mass spectrometry-based metabolomics techniques.

impel it to be an ideal instrument for metabolomics applications. At present, MS not only allows chemical information analysis of the extract content by LC–MS or DI–MS methods but can also use imaging methods to intact analyzed tissue or cells to provide spatial information for the location of given metabolites. In order to better match with clinical phenotype, sample processing, data analysis, method validation, and other details need to be standardized. In the future, it is still necessary to continually develop new technologies to expand the coverage of metabolites and improve data quality. Data processing technology and multivariate statistical analysis technology should also be combined in order to further nourish the development of metabolomics.

Acknowledgments

This study was supported by the National Natural Science Foundation of China (Grant No. 81903818, 81973745, 81302905), the University Nursing Program for Young Scholars with Creative Talents in Heilongjiang Province (UNPYSCT-2020224, UNPYSCT-2015118), and Foundation of Heilongjiang University of Chinese Medicine (2018bs02, 2018jc01).

References

1 Jang, C., Chen, L., and Rabinowitz, J.D. (2018). Metabolomics and isotope tracing. *Cell* 173 (4): 822–837.
2 Zhang, A., Sun, H., and Wang, X. (2014). Urinary metabolic profiling of rat models revealed protective function of scoparone against alcohol induced hepatotoxicity. *Scientific Reports* 4: 6768.
3 Zhao, Y.Y., Liu, J., Cheng, X.L. et al. (2012). Urinary metabonomics study on biochemical changes in an experimental model of chronic renal failure by adenine based on UPLC Q-TOF/MS. *Clinica Chimica Acta; International Journal of Clinical Chemistry* 413 (5–6): 642–649.
4 Wishart, D.S. (2019). Metabolomics for investigating physiological and pathophysiological processes. *Physiological Reviews* 99 (4): 1819–1875.
5 Zhang, A., Liu, Q., Zhao, H. et al. (2016). Phenotypic characterization of nanshi oral liquid alters metabolic signatures during disease prevention. *Scientific Reports* 6: 19333.
6 Emwas, A.H., Roy, R., McKay, R.T. et al. (2019). NMR spectroscopy for metabolomics research. *Metabolites* 9 (7): 123–161.
7 Yang, H., Lau, W.B., Lau, B. et al. (2017). A mass spectrometric insight into the origins of benign gynecological disorders. *Mass Spectrometry Reviews* 36 (3): 450–470.
8 Papadimitropoulos, M.P., Vasilopoulou, C.G., Maga-Nteve, C., and Klapa, M.I. (2018). Untargeted GC-MS metabolomics. *Methods in Molecular Biology* 1738: 133–147.

9 Fiehn, O. (2016). Metabolomics by gas chromatography-mass spectrometry: combined targeted and untargeted profiling. *Current Protocols in Molecular Biology* 114: 30.34.31–30.34.32.

10 Dunn, W.B., Broadhurst, D., Begley, P. et al. (2011). Procedures for large-scale metabolic profiling of serum and plasma using gas chromatography and liquid chromatography coupled to mass spectrometry. *Nature Protocols* 6 (7): 1060–1083.

11 Kanani, H., Chrysanthopoulos, P.K., and Klapa, M.I. (2008). Standardizing GC-MS metabolomics. *Journal of Chromatography B, Analytical Technologies in the Biomedical and Life Sciences* 871 (2): 191–201.

12 Wong, Y.F., Hartmann, C., and Marriott, P.J. (2014). Multidimensional gas chromatography methods for bioanalytical research. *Bioanalysis* 6 (18): 2461–2479.

13 Marriott, P.J., Chin, S.T., and Nolvachai, Y. (1636). Techniques and application in comprehensive multidimensional gas chromatography–mass spectrometry. *Journal of Chromatography A* 2021: 461788.

14 Cortes, H.J., Winniford, B., Luong, J., and Pursch, M. (2009). Comprehensive two dimensional gas chromatography review. *Journal of Separation Science* 32 (5–6): 883–904.

15 Ryan, D., Morrison, P., and Marriott, P. (2005). Orthogonality considerations in comprehensive two-dimensional gas chromatography. *Journal of Chromatography A* 1071 (1–2): 47–53.

16 Miyazaki, T., Okada, K., Yamashita, T., and Miyazaki, M. (2017). Two-dimensional gas chromatography time-of-flight mass spectrometry-based serum metabolic fingerprints of neonatal calves before and after first colostrum ingestion. *Journal of Dairy Science* 100 (6): 4354–4364.

17 Plumb, R., Castro-Perez, J., Granger, J. et al. (2004). Ultra-performance liquid chromatography coupled to quadrupole-orthogonal time-of-flight mass spectrometry. *Rapid Communications in Mass Spectrometry: RCM* 18 (19): 2331–2337.

18 Zhang, A.H., Sun, H., Yan, G.L., and Wang, X.J. (2017). Recent developments and emerging trends of mass spectrometry for herbal ingredients analysis. *TrAC Trends in Analytical Chemistry* 94: 70–76.

19 Nordstrom, A., O'Maille, G., Qin, C., and Siuzdak, G. (2006). Nonlinear data alignment for UPLC-MS and HPLC-MS based metabolomics: quantitative analysis of endogenous and exogenous metabolites in human serum. *Analytical Chemistry* 78 (10): 3289–3295.

20 Wang, X.J., Li, J., and Zhang, A.H. (2016). Urine metabolic phenotypes analysis of extrahepatic cholangiocarcinoma disease using ultra-high performance liquid chromatography-mass spectrometry. *RSC Advances* 6 (67): 63049–63057.

21 Liang, Q., Liu, H., Xing, H.T. et al. (2016). UPLC-QTOF/MS based metabolomics reveals metabolic alterations associated with severe sepsis. *RSC Advances* 49: 43293–43298.

22 Want, E.J., Wilson, I.D., Gika, H. et al. (2010). Global metabolic profiling procedures for urine using UPLC-MS. *Nature Protocols* 5 (6): 1005–1018.

23 Liang, Q., Liu, H., Xing, H.T. et al. (2016). Urinary UPLC-MS metabolomics dissecting the underlying mechanisms of Huaxian capsule protects against sepsis. *RSC Advances* 46: 40436–40441.

24 Lu, W., Bennett, B.D., and Rabinowitz, J.D. (2008). Analytical strategies for LC-MS-based targeted metabolomics. *Journal of Chromatography B, Analytical Technologies in the Biomedical and Life Sciences* 871 (2): 236–242.

25 Putri, S.P., Yamamoto, S., Tsugawa, H., and Fukusaki, E. (2013). Current metabolomics: technological advances. *Journal of Bioscience and Bioengineering* 116 (1): 9–16.

26 Joachim, W. (2004). *Handbook of Ion Chromatography*. Wiley.

27 Guo, L., Worth, A.J., Mesaros, C. et al. (2016). Diisopropylethylamine /hexafluoroisopropanol-mediated ion-pairing ultra-high-performance liquid chromatography/mass spectrometry for phosphate and carboxylate metabolite analysis: utility for studying cellular metabolism. *Rapid Communications in Mass Spectrometry: RCM* 30 (16): 1835–1845.

28 Buscher, J.M., Czernik, D., Ewald, J.C. et al. (2009). Cross-platform comparison of methods for quantitative metabolomics of primary metabolism. *Analytical Chemistry* 81 (6): 2135–2143.

29 Tang, D.Q., Zou, L., Yin, X.X., and Ong, C.N. (2016). HILIC-MS for metabolomics: an attractive and complementary approach to RPLC-MS. *Mass Spectrometry Reviews* 35 (5): 574–600.

30 Cífková, E., Hajek, R., Lisa, M., and HolLapek, M. (2016). Hydrophilic interaction liquid chromatography-mass spectrometry of (lyso)phosphatidic acids, (lyso)phosphatidylserines and other lipid classes. *Journal of Chromatography A* 1439: 65–73.

31 Gordillo, R. (2021). Supercritical fluid chromatography hyphenated to mass spectrometry for metabolomics applications. *Journal of Separation Science* 44 (1): 448–463.

32 van de Velde, B., Guillarme, D., and Kohler, I. (2020). Supercritical fluid chromatography–mass spectrometry in metabolomics: past, present, and future perspectives. *Journal of Chromatography B, Analytical Technologies in the Biomedical and Life Sciences* 1161: 122444.

33 Shulaev, V. and Isaac, G. (2018). Supercritical fluid chromatography coupled to mass spectrometry – a metabolomics perspective. *Journal of Chromatography B, Analytical Technologies in the Biomedical and Life Sciences* 1092: 499–505.

34 Pinkston, J.D., Wen, D., Morand, K.L. et al. (2006). Comparison of LC/MS and SFC/MS for screening of a large and diverse library of pharmaceutically relevant compounds. *Analytical Chemistry* 78 (21): 7467–7472.

35 Taguchi, K., Fukusaki, E., and Bamba, T. (2014). Supercritical fluid chromatography/mass spectrometry in metabolite analysis. *Bioanalysis* 6 (12): 1679–1689.

36 Lisa, M. and Holcapek, M. (2015). High-throughput and comprehensive lipidomic analysis using ultrahigh-performance supercritical fluid chromatography-mass spectrometry. *Analytical Chemistry* 87 (14): 7187–7195.

37 Taguchi, K., Fukusaki, E., and Bamba, T. (2013). Simultaneous and rapid analysis of bile acids including conjugates by supercritical fluid chromatography coupled to tandem mass spectrometry. *Journal of Chromatography A* 1299: 103–109.

38 Masuo, Y., Imai, T., Shibato, J. et al. (2009). Omic analyses unravels global molecular changes in the brain and liver of a rat model for chronic sake (Japanese alcoholic beverage) intake. *Electrophoresis* 30 (8): 1259–1275.

39 Beale, D.J., Jones, O.A., Karpe, A.V. et al. (2016). Review of analytical techniques and their application in disease diagnosis in breathomics and salivaomics research. *International Journal of Molecular Sciences* 18 (1).

40 Wolfender, J.L., Marti, G., Thomas, A., and Bertrand, S. (2015). Current approaches and challenges for the metabolite profiling of complex natural extracts. *Journal of Chromatography A* 1382: 136–164.

41 Zhang, W., Hankemeier, T., and Ramautar, R. (1972). Capillary electrophoresis-mass spectrometry for metabolic profiling of biomass-limited samples. *Methods in Molecular Biology* 2019: 165–172.

42 Garcia, A., Godzien, J., Lopez-Gonzalvez, A., and Barbas, C. (2017). Capillary electrophoresis mass spectrometry as a tool for untargeted metabolomics. *Bioanalysis* 9 (1): 99–130.

43 Kleparnik, K. (2015). Recent advances in combination of capillary electrophoresis with mass spectrometry: methodology and theory. *Electrophoresis* 36 (1): 159–178.

44 Wenz, C., Barbas, C., Lopez-Gonzalvez, A. et al. (2015). Interlaboratory study to evaluate the robustness of capillary electrophoresis-mass spectrometry for peptide mapping. *Journal of Separation Science* 38 (18): 3262–3270.

45 Stalmach, A., Albalat, A., Mullen, W., and Mischak, H. (2013). Recent advances in capillary electrophoresis coupled to mass spectrometry for clinical proteomic applications. *Electrophoresis* 34 (11): 1452–1464.

46 Onjiko, R.M., Portero, E.P., Moody, S.A., and Nemes, P. (2017). In situ microprobe single-cell capillary electrophoresis mass spectrometry: metabolic reorganization in single differentiating cells in the live vertebrate (*Xenopus laevis*) embryo. *Analytical Chemistry* 89 (13): 7069–7076.

47 Zhang, X., Quinn, K., Cruickshank-Quinn, C. et al. (2018). The application of ion mobility mass spectrometry to metabolomics. *Current Opinion in Chemical Biology* 42: 60–66.

48 Paglia, G. and Astarita, G. (2017). Metabolomics and lipidomics using traveling-wave ion mobility mass spectrometry. *Nature Protocols* 12 (4): 797–813.

49 Zhang, X., Chiu, V.M., Stoica, G. et al. (2014). Metabolic analysis of striatal tissues from Parkinson's disease-like rats by electrospray ionization ion mobility mass spectrometry. *Analytical Chemistry* 86 (6): 3075–3083.

50 Bowman, A.P., Abzalimov, R.R., and Shvartsburg, A.A. (2017). Broad separation of isomeric lipids by high-resolution differential ion mobility spectrometry with tandem mass spectrometry. *Journal of the American Society for Mass Spectrometry* 28 (8): 1552–1561.

51 Gonzalez-Dominguez, R., Sayago, A., and Fernandez-Recamales, A. (2017). Direct infusion mass spectrometry for metabolomic phenotyping of diseases. *Bioanalysis* 9 (1): 131–148.

52 Lin, L., Yu, Q., Yan, X. et al. (2010). Direct infusion mass spectrometry or liquid chromatography mass spectrometry for human metabonomics? A serum metabonomic study of kidney cancer. *The Analyst* 135 (11): 2970–2978.

53 Kapron, J.T., Pace, E., Van Pelt, C.K., and Henion, J. (2003). Quantitation of midazolam in human plasma by automated chip-based infusion nanoelectrospray tandem mass spectrometry. *Rapid Communications in Mass Spectrometry: RCM* 17 (18): 2019–2026.

54 Hop, C.E., Chen, Y., and Yu, L.J. (2005). Uniformity of ionization response of structurally diverse analytes using a chip-based nanoelectrospray ionization source. *Rapid Communications in Mass Spectrometry: RCM* 19 (21): 3139–3142.

55 Valaskovic, G.A., Utley, L., Lee, M.S., and Wu, J.T. (2006). Ultra-low flow nanospray for the normalization of conventional liquid chromatography/mass spectrometry through equimolar response: standard-free quantitative estimation of metabolite levels in drug discovery. *Rapid Communications in Mass Spectrometry: RCM* 20 (7): 1087–1096.

56 Zhang, Y., He, C., Qiu, L. et al. (2016). Serum unsaturated free fatty acids: a potential biomarker panel for early-stage detection of colorectal cancer. *Journal of Cancer* 7 (4): 477–483.

57 Zhang, Y., Qiu, L., He, C. et al. (2015). Serum unsaturated free fatty acids: a potential biomarker panel for differentiating benign thyroid diseases from thyroid cancer. *Journal of Cancer* 6 (12): 1276–1281.

58 Zhang, Y., Qiu, L., Wang, Y. et al. (2014). Unsaturated free fatty acids: a potential biomarker panel for early detection of gastric cancer. *Biomarkers: Biochemical Indicators of Exposure, Response, and Susceptibility to Chemicals* 19 (8): 667–673.

59 Tsuchida, S., Umemura, H., and Nakayama, T. (2020). Current status of matrix-assisted laser desorption/ionization-time-of-flight mass spectrometry (MALDI-TOF MS) in clinical diagnostic microbiology. *Molecules* 25 (20): 4775–4788.

60 Boughton, B.A. and Hamilton, B. (2017). Spatial metabolite profiling by matrix-assisted laser desorption ionization mass spectrometry imaging. *Advances in Experimental Medicine and Biology* 965: 291–321.

61 Sleno, L. and Volmer, D.A. (2005). Some fundamental and technical aspects of the quantitative analysis of pharmaceutical drugs by matrix-assisted laser desorption/ionization mass spectrometry. *Rapid Communications in Mass Spectrometry: RCM* 19 (14): 1928–1936.

62 Baker, S.N. and Baker, G.A. (2010). Luminescent carbon nanodots: emergent nanolights. *Angewandte Chemie* 49 (38): 6726–6744.

63 Novoselov, K.S., Geim, A.K., Morozov, S.V. et al. (2004). Electric field effect in atomically thin carbon films. *Science* 306 (5696): 666–669.

64 Chughtai, K. and Heeren, R.M. (2010). Mass spectrometric imaging for biomedical tissue analysis. *Chemical Reviews* 110 (5): 3237–3277.

65 Angel, P.M. and Caprioli, R.M. (2013). Matrix-assisted laser desorption ionization imaging mass spectrometry: in situ molecular mapping. *Biochemistry* 52 (22): 3818–3828.

66 Pareek, V., Tian, H., Winograd, N., and Benkovic, S.J. (2020). Metabolomics and mass spectrometry imaging reveal channeled de novo purine synthesis in cells. *Science* 368 (6488): 283–290.

67 Weaver, E.M. and Hummon, A.B. (2013). Imaging mass spectrometry: from tissue sections to cell cultures. *Advanced Drug Delivery Reviews* 65 (8): 1039–1055.

68 Ly, A., Buck, A., Balluff, B. et al. (2016). High-mass-resolution MALDI mass spectrometry imaging of metabolites from formalin-fixed paraffin-embedded tissue. *Nature Protocols* 11 (8): 1428–1443.

69 Cody, R.B., Laramee, J.A., and Durst, H.D. (2005). Versatile new ion source for the analysis of materials in open air under ambient conditions. *Analytical Chemistry* 77 (8): 2297–2302.

70 Calvano, C.D., van der Werf, I.D., Palmisano, F., and Sabbatini, L. (2016). Revealing the composition of organic materials in polychrome works of art: the role of mass spectrometry-based techniques. *Analytical and Bioanalytical Chemistry* 408 (25): 6957–6981.

71 Dunn, W.B., Bailey, N.J., and Johnson, H.E. (2005). Measuring the metabolome: current analytical technologies. *The Analyst* 130 (5): 606–625.

72 Gurdeniz, G., Kristensen, M., Skov, T., and Dragsted, L.O. (2012). The effect of LC-MS data preprocessing methods on the selection of plasma biomarkers in fed vs. fasted rats. *Metabolites* 2 (1): 77–99.

73 Coble, J.B. and Fraga, C.G. (2014). Comparative evaluation of preprocessing freeware on chromatography/mass spectrometry data for signature discovery. *Journal of Chromatography A* 1358: 155–164.

74 Scalbert, A., Brennan, L., Fiehn, O. et al. (2009). Mass-spectrometry-based metabolomics: limitations and recommendations for future progress with particular focus on nutrition research. *Metabolomics: Official Journal of the Metabolomic Society* 5 (4): 435–458.

75 Bylesjo, M. (2015). Extracting meaningful information from metabonomic data using multivariate statistics. *Methods in Molecular Biology* 1277: 137–146.

76 Jolliffe, I.T. (2002). *Principal Component Analysis*, 2e. Springer.

77 Wold, S., Sjöström, M., and Eriksson, L. (2001). PLS-regression: a basic tool of chemometrics. *Chemometrics and Intelligent Laboratory Systems* 58 (2): 109–130.

78 Westerhuis, J.A., van Velzen, E.J., Hoefsloot, H.C., and Smilde, A.K. (2010). Multivariate paired data analysis: multilevel PLSDA versus OPLSDA. *Metabolomics: Official Journal of the Metabolomic Society* 6 (1): 119–128.

79 Bartel, J., Krumsiek, J., and Theis, F.J. (2013). Statistical methods for the analysis of high-throughput metabolomics data. *Computational and Structural Biotechnology Journal* 4: e201301009.

80 Boccard, J., Veuthey, J.L., and Rudaz, S. (2010). Knowledge discovery in metabolomics: an overview of MS data handling. *Journal of Separation Science* 33 (3): 290–304.

81 Eisen, M.B., Spellman, P.T., Brown, P.O., and Botstein, D. (1998). Cluster analysis and display of genome-wide expression patterns. *Proceedings of the National Academy of Sciences of the United States of America* 95 (25): 14863–14868.

82 Heinemann, J. (1859). Cluster analysis of untargeted metabolomic experiments. *Methods in Molecular Biology* 2019: 275–285.

83 Breiman, L. (2001). Random forests. *Machine Learning* 45: 5–12.

84 Amaratunga, D., Cabrera, J., and Lee, Y.S. (2008). Enriched random forests. *Bioinformatics* 24 (18): 2010–2014.

85 Chen, T., Cao, Y., Zhang, Y. et al. (2013). Random forest in clinical metabolomics for phenotypic discrimination and biomarker selection. *Evidence-based Complementary and Alternative Medicine: eCAM* 2013: 298183.

86 Sumner, L.W., Amberg, A., Barrett, D. et al. (2007). Proposed minimum reporting standards for chemical analysis chemical analysis working group (CAWG) metabolomics standards initiative (MSI). *Metabolomics: Official Journal of the Metabolomic Society* 3 (3): 211–221.

87 Salek, R.M., Steinbeck, C., Viant, M.R. et al. (2013). The role of reporting standards for metabolite annotation and identification in metabolomic studies. *GigaScience* 2 (1): 13.

5

Mass Spectrometry-Driven Metabolomics for Metabolites and Metabolic Pathway Analysis

Aihua Zhang, Hui Sun, Shi Qiu, and Xijun Wang

Heilongjiang University of Chinese Medicine, National Chinmedomics Research Center, Functional Metabolomics Laboratory, Department of Pharmaceutical Analysis, Heping Road 24, Harbin 150040, China

5.1 Introduction

Metabolomics could lead to the identification of all the metabolites (molecules <1 kDa) in a biological system. It offers the potential for a holistic approach to clinical medicine, thus improving disease diagnosis and understanding of the pathological mechanisms [1–3]. Metabolites are important indicators of physiological or pathological states that can provide information for the identification of biomarkers for diseases and can help to understand its occurrence and progression [4]. Metabolomics thus has the potential not only to be an effective tool for early diagnosis of disease but also to be a predictor of treatment response and survival in medical studies [5–8].

In recent years, a wide variety of methods have been used to separate and quantitate components of the metabolome, and no single analytical platform can capture all metabolic information in a sample [9–11]. Mass spectrometry (MS, Figure 5.1) represents a universal, sensitive tool that can be used to characterize and quantify large number of compounds in a biological sample where metabolite concentrations might cover a broad range [12–16]. The advances in MS and chromatography have particularly improved the efficiency and sensitivity as well as reliability of targeted and untargeted analyses [17–20]. Metabolomics requires proper separation of the compounds to be assayed, and chemical separation techniques such as GC and LC or capillary electrophoresis can all be combined to MS detection [21–23]. Therefore, the hyphenated analytical techniques, such as liquid chromatography–mass spectrometry (LC–MS), gas chromatography–mass spectrometry (GC–MS), capillary electrophoresis–mass spectrometry (CE–MS), and matrix-assisted laser desorption ionization–MS, etc., often lead to better analyte coverage of the metabolome [24–26]. Bio-information derived from these techniques could provide accurate and clinically useful diagnostic capability for the management of diseases at

Mass Spectrometry-Based Metabolomics in Clinical and Herbal Medicines: Strategies, Technologies, and Applications, First Edition. Edited by Aihua Zhang and Wanying Wang.

Figure 5.1 High-definition mass spectrometry system for targeted and untargeted metabolomics analyses.

the metabolic level [27–30]. MS-based metabolomics will aid the discovery of novel biomarkers, hopefully leading to the early detection of various diseases.

Metabolomics promises to have broad implications for medical practice because it can capture information with regard to mechanisms of disease, making it possible to map disease risk to metabolic pathways [31, 32]. Metabolomic analysis using MS can be performed using biofluids such as urine, blood, or cerebrospinal fluid and may permit the identification of disease-specific metabolite signatures that may be useful as disease biomarkers [33–35]. MS-metabolomics offers the opportunity to generate novel noninvasive diagnostic tests and provides a unique insight into established and novel metabolic pathways, which are simple and cost effective yet can retain high sensitivity and specificity characteristics. This chapter will highlight the importance and benefits with certain examples of MS-metabolomics for identifying potential biomarker metabolites and metabolic pathways and for gaining mechanistic insight into disease processes.

5.2 Mass Spectrometry-Driven Metabolomics for Biomarker Discovery

The burden of diseases is growing worldwide, and a more desperate need for better tools to detect, diagnose, and monitor the disease is required. Most clinical chemistry tests that rely on the outdated technologies are neither sensitive nor specific for diseases. A fundamental task in biomedicine is to detect the disease as early as possible. In order to meet this, MS-metabolomics offers new opportunities for biomarker discovery in complex diseases and may provide physiologic understanding of diseases beyond traditional technologies. Biomarkers are measurable internal indicators of molecular and/or cellular alterations that may appear in an organism after or during exposure to disease or therapeutic intervention. They are compounds or a set of compounds that must be quantitatively, sensitively, specifically, and easily measure on noninvasively collected biological media. Metabolomics allows the simultaneous and relative quantification of thousands of different metabolites within a given sample using sensitive and specific MS methodology. MS platforms are frequently used to provide the sensitive and reproducible detection of hundreds to thousands of metabolites in biofluids [36–40]. Lots of studies have demonstrated that the possibilities of MS-metabolomics as a diagnostic tool in diseases provided further insight into pathophysiologic mechanisms [41–43].

5.3 Data Extraction and Analysis for Metabolites and Metabolic Pathway

In fact, MS has already become the main analytical platform for metabolomics. It aims to analyze a sample based on the mass-to-charge ratio (m/z) of its ionized components. Original data of metabolome were firstly processed for peak extraction, noise reduction, matching, alignment, and detection and then obtained the information parameters of ion identity (ID), m/z, peak area, retention time (Rt), and ionic strength (Figure 5.2). Due to the complexity of the data, pattern recognition approach has become predominant in the metabolomics applications [28, 44, 45]. By applying advanced analytical and statistical tools, the metabolome is mined for biomarker metabolites and metabolic pathway that are associated with the diseases [46–49]. Multivariate statistical methods were developed to handle the complexity data acquired from health and disease [50–53]. Using pattern recognition approaches such as "unsupervised" principal component analysis (PCA) and "supervised" partial least squares-discriminant analysis (PLS-DA), metabolomics could provide a comprehensive assessment of the metabolite. The commonly used unsupervised technique for identifying patterns and trends in metabolomics data is PCA that attempts to reduce multidimensional data. The PCA are displayed as score plots indicating the scatter of the samples, which indicate similar metabolomic compositions when clustered together and compositionally different metabolome when dispersed. PCA is followed by a supervised PLS-DA that attempts to maximize the covariance between the independent and dependent variables to discriminate among samples. Potential biomarkers were extracted from loading plots constructed following the PLS-DA, and biomarkers were chosen depending on their contribution to the correlation within the data set. Numerous other supervised forms such as orthogonal projections to latent structures, artificial neural networks, support vector machines, and random forest methods have also been used to analyze metabolic data [54–57].

5.4 Biomarker Identification

Biomarkers are of great importance for diagnosing and monitoring disease progression, aiding for patient treatment [58–62]. The use of high-resolution MS tools to explore the metabolism could insight into disease progression [63–67]. There is an increasing need for diagnostic biomarkers to ensure the therapies are targeted at the right patient. One of the challenges is the validation of biomarkers in metabolomics to identify the perturbed pathways. The identification of metabolic biomarkers is a challenging work. Another one is the lab-to-lab reproducibility. Most of the biomarkers have failed to replace the clinical tests. It has seen that advancement in metabolomics study has been producing potential biomarkers, to become a clinically approved test. The potential biomarker should be validated using hundreds of specimens. We can apply the MS-metabolomics to aid the discovery of potential biomarkers and combine the different molecules that can provide information compensating for the shortcoming of individual tests, hopefully leading to the early detection of various diseases. In future work, the potential biomarkers should be further validated with a large enough patient cohort to achieve earlier diagnosis.

5.5 Applications

Powerful analytical techniques like liquid chromatography (LC) coupled to tandem MS offer a rapid, effective, and economical way to analyze metabolic alterations of pre-defined target metabolites in biological samples. Novel hyphenated technical method like the combination of tandem MS combines both identification and quantification of known and unknown metabolic targets. In a study, serum metabolomics based on ultra-performance liquid chromatography–quadrupole time of flight–tandem MS (UPLC–Q-TOF–MS) combined with ingenuity pathway analysis was used to explore the therapeutic mechanisms of geniposide [15]. A total of 33 metabolites were identified and related with the pathogenesis of alcoholic liver disease (ALD), which 21 of them are regulated by geniposide, involving the relevant metabolic pathways such as arachidonic acid metabolism, amino acid metabolism, pyruvate metabolism, and tricarboxylic acid (TCA) cycle. Zhong et al. had investigated the serum metabolic profiling of gout patients by UPLC–Q-TOF–MS. Principal components analysis was performed to detect different compounds between the

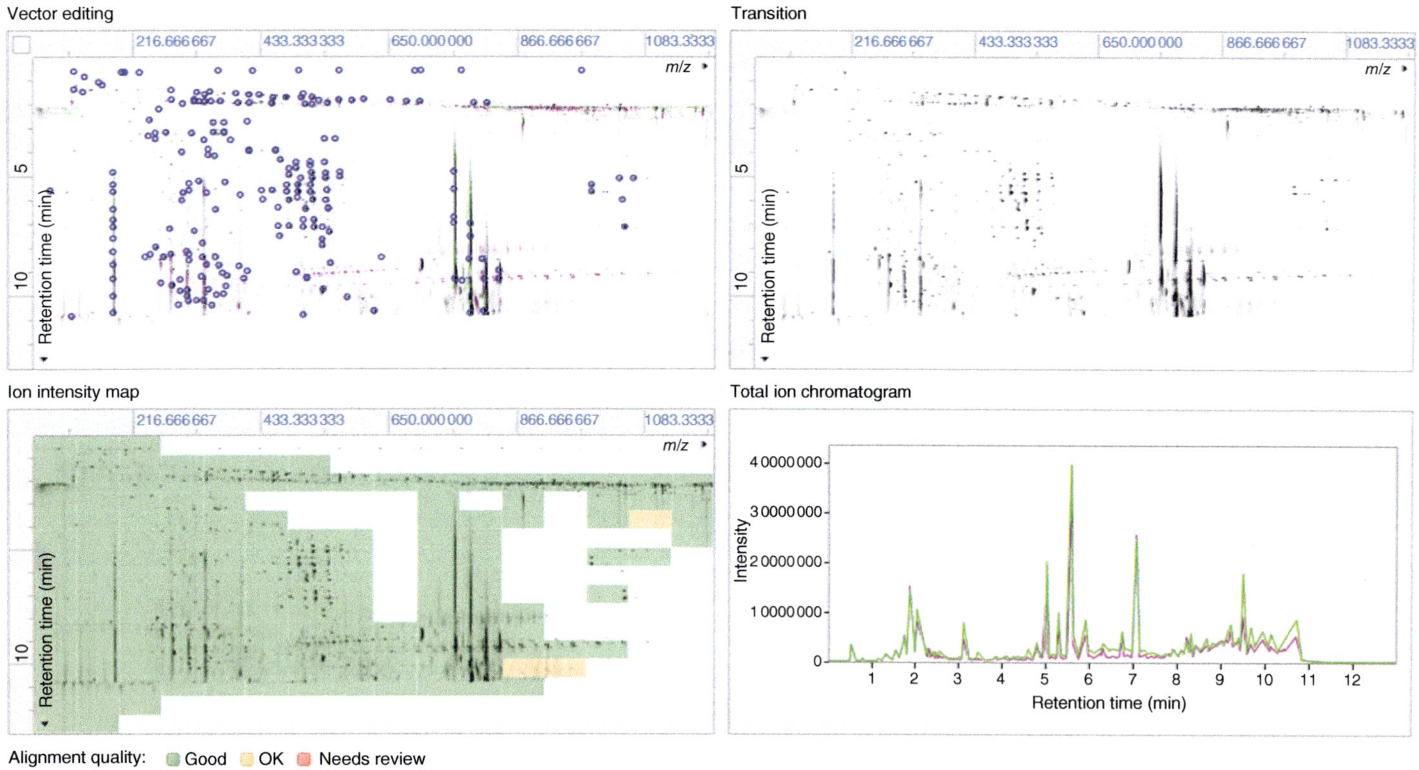

Figure 5.2 Data extraction for liquid chromatography–mass spectrometry metabolomics analyses. The data were processed for peak extraction, noise reduction, matching, alignment, and detection and then obtained the information parameters of ion identity, *m/z*, peak area, and retention time.

gout patients and healthy controls [68]. A total of 138 significantly different compounds were selected and mainly involved in primary bile acid biosynthesis, purine metabolism, and glycerophospholipid metabolism. Receiver operating characteristic curve (ROC) curve analysis revealed that 4-hydroxytriazolam, urate, and bilirubin exerted higher area under the curve (AUC) values. The metabolite profiling of myocardial injury rats was performed through ultra-high performance liquid chromatography combined with MS combined with pathway analysis [69]. A total of 15 metabolites that were considered as potential biomarkers of myocardial injury are closely related to glycerophospholipid metabolism, sphingolipid metabolism, purine metabolism, etc.

LC–MS was performed to analyze the metabolic profile before and after transcatheter aortic valve replacement (TAVR) [70]. The early changed metabolites in the peripheral vein were involved in primary bile acid biosynthesis, glycerolipid metabolism, purine metabolism, amino sugar and nucleotide sugar metabolism, etc. The cardiac-specific changed metabolites were enriched to the sphingolipid metabolism pathway after TAVR. LC–MS was also used to identify 68 endogenous differential metabolites in the urine of rats with chronic atrophic gastritis [71]. The main metabolic pathways were D-glutamine and D-glutamic acid metabolism, histidine metabolism, and purine metabolism. Four metabolites, kynurenine, 1-myristoyl-glycero-3-phosphocholine, 2-piperidinone, and hippuric acid, were identified using liquid chromatography/MS metabolomics as prognostic factors in the preoperative plasma from esophageal squamous cell carcinoma (ESCC) patients [72]. It revealed dysregulated kynurenine pathway as an important metabolic feature of ESCC, including accumulation of tryptophan, formylkynurenine, and kynurenine, as well as upregulated indoleamine 2,3-dioxygenase.

Changes of serum metabolites and pathways in chronic glomerulonephritis were analyzed by non-targeted metabolomic method based on LC/MS [73]. A total of 25 potential biomarkers were identified to express serum metabolic turbulence in chronic glomerulonephritis, and then 16 biomarkers were regulated by rhein trending to the normal states. From metabolite enrichment and pathway analysis, pharmacological effect was mainly involved in phenylalanine, phenylalanine metabolism, tyrosine and tryptophan biosynthesis, arachidonic acid metabolism, tricarboxylic acid cycle, and proline metabolism. Untargeted plasma metabolomic profiling of subjects undergoing coronary angiography was performed using LC/MS [74]. Metabolic features and pathways associated with mortality were identified, and differential regulation of the metabolic pathways involved in myocardial energetics and systemic inflammation is independently associated with mortality in patients with coronary artery disease. An LC/MS-based metabonomic approach was established to discriminate the metabolic modifications between the cohort of autism patients and the healthy subjects [75]. It revealed that 24 of the identified potential biomarkers were involved in the tryptophan kynurenine, amino acid, or lipid metabolism pathway. The combination of 7,8-dihydroneopterin, nicotinamide, D-neopterin, and anthranilic acid allows for discrimination between autism spectrum disorder patients and controls. It indicated that UPLC/MS-based metabolomics is capable of rapidly profiling autism metabolites and is a promising technique for the discovery of potential biomarkers related to autism.

Chronic pancreatitis (CP), a fibroinflammatory syndrome, leads to an increased risk for pancreatic cancer. Owing to the lack of specific biomarkers, the diagnosis is based on symptoms and insensitive imaging features, preventing an early appropriate treatment. A total of eight metabolites were identified in large and prospective cohorts and produced optimal classification for CP according to these metabolic signature with an area under the curve of 0.85 [76]. Untargeted plasma metabolomics was used to identify metabolite signature associated with cancer-related fatigue [77]. It yielded a total of 1120 compounds, and metabolic pathways unique to fatigued cancer included histidine metabolism, sphingolipid metabolism, and cysteine and methionine metabolism and provided a global metabolic phenotype for cancer-related fatigue. Metabolite signature associated with the prediction of early allograft dysfunction were identified, and a combination of four metabolites including lysoPC (16 : 0), betaine, PC (18 : 0/20 : 5), and palmitic acid (C16 : 0) showed an AUC of 0.821 and further improved to 0.846 by the addition of total bilirubin parameter [78]. It showed the combination of clinical parameters and metabolomics-derived biomarkers exhibited the power of panels in diagnostic evaluation of disease.

5.6 Conclusions

At the global omics level, metabolomics has the potential to contribute significantly to biomedical research and to clinical practice. With technological advances, high-definition MS, coupled with pattern recognition approach, could obtain the comprehensive metabolic profiling and metabolic pathways from large biological data sets. MS-based metabolomics enables a great impact on discovering biomarkers and identifying perturbed pathways for diagnosing diseases and the response to treatment. It can be expected to result in enhanced understanding of disease mechanisms and has a strong potential for improving the early diagnosis of diseases.

Acknowledgments

This work was supported by grants from the Key Program of Natural Science Foundation of State (Grant No. 81973745, 81302905), Natural Science Foundation of Heilongjiang Province (YQ2019H030), and Heilongjiang Touyan Innovation Team Program.

References

1 Bar, N., Korem, T., Weissbrod, O. et al. (2020). A reference map of potential determinants for the human serum metabolome. *Nature* 588 (7836): 135–140.

2 Wang, X., Zhang, A., and Sun, H. (2013). Power of metabolomics in diagnosis and biomarker discovery of hepatocellular carcinoma. *Hepatology* 57 (5): 2072–2077.

3 Zhang, H.L., Zhang, A.H., Miao, J.H. et al. (2019). Targeting regulation of tryptophan metabolism for colorectal cancer therapy: a systematic review. *RSC Advances* 9 (6): 3072–3080.

4 Fessenden, M. (2016). Metabolomics: small molecules, single cells. *Nature* 540 (7631): 153–155.

5 Pedersen, H.K., Gudmundsdottir, V., Nielsen, H.B. et al. (2016). Human gut microbes impact host serum metabolome and insulin sensitivity. *Nature* 535 (7612): 376–381.

6 Arakaki, A.K., Skolnick, J., and McDonald, J.F. (2008). Marker metabolites can be therapeutic targets as well. *Nature* 456: 443.

7 Nicholson, J.K. and Wilson, I.D. (2003). Understanding 'global' systems biology: metabonomics and the continuum of metabolism. *Nature Reviews Drug Discovery* 2: 668–676.

8 Garcia-Perez, I., Posma, J.M., Serrano-Contreras, J.I. et al. (2020). Identifying unknown metabolites using NMR-based metabolic profiling techniques. *Nature Protocols* 15 (8): 2538–2567.

9 Everett, J.R., Holmes, E., Veselkov, K.A. et al. (2019). A unified conceptual framework for metabolic phenotyping in diagnosis and prognosis. *Trends in Pharmacological Sciences* 40 (10): 763–773.

10 Domingo-Almenara, X., Montenegro-Burke, J.R., Ivanisevic, J. et al. (2018). XCMS-MRM and METLIN-MRM: a cloud library and public resource for targeted analysis of small molecules. *Nature Methods* 15 (9): 681–684.

11 Garcia-Perez, I., Posma, J.M., Gibson, R. et al. (2017). Objective assessment of dietary patterns by use of metabolic phenotyping: a randomised, controlled, crossover trial. *The Lancet Diabetes & Endocrinology* 5 (3): 184–195.

12 Want, E.J., Masson, P., Michopoulos, F. et al. (2013). Global metabolic profiling of animal and human tissues via UPLC-MS. *Nature Protocols* 8 (1): 17–32.

13 Nicholson, J.K. and Wilson, I.D. (2003). Understanding 'global' systems biology: metabonomics and the continuum of metabolism. *Nature Reviews Drug Discovery* 2 (8): 668–676.

14 Zhang, A., Sun, H., Yan, G. et al. (2014). Metabolomics in diagnosis and biomarker discovery of colorectal cancer. *Cancer Letters* 345 (1): 17–20.

15 Qiu, S., Zhang, A.H., Guan, Y. et al. (2020). Functional metabolomics using UPLC-Q/TOF-MS combined with ingenuity pathway analysis as a promising strategy for evaluating the efficacy and discovering amino acid metabolism as a potential therapeutic mechanism-related target for geniposide against alcoholic liver disease. *RSC Advances* 10 (5): 2677–2690.

16 Zhang, A., Sun, H., Wang, P. et al. (2012). Recent and potential developments of biofluid analyses in metabolomics. *Journal of Proteomics* 75 (4): 1079–1088.

17 Sun, H., Zhang, A., Yan, G. et al. (2013). Metabolomic analysis of key regulatory metabolites in hepatitis C virus-infected tree shrews. *Molecular & Cellular Proteomics* 12 (3): 710–719.

18 Wang, X., Zhang, A., Han, Y. et al. (2012). Urine metabolomics analysis for biomarker discovery and detection of jaundice syndrome in patients with liver disease. *Molecular & Cellular Proteomics* 11 (8): 370–380.

19 Wang, X., Yang, B., Sun, H. et al. (2012). Pattern recognition approaches and computational systems tools for ultra performance liquid chromatography–mass spectrometry-based comprehensive metabolomic profiling and pathways analysis of biological data sets. *Analytical Chemistry* 84 (1): 428–439.

20 Zhang, A., Sun, H., Yan, G. et al. (2016). Mass spectrometry-based metabolomics: applications to biomarker and metabolic pathway research. *Biomedical Chromatography* 30 (1): 7–12.

21 Yang, Q., Zhang, A.H., Miao, J.H. et al. (2019). Metabolomics biotechnology, applications, and future trends: a systematic review. *RSC Advances* 9 (64): 37245–37257.

22 Zhang, A.H., Ma, Z.M., Kong, L. et al. (2020). High-throughput lipidomics analysis to discover lipid biomarkers and profiles as potential targets for evaluating efficacy of Kai–Xin–San against APP/PS1 transgenic mice based on UPLC–Q/TOF–MS. *Biomedical Chromatography* 34 (2): e4724.

23 Zhang, A., Sun, H., and Wang, X. (2017). Emerging role and recent applications of metabolomics biomarkers in obesity disease research. *RSC Advances* 7 (25): 14966–14973.

24 Zhang, A., Zhou, X., Zhao, H. et al. (2017). Metabolomics and proteomics technologies to explore the herbal preparation affecting metabolic disorders using high resolution mass spectrometry. *Molecular BioSystems* 13 (2): 320–329.

25 Wu, X.H., Sun, X.L., Zhao, C. et al. (2019). Exploring the pharmacological effects and potential targets of paeoniflorin on the endometriosis of cold coagulation and blood stasis model rats by ultra-performance liquid chromatography tandem mass spectrometry with a pattern recognition approach. *RSC Advances* 9 (36): 20796–20805.

26 Zhang, A., Sun, H., Wu, X. et al. (2012). Urine metabolomics. *Clinica Chimica Acta* 414: 65–69.

27 Liang, Q., Liu, H., Xie, L.X. et al. (2017). High-throughput metabolomics enables biomarker discovery in prostate cancer. *RSC Advances* 7 (5): 2587–2593.

28 Zhang, A., Liu, Q., Zhao, H. et al. (2016). Phenotypic characterization of nanshi oral liquid alters metabolic signatures during disease prevention. *Scientific Reports* 6: 19333.

29 Liang, Q., Liu, H., Xing, H. et al. (2016). High-resolution mass spectrometry for exploring metabolic signatures of sepsis-induced acute kidney injury. *RSC Advances* 6 (36): 29863–29868.

30 Zhang, A.H., Ma, Z.M., Sun, H. et al. (2019). High-throughput metabolomics evaluate the efficacy of total lignans from *Acanthophanax senticosus* stem against ovariectomized osteoporosis rat. *Frontiers in Pharmacology* 10: 553.

31 Liang, Q., Liu, H., Zhang, T. et al. (2016). Discovery of serum metabolites for diagnosis of mild cognitive impairment to Alzheimer's disease progression using an optimized metabolomics method. *RSC Advances* 6: 3586–3591.

32 Zhang, A., Fang, H., Wang, Y. et al. (2017). Discovery and verification of the potential targets from bioactive molecules by network pharmacology-based target prediction combined with high-throughput metabolomics. *RSC Advances* 7 (81): 51069–51078.

33 Liang, Q., Zhu, Y., Liu, H. et al. (2017). High-throughput lipidomics enables discovery of the mode of action of huaxian capsule impacting the metabolism of sepsis. *RSC Advances* 7 (71): 44990–44996.

34 Zhang, A., Wang, H., Sun, H. et al. (2015). Metabolomics strategy reveals therapeutical assessment of limonin on nonbacterial prostatitis. *Food & Function* 6 (11): 3540–3549.

35 Fang, H., Zhang, A.H., Sun, H. et al. (2019). High-throughput metabolomics screen coupled with multivariate statistical analysis identifies therapeutic targets in alcoholic liver disease rats using liquid chromatography–mass spectrometry. *Journal of Chromatography B* 1109: 112–120.

36 Zhang, Y., Liu, P., Li, Y. et al. (2017). Exploration of metabolite signatures using high-throughput mass spectrometry coupled with multivariate data analysis. *RSC Advances* 7 (11): 6780–6787.

37 Zhang, A., Sun, H., and Wang, X. (2012). Saliva metabolomics opens door to biomarker discovery, disease diagnosis, and treatment. *Applied Biochemistry and Biotechnology* 168 (6): 1718–1727.

38 Qiu, S., Zhang, A., Zhang, T. et al. (2017). Dissect new mechanistic insights for geniposide efficacy on the hepatoprotection using multiomics approach. *Oncotarget* 8 (65): 108760.

39 Zhang, A., Sun, H., Wang, P. et al. (2012). Modern analytical techniques in metabolomics analysis. *Analyst* 137 (2): 293–300.

40 Zhang, A., Sun, H., Xu, H. et al. (2013). Cell metabolomics. *Omics: A Journal of Integrative Biology* 17 (10): 495–501.

41 Liang, Q., Liu, H., Xing, H. et al. (2016). UPLC-QTOF/MS based metabolomics reveals metabolic alterations associated with severe sepsis. *RSC Advances* 6 (49): 43293–43298.

42 Zhang, A.-H., Qiu, S., Xu, H.-Y. et al. (2014). Metabolomics in diabetes. *Clinica Chimica Acta* 429: 106–110.

43 Liang, Q., Liu, H., Jiang, Y. et al. (2016). Novel liquid chromatography-mass spectrometry for metabolite biomarkers of acute lung injury disease. *Analytical Methods* 8 (31): 6017–6022.

44 Zhang, A.H., Sun, H., Yan, G.L. et al. (2019). Chinmedomics: a powerful approach integrating metabolomics with serum pharmacochemistry to evaluate the efficacy of traditional Chinese medicine. *Engineering* 5 (1): 60–68.

45 Wang, X., Wang, Q., Zhang, A. et al. (2013). Metabolomics study of intervention effects of wen-Xin-formula using ultra high-performance liquid chromatography/mass spectrometry coupled with pattern recognition approach. *Journal of Pharmaceutical and Biomedical Analysis* 74: 22–30.

46 Zhang, A., Sun, H., and Wang, X. (2014). Urinary metabolic profiling of rat models revealed protective function of scoparone against alcohol induced hepatotoxicity. *Scientific Reports* 4: 6768.

47 Wang, X., Yang, B., Zhang, A. et al. (2012). Potential drug targets on insomnia and intervention effects of Jujuboside a through metabolic pathway analysis as revealed by UPLC/ESI-SYNAPT-HDMS coupled with pattern recognition approach. *Journal of Proteomics* 75 (4): 1411–1427.

48 Zhang, A.H., Sun, H., Yan, G.L. et al. (2014). Metabolomics study of type 2 diabetes using ultra-performance LC-ESI/quadrupole-TOF high-definition MS coupled with pattern recognition methods. *Journal of Physiology and Biochemistry* 70 (1): 117–128.

49 Zhang, A., Sun, H., and Wang, X. (2012). Serum metabolomics as a novel diagnostic approach for disease: a systematic review. *Analytical and Bioanalytical Chemistry* 404 (4): 1239–1245.

50 Fang, H., Zhang, A., Yu, J. et al. (2016). Insight into the metabolic mechanism of scoparone on biomarkers for inhibiting Yanghuang syndrome. *Scientific Reports* 6 (1): 1–9.

51 Zhang, A., Sun, H., Han, Y. et al. (2012). Exploratory urinary metabolic biomarkers and pathways using UPLC-Q-TOF-HDMS coupled with pattern recognition approach. *Analyst* 137: 4200–4208.

52 Zhang, A., Sun, H., Dou, S. et al. (2013). Metabolomics study on the hepatoprotective effect of scoparone using ultra-performance liquid chromatography/electrospray ionization quadruple time-of-flight mass spectrometry. *Analyst* 138: 353–361.

53 Wang, X., Sun, H., Zhang, A. et al. (2011). Ultra-performance liquid chromatography coupled to mass spectrometry as a sensitive and powerful technology for metabolomic studies. *Journal of Separation Science* 34 (24): 3451–3459.

54 Dong, H., Zhang, A., Sun, H. et al. (2012). Ingenuity pathways analysis of urine metabolomics phenotypes toxicity of Chuanwu in Wistar rats by UPLC-Q-TOF-HDMS coupled with pattern recognition methods. *Molecular BioSystems* 8 (4): 1206–1221.

55 Zhang, A., Sun, H., Han, Y. et al. (2013). Urinary metabolic biomarker and pathway study of hepatitis B virus infected patients based on UPLC-MS system. *PLoS One* 8 (5): e64381.

56 Wang, X., Zhang, A., Sun, H. et al. (2016). Discovery and development of innovative drug from traditional medicine by integrated chinmedomics strategies in the post-genomic era. *TrAC Trends in Analytical Chemistry* 76: 86–94.

57 Zhang, A., Sun, H., Yan, G. et al. (2017). Recent developments and emerging trends of mass spectrometry for herbal ingredients analysis. *TrAC Trends in Analytical Chemistry* 94: 70–76.

58 Wang, X., Zhang, A., and Sun, H. (2012). Future perspectives of Chinese medical formulae: chinmedomics as an effector. *Omics: A Journal of Integrative Biology* 16 (7–8): 414–421.

59 Wang, X.J., Ren, J.L., Zhang, A.H. et al. (2019). Novel applications of mass spectrometry-based metabolomics in herbal medicines and its active ingredients: current evidence. *Mass Spectrometry Reviews* 38 (4–5): 380–402.

60 Wang, X., Sun, H., Zhang, A. et al. (2011). Potential role of metabolomics apporoaches in the area of traditional Chinese medicine: as pillars of the bridge between Chinese and Western medicine. *Journal of Pharmaceutical and Biomedical Analysis* 55 (5): 859–868.

61 Wang, X.J., Zhang, A.H., Kong, L. et al. (2019). Rapid discovery of quality-markers from Kaixin san using chinmedomics analysis approach. *Phytomedicine* 54: 371–381.

62 Chu, H., Zhang, A., Han, Y. et al. (2016). Metabolomics approach to explore the effects of Kai-Xin-San on Alzheimer's disease using UPLC/ESI-Q-TOF mass spectrometry. *Journal of Chromatography B* 1015: 50–61.

63 Nan, Y., Zhou, X., Liu, Q. et al. (2016). Serum metabolomics strategy for understanding pharmacological effects of ShenQi pill acting on kidney yang deficiency syndrome. *Journal of Chromatography B* 1026: 217–226.

64 Yang, B., Zhang, A., Sun, H. et al. (2012). Metabolomic study of insomnia and intervention effects of Suanzaoren decoction using ultra-performance liquid-chromatography/electrospray-ionization synapt high-definition mass spectrometry. *Journal of Pharmaceutical and Biomedical Analysis* 58: 113–124.

65 Zhang, A., Sun, H., Yan, G. et al. (2013). Urinary metabolic profiling identifies a key role for glycocholic acid in human liver cancer by ultra-performance liquid-chromatography coupled with high-definition mass spectrometry. *Clinica Chimica Acta* 418: 86–90.

66 Li, X., Zhang, A., Sun, H. et al. (2017). Metabolic characterization and pathway analysis of berberine protects against prostate cancer. *Oncotarget* 8 (39): 65022.

67 Li, X.N., Zhang, A., Wang, M. et al. (2017). Screening the active compounds of Phellodendri Amurensis cortex for treating prostate cancer by high-throughput chinmedomics. *Scientific Reports* 7 (1): 1–15.

68 Zhong, Z., Huang, Y., Huang, Q. et al. (2020). Serum metabolic profiling analysis of gout patients based on UPLC-Q-TOF/MS. *Clinica Chimica Acta* 515: 52–60.

69 Zhao, L.K., Zhao, Y.B., and Zhang, P.X. (2020). High-throughput metabolomics discovers metabolite biomarkers and insights the protective mechanism of schisandrin B on myocardial injury rats. *Journal of Separation Science* https://doi.org/10.1002/jssc.202000875.

70 Liao, Y., Liu, C., Xiong, T. et al. (2020). Metabolic modulation and potential biomarkers of the prognosis identification for severe aortic stenosis after TAVR by a metabolomics study. *Cardiology Research and Practice* 2020: 3946913.

71 Zu, G.X., Sun, Q.Q., Chen, J. et al. (2020). Urine metabolomics of rats with chronic atrophic gastritis. *PLoS One* 15 (11): e0236203.

72 Chen, Z., Dai, Y., Huang, X. et al. (2020). Combined metabolomic analysis of plasma and tissue reveals a prognostic risk score system and metabolic dysregulation in esophageal squamous cell carcinoma. *Frontiers in Oncology* 10: 1545.

73 Yu, W., Yang, W., Zhao, M.Y. et al. (2020). Functional metabolomics analysis elucidating the metabolic biomarker and key pathway change associated with the chronic glomerulonephritis and revealing action mechanism of Rhein. *Frontiers in Pharmacology* 11: 554783.

74 Mehta, A., Liu, C., Nayak, A. et al. (2020). Untargeted high-resolution plasma metabolomic profiling predicts outcomes in patients with coronary artery disease. *PLoS One* 15 (8): e0237579.

75 Liang, Y., Ke, X., Xiao, Z. et al. (2020). Untargeted metabolomic profiling using UHPLC-QTOF/MS reveals metabolic alterations associated with autism. *BioMed Research International* 2020: 6105608.

76 Adam, M.G., Beyer, G., Christiansen, N. et al. (2021). Identification and validation of a multivariable prediction model based on blood plasma and serum metabolomics for the distinction of chronic pancreatitis subjects from non-pancreas disease control subjects. *Gut* 4: 320723–320729.

77 Feng, L.R., Barb, J.J., Regan, J. et al. (2021). Plasma metabolomic profile associated with fatigue in cancer patients. *Cancer Medicine* https://doi.org/10.1002/cam4.3749.

78 Tsai, H.I., Lo, C.J., Lee, C.W. et al. (2021). A panel of biomarkers in the prediction for early allograft dysfunction and mortality after living donor liver transplantation. *American Journal of Translational Research* 13 (1): 372–382.

6

Metabolomics as Drivers for Biomarker Discovery and Mechanism Interpretation

Wanying Wang[1], *Aihua Zhang*[2], *Hui Sun*[2], *Guangli Yan*[2], *and Xijun Wang*[2]

[1] *Macau University of Science and Technology, State Key Laboratory of Quality Research in Chinese Medicine, Avenida Wai Long, Taipa, Macau, China*
[2] *Heilongjiang University of Chinese Medicine, National Chinmedomics Research Center, Functional Metabolomics Laboratory, Heping Road 24, Harbin 150040, China*

6.1 Introduction

Metabolomics, an omic science, deals with the global assessment of the metabolites present in a biological system to evaluate the progress of the disease, select potential biomarkers, and provide insights into the underlying pathophysiology [1–3]. Recent improvements in metabolomics technologies reveal the important value of biomarker discovery and mechanism analysis [4–7]. Because metabolites represent the downstream expression of genome, transcriptome, and proteome, they can closely reflect the phenotype of an organism [8–12]. Analyzing metabolic differences could provide insight into the underlying pathology, diagnosis, and prognosis of disease [13–15]. Additionally, small-molecule metabolites can provide mechanistic insights into novel biomarkers for diseases [16]. Metabolomics may provide useful biomarkers for early-stage disease diagnosis and prognosis [17].

The development of metabolomic technologies holds the promise to significantly improve diagnosis, unravel more appropriate therapeutic targets, and enable more precise prognosis of disease development [18–23]. The potential of this approach for clinical diagnostics is huge, since only minimal biological preparation is necessary. In recent years, the development of instrumental systems including the ultra-performance liquid chromatography, high-resolution mass spectrometry (MS), and nuclear magnetic resonance (NMR) spectroscopy has enabled more comprehensive coverage of the metabolome [13, 24–28]. Advances in chromatography coupled with mass spectrometry have particularly improved both the efficiency and reliability of metabolic profiling, representing one of the major platforms and showing great potential and value in the clinic [29–34].

Early detection is the most effective way to improve the clinical outcome of diseases. Metabolomics provides a powerful platform for discovering novel biomarkers and metabolic pathways for mechanism interpretation [35–40]. Metabolites as important indicators of physiological or pathological states could provide

Mass Spectrometry-Based Metabolomics in Clinical and Herbal Medicines: Strategies, Technologies, and Applications, First Edition. Edited by Aihua Zhang and Wanying Wang.

key information for the identification of early biomarkers for disease [41–46]. Metabolomics has the potential to generate novel noninvasive diagnostic tests and provides a unique insight into the novel metabolic pathways, which are simple and cost effective yet can retain high sensitivity and specificity characteristics [47–50]. It has provided a significant contribution in understanding disease pathophysiology, in promoting the discovery of aberrant biochemical pathways, and in contributing to the development of targeted therapies. Because of this potential, the objective of this chapter is to highlight the clinical applications of metabolomic techniques used for bioanalytical data interpretation.

6.2 Metabolomics for Biomarker Discovery

Early diagnosis is the key to the effective treatment of disease. The detection of disease biomarkers plays a critical role not only in disease early diagnosis but also in classification and disease progression, or assessment prognosis and treatment response [51–55]. Traditional diagnosis of disease is based on clinical and radiological findings, which, however, are non-specific for diseases. In addition, by the time clinical and radiological findings become apparent, some diseases have already progressed to the advanced stage. Recently, a lot of research has focused on the discovery of biomarkers, which could accurately predict and make an early diagnosis. More recently applied metabolomic techniques can not only provide important insight into the pathophysiological mechanisms but can also aid the detection of potentially predictive, early diagnostic, and prognostic biomarkers [56–61]. Progress in metabolomics technology has allowed for the simultaneous measurement of a large number of metabolic products, using serum/plasma, urine, feces, tissues, and other biological specimens [62–64].

The preclinical biomarkers of disease will be key to the development of disease preventative therapies. The early detection of diseases is pivotal for successful patient treatment and management. With the recent emergence of new technologies, the biomarker discovery has been the subject of intense research [65–68]. The alterations of metabolites in biofluids are indicators of variations in physiology or pathology. Metabolomics is a well-established rapidly developing research field involving quantitative and qualitative metabolite assessment within biological systems [69–71]. The metabolome represents the endpoint of the omics cascade and is also the closest point in the cascade to the phenotype [72–74]. Therefore, metabolome analysis can be a useful approach for finding effective diagnostic markers and examining unknown pathological conditions. Metabolomics involves the establishment of relationships between phenotype and metabolic signatures, which are key aspects of biological function, and is a powerful technique for the elucidation of biochemical pathways to improve diagnosis and therapy [75–78]. It has the potential to provide insight into the pathogenesis of disease states and discover diagnostic markers for therapeutic targets. An advantage or the predictive power of metabolomics performed better in both sensitivity and specificity and may be of advantage in the future for the determination of biomarkers. Random forest (Figure 6.1)

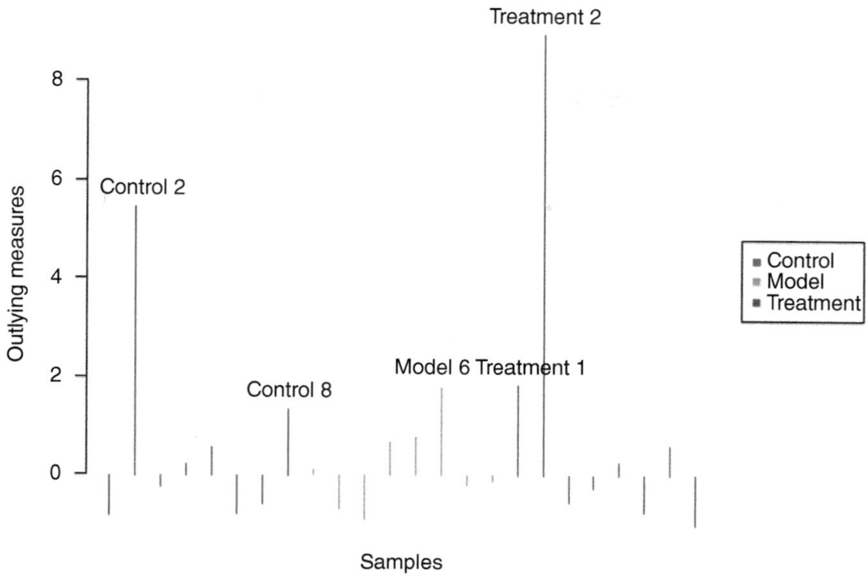

Figure 6.1 The random forest for evaluating the performance for potential outliers of bio-samples.

and heatmap (Figure 6.2) can be used to evaluate the performance for potential outliers of bio-samples in the corresponding group. The high-dimensional feature selection-significance analysis of microarrays (SAM) can be used to select significant metabolites (Figure 6.3). The classical univariate receiver operating characteristic (ROC) (Figure 6.4) and the multivariate ROC curve (Figure 6.5) can be used to evaluate the performance for sensitivity and specificity of metabolomics-derived biomarkers (Figure 6.4). Moreover, metabolic profiling is highly simple, accurate, and specific and should prove equally valuable in metabolomic research applications.

6.3 Metabolomics for Mechanism Interpretation

Dissecting the metabolic changes through metabolomics has shown the potential to improve understanding of the pathophysiological mechanisms. High-throughput metabolite profiling provides an opportunity to identify biomarkers and to reveal metabolic mechanisms. Data coming from metabolomics contribute to better understanding of the mechanisms as well as the development of biomarkers helping to move forward to take actions. Research has reached a stage in which sufficiently robust data call for new initiatives focused on early stage of diseases [47, 79] of microarrays. Prevention starting at an early-stage diagnosis is likely to dispose of a very large impact on reducing disease incidence and its associated effects at the personal, economic, and social levels. The recent application of metabolomics to clinical research showed its potential role for a better understanding of the pathophysiology of diseases [80]. This technology offers the clear opportunity to identify

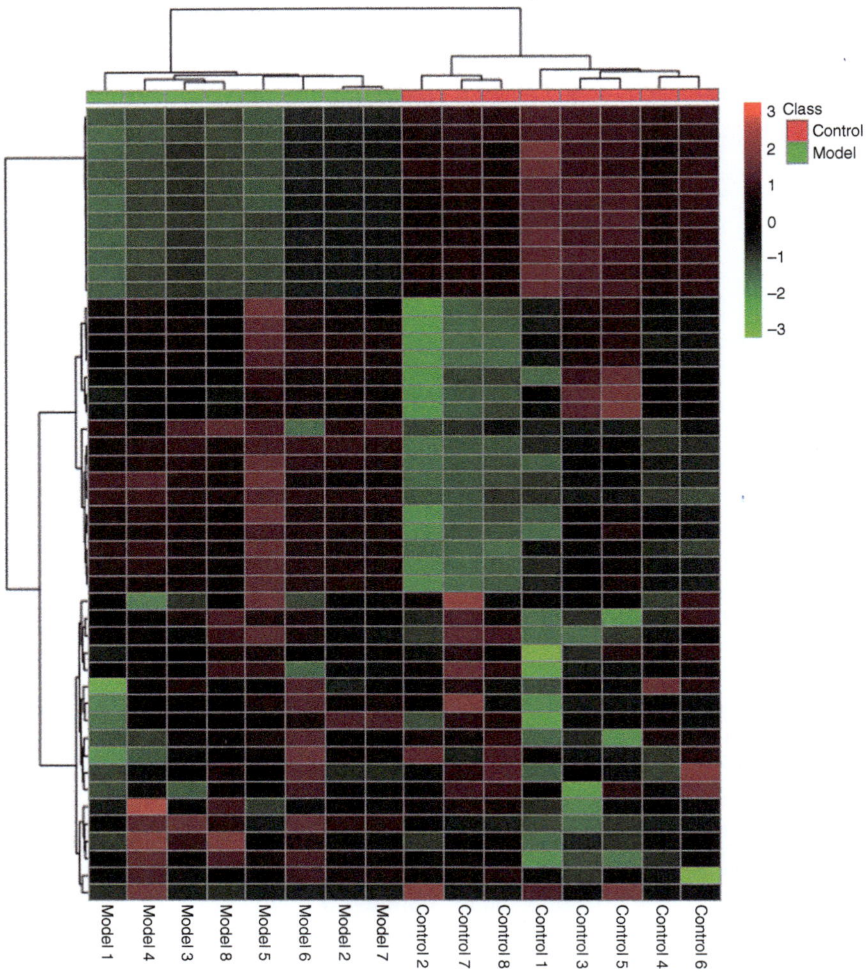

Figure 6.2 The cluster analysis heatmap for evaluating the performance for potential outliers of bio-samples.

risk factors, disease-specific or stage-specific biomarkers, and to predict therapeutic response. Metabolomic-based studies have been recently applied to assess biological system disturbances caused by specific genetic mutations resulting in disorders [81]. Analysis of the key metabolites has become an important role to monitor the state of biological organisms, and metabolomics is useful for understanding the biochemical changes in related diseases. It is increasingly being applied toward the identification of biomarkers for disease diagnosis and risk prediction, leading to the improved treatment.

Recent developments have suggested that metabolomics is the relatively new field in bioinformatics that uses measurements on metabolite abundance as a tool for disease diagnosis and other medical purposes. It holds promise for early diagnosis, increases choice of therapy, and identifies new metabolic pathways that could

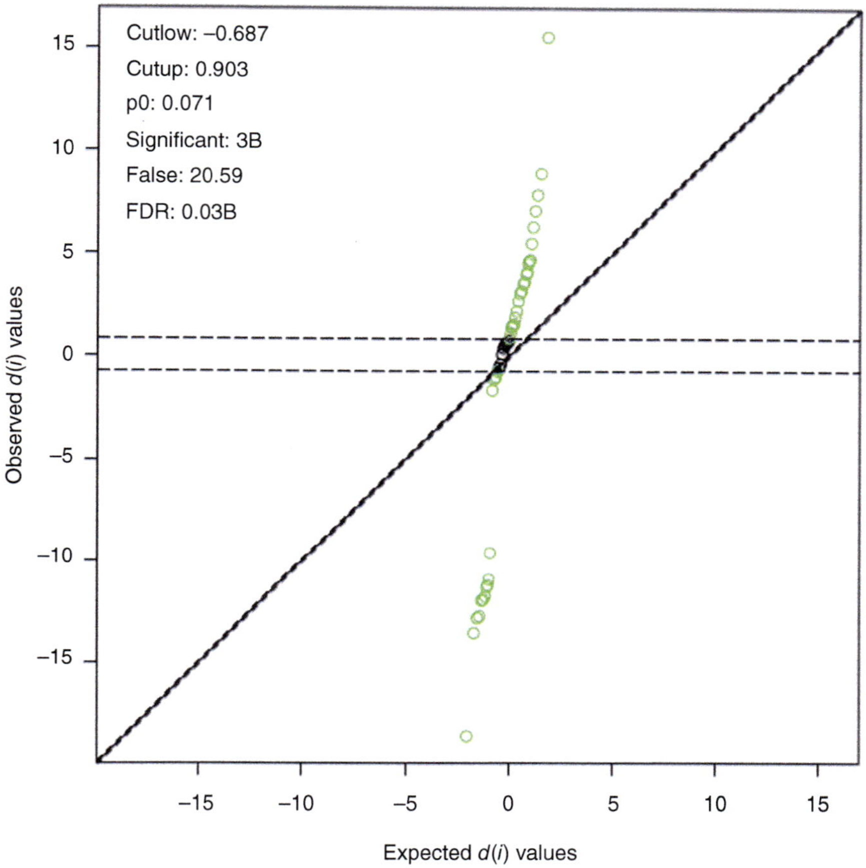

Figure 6.3 The high-dimensional feature selection-significance analysis of microarrays for selecting the significance metabolites.

potentially be targeted in diseases. Improvements in analytical technologies have made it possible to rapidly determine the concentrations of thousands of metabolites in any biological sample, which has resulted in metabolome analysis being applied to various clinical studies. The advent of the advanced metabolomics technologies and associated bioinformatics development is bringing these goals into focus. Metabolomics technology is both still in development and is a promising tool for elucidating biological pathways and discovering clinical biomarkers, strengthening the efforts for optimizing both the prevention and treatment of disease.

6.4 Application

Over recent years, metabolomics has been featured as the state-of-the-art technology that successfully opens the paths to facilitating biomarker discovery and understanding biological mechanisms. Metabolites are closely related to disease

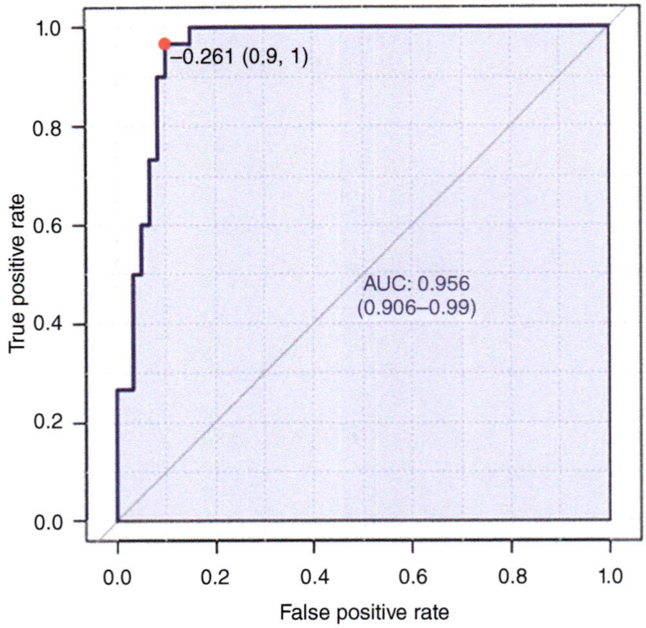

Figure 6.4 The classical univariate receiver operating characteristic (ROC) curve evaluating the performance of metabolomics-derived biomarkers. Performed by MetaboAnalyst online tool (www.metaboanalyst.ca) that generate ROC curve, to calculate AUC and their 95% confidence intervals, to compute optimal cutoffs for any given feature, and to generate performance for sensitivity, specificity, and confidence intervals at different cutoffs.

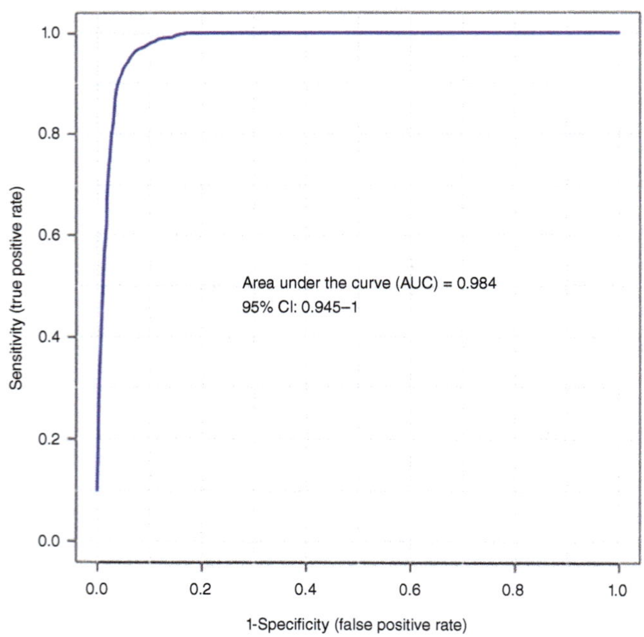

Figure 6.5 The multivariate receiver operating characteristic (ROC) curve evaluating the performance of metabolomics-derived biomarkers. Performed by MetaboAnalyst online tool (www.metaboanalyst.ca).

phenotypes, and the metabolomics has provided information regarding metabolite changes. Metabolomics has great potential for clinical diagnosis biomarkers development, for prediction of the efficacy, for assessing the therapeutic effects, for exploring the potential mechanisms of diverse diseases, and for the discovery of metabolic metabolism-based targets. It can clarify the underlying molecular mechanism of diseases via the qualitative and quantitative analysis of metabolites. High-throughput metabolomics has revealed the perturbed metabolic pathways and biomarkers of Yang-Huang syndrome (YHS), as potential targets to evaluate therapeutic effects and mechanism of geniposide [82]. A total of 33 metabolites, including tyramine glucuronide, taurine, and L-cysteine, were identified relating to YHS. These differential metabolites directly participated in the disturbance of phase I reaction and hydrophilic transformation of bilirubin and completely reversed by geniposide. The key metabolites were significantly associated with YHS to improve the understanding of the precise pathogenesis. It reported that geniposide indirectly regulates the metabolite "tyramine glucuronide" through its direct effect on the target heme oxygenase *in vivo*. A study has employed metabolomics for the detection of disease-related changes in autoimmune diseases [83]. Quantitative analysis of 28 urine organic acids was performed using MS. Methylmalonate, 2-hydroxyglutarate, and 2-hydroxybutyrate were suggested as potential biomarkers for autoimmune diseases. Urine organic acid levels related to the mechanisms of energy production and detoxification were associated with the presence of autoimmune diseases and could be an adjunct tool for early diagnosis and prediction.

Alzheimer's disease (AD) leads to the progressive loss of memory and other cognitive functions. It is the most common form of dementia in the elderly and has become a major public health problem due to the increase in life expectancy. Although the detection of AD is based on several neuropsychological tests and imaging and biological analyses, none of these biomarkers allows a clear understanding of the pathophysiological mechanisms of the disease. Inadequate understanding of the mechanisms underlying AD has hampered the development of efficient tools for diagnosis and curative interventions. In a study, ultra-performance liquid chromatography–mass spectrometry-based untargeted metabolomics were used to measure the concentrations of plasma metabolites in a cohort of subjects with AD [84]. It showed marked reductions in the levels of polyunsaturated fatty acids, acyl-carnitines, degradation products of tryptophan, and elevated levels of bile acids, and the concentrations of these metabolites were significantly correlated with disease severity.

Metabolomic analysis was conducted for comparison of urine metabolic profiles between asthmatic children and healthy controls [85]. It revealed significant changes of 20 differential metabolites. Pathway-enrichment analysis based on these differential metabolites indicated that several perturbed pathways (e.g. sphingolipid metabolism and citric acid cycle) may be related to asthma pathogenesis and 2-hydroxybutanoic acid could show excellent discriminatory performance that can serve as a potential biomarker for asthma diagnosis. Early lung cancer diagnostic biomarker discovery was used by combining metabolomics and machine learning methods [86]. Levels of 61 plasma metabolites were from the targeted metabolomics analysis, and the top five relative importance metabolic biomarkers could be potential screening biomarkers for early detection of lung cancer. Targeted metabolomics were used to identify potential biomarkers to analyze and better understand the mechanism of hyperlipidemia by a liquid chromatography–mass spectrometry method [87]. The disordered glucose metabolism, energy metabolism, and lipid metabolism were important characteristics of metabolic perturbations in hyperlipidemia. Global metabolomics profiling of cerebrospinal fluid (CSF) samples was conducted to provide insight into the underlying mechanisms of cancer-related fatigue [88]. Eight metabolites were significantly associated with fatigue in the discovery cohort. Notably, the abundance of γ-glutamylglutamine in CSF samples was also significantly associated with fatigue, suggesting that glutamatergic pathways or oxidative stress may contribute to acute lymphoblastic leukemia-associated cancer-related fatigue. Serum metabolome has provided useful biomarkers for the diagnosis of intrahepatic cholangiocarcinoma and hepatocellular carcinoma. The proposed model of combining aspartic acid, SM(42 : 3), glycine, and SM(43 : 2) could accurately differentiate these types, and it exhibited 0.890 AUC, 75% sensitivity, and 90% specificity. Specific changes in these metabolites were beneficial to differentiate intrahepatic cholangiocarcinoma from hepatocellular carcinoma and helped in the early diagnosis of these diseases [89]. These studies provide researchers the opportunity to gain new insights into the potential of these metabolites in the early diagnosis of patients and into the pathophysiological mechanisms.

6.5 Current Trends and Future Perspectives

Disease diagnosis with current biomarkers is difficult and often results in unnecessary invasive procedures as well as over-diagnosis or over-treatment, highlighting the need for novel biomarkers. Identifying biomarkers for the early detection of diseases will result in more efficient treatments, reduction in suffering, and lower mortality rates. There is a dire need for sensitive and more affordable diagnostic tools to develop effective therapies. Metabolomics appears to be an important tool to gain information on beneficial metabolites and to analyze, understand, and construct the metabolic pathways. One of the major challenges in metabolomics is the validation of metabolic biomarkers for clinical use. The identification and interpretation of metabolic biomarkers are also a challenging task. To become a clinically approved test, a potential biomarker should be confirmed and validated using hundreds of specimens and should be reproducible, specific, and sensitive. Since at present it does not apply the application of the perfect marker, maybe combining different candidate biomarker molecules will provide information compensating for the shortcoming of individual tests. As metabolomics identify the modified metabolic pathways by disease, the treatment should be used to target these pathways based on metabolomics biomarkers. Currently, metabolomics has become an important tool in clinical research and is a hot topic. In future work, the potential biomarkers should be further validated with a large enough patient cohort to achieve earlier diagnosis. Moreover, the improvement of patient prognosis and treatment will be another aim of such investigations. In the era of precise medicine, the reliable and sensitive biomarkers during the course of disease and treatment will be needed.

6.6 Conclusions

To improve the clinical course of diseases, more accurate diagnostic and assessment methods are required as early as possible. In order to achieve this, metabolomics offers new opportunities for biomarker discovery and identifying the perturbed pathways in complex diseases, and it may provide pathological understanding of diseases beyond traditional technologies that helps to identify patients at risk for disease and predicts the response to specific treatments. In short, metabolomics is a powerful tool for discovering new diagnostic biomarkers and biochemical pathways, thereby advancing the understanding of the complex mechanisms of disease to provide evidence for treatment.

Acknowledgments

This work was supported by grants from the National Key Research and Development Program of China (2018YFC1706103), Key Program of Natural Science Foundation of State (Grant No. 81830110, 81861168037, 81973745, 81302905), Natural Science Foundation of Heilongjiang Province (YQ2019H030), and Heilongjiang Touyan Innovation Team Program.

References

1 Pareek, V., Tian, H., Winograd, N. et al. (2020). Metabolomics and mass spectrometry imaging reveal channeled de novo purine synthesis in cells. *Science* 368 (6488): 283–290.

2 Park, T.J., Reznick, J., Peterson, B.L. et al. (2017). Fructose-driven glycolysis supports anoxia resistance in the naked mole-rat. *Science* 356 (6335): 307–311.

3 Guo, H., Chou, W.C., Lai, Y. et al. (2020). Multi-omics analyses of radiation survivors identify radioprotective microbes and metabolites. *Science* 370 (6516): 9097.

4 Zhang, A., Sun, H., Wang, P. et al. (2012). Recent and potential developments of biofluid analyses in metabolomics. *Journal of Proteomics* 75 (4): 1079–1088.

5 Anyfanti, P., Nikolaidou, B., and Gkaliagkousi, E. (2020). Urine metabolomic phenotyping for detection of adrenocortical carcinoma: still a long way to go. *The Lancet Diabetes & Endocrinology* 8 (11): 876–877.

6 Zhang, A., Sun, H., Yan, G. et al. (2014). Metabolomics in diagnosis and biomarker discovery of colorectal cancer. *Cancer Letters* 345 (1): 17–20.

7 Sun, H., Zhang, A., Yan, G. et al. (2013). Metabolomic analysis of key regulatory metabolites in hepatitis C virus-infected tree shrews. *Molecular & Cellular Proteomics* 12 (3): 710–719.

8 Wang, X., Zhang, A., and Sun, H. (2013). Power of metabolomics in diagnosis and biomarker discovery of hepatocellular carcinoma. *Hepatology* 57 (5): 2072–2077.

9 Bancos, I., Taylor, A.E., Chortis, V. et al. (2020). Urine steroid metabolomics for the differential diagnosis of adrenal incidentalomas in the EURINE-ACT study: a prospective test validation study. *The Lancet Diabetes & Endocrinology* 8 (9): 773–781.

10 Zhang, H.L., Zhang, A.H., Miao, J.H. et al. (2019). Targeting regulation of tryptophan metabolism for colorectal cancer therapy: a systematic review. *RSC Advances* 9 (6): 3072–3080.

11 Qiu, S., Zhang, A.H., Guan, Y. et al. (2020). Functional metabolomics using UPLC-Q/TOF-MS combined with ingenuity pathway analysis as a promising strategy for evaluating the efficacy and discovering amino acid metabolism as a potential therapeutic mechanism-related target for geniposide against alcoholic liver disease. *RSC Advances* 10 (5): 2677–2690.

12 van Laarhoven, A., Dian, S., Aguirre-Gamboa, R. et al. (2018). Cerebral tryptophan metabolism and outcome of tuberculous meningitis: an observational cohort study. *The Lancet Infectious Diseases* 18 (5): 526–535.

13 Zhang, A., Liu, Q., Zhao, H. et al. (2016). Phenotypic characterization of nanshi oral liquid alters metabolic signatures during disease prevention. *Scientific Reports* 6: 19333.

14 Liang, Q., Liu, H., Xing, H. et al. (2016). High-resolution mass spectrometry for exploring metabolic signatures of sepsis-induced acute kidney injury. *RSC Advances* 6 (36): 29863–29868.

15 Harder, J.M., Guymer, C., Wood, J. et al. (2020). Disturbed glucose and pyruvate metabolism in glaucoma with neuroprotection by pyruvate or rapamycin. *Proceedings of the National Academy of Sciences of the United States of America* 117 (52): 33619–33627.

16 Zhang, A., Fang, H., Wang, Y. et al. (2017). Discovery and verification of the potential targets from bioactive molecules by network pharmacology-based target prediction combined with high-throughput metabolomics. *RSC Advances* 7 (81): 51069–51078.

17 Liang, Q., Liu, H., Zhang, T. et al. (2016). Discovery of serum metabolites for diagnosis of mild cognitive impairment to Alzheimer's disease progression using an optimized metabolomics method. *RSC Advances* 6: 3586–3591.

18 Vial, T., Tan, W.L., Deharo, E. et al. (2020). Mosquito metabolomics reveal that dengue virus replication requires phospholipid reconfiguration via the remodeling cycle. *Proceedings of the National Academy of Sciences of the United States of America* 117 (44): 27627–27636.

19 Zhang, A., Sun, H., and Wang, X. (2012). Saliva metabolomics opens door to biomarker discovery, disease diagnosis, and treatment. *Applied Biochemistry and Biotechnology* 168 (6): 1718–1727.

20 Liang, Q., Liu, H., Xing, H. et al. (2016). UPLC-QTOF/MS based metabolomics reveals metabolic alterations associated with severe sepsis. *RSC Advances* 6 (49): 43293–43298.

21 Zhang, A.-H., Qiu, S., Xu, H.-Y. et al. (2014). Metabolomics in diabetes. *Clinica Chimica Acta* 429: 106–110.

22 Ryabova, A., Cornette, R., Cherkasov, A. et al. (2020). Combined metabolome and transcriptome analysis reveals key components of complete desiccation tolerance in an anhydrobiotic insect. *Proceedings of the National Academy of Sciences of the United States of America* 117 (32): 19209–19220.

23 Zhang, A., Sun, H., Xu, H. et al. (2013). Cell metabolomics. *Omics: A Journal of Integrative Biology* 17 (10): 495–501.

24 Hou, Y.C., Yu, H.C., Martin, R. et al. (2020). Precision medicine integrating whole-genome sequencing, comprehensive metabolomics, and advanced imaging. *Proceedings of the National Academy of Sciences of the United States of America* 117 (6): 3053–3062.

25 Zhang, A.H., Sun, H., Yan, G.L. et al. (2019). Chinmedomics: a powerful approach integrating metabolomics with serum pharmacochemistry to evaluate the efficacy of traditional Chinese medicine. *Engineering* 5 (1): 60–68.

26 Liang, Q., Liu, H., Jiang, Y. et al. (2016). Novel liquid chromatography–mass spectrometry for metabolite biomarkers of acute lung injury disease. *Analytical Methods* 8 (31): 6017–6022.

27 Sovio, U., Goulding, N., McBride, N. et al. (2020). A maternal serum metabolite ratio predicts fetal growth restriction at term. *Nature Medicine* 26 (3): 348–353.

28 Wang, X., Wang, Q., Zhang, A. et al. (2013). Metabolomics study of intervention effects of wen-Xin-formula using ultra high-performance liquid chromatography/mass spectrometry coupled with pattern recognition approach. *Journal of Pharmaceutical and Biomedical Analysis* 74: 22–30.

29 Zhang, A.H., Sun, H., Yan, G.L. et al. (2014). Metabolomics study of type 2 diabetes using ultra-performance LC-ESI/quadrupole-TOF high-definition MS coupled with pattern recognition methods. *Journal of Physiology and Biochemistry* 70 (1): 117–128.

30 Yachida, S., Mizutani, S., Shiroma, H. et al. (2019). Metagenomic and metabolomic analyses reveal distinct stage-specific phenotypes of the gut microbiota in colorectal cancer. *Nature Medicine* 25 (6): 968–976.

31 Zhang, A., Sun, H., and Wang, X. (2014). Urinary metabolic profiling of rat models revealed protective function of scoparone against alcohol induced hepatotoxicity. *Scientific Reports* 4: 6768.

32 Liu, R., Hong, J., Xu, X. et al. (2017). Gut microbiome and serum metabolome alterations in obesity and after weight-loss intervention. *Nature Medicine* 23 (7): 859–868.

33 Zhang, A.H., Ma, Z.M., Kong, L. et al. (2020). High-throughput lipidomics analysis to discover lipid biomarkers and profiles as potential targets for evaluating efficacy of Kai–Xin–san against APP/PS1 transgenic mice based on UPLC–Q/TOF–MS. *Biomedical Chromatography* 34 (2): e4724.

34 Wang, X., Zhang, A., Han, Y. et al. (2012). Urine metabolomics analysis for biomarker discovery and detection of jaundice syndrome in patients with liver disease. *Molecular & Cellular Proteomics* 11 (8): 370–380.

35 Zhang, A., Sun, H., Yan, G. et al. (2016). Mass spectrometry-based metabolomics: applications to biomarker and metabolic pathway research. *Biomedical Chromatography* 30 (1): 7–12.

36 Yang, Q., Zhang, A.H., Miao, J.H. et al. (2019). Metabolomics biotechnology, applications, and future trends: a systematic review. *RSC Advances* 9 (64): 37245–37257.

37 Wang, X., Yang, B., Sun, H. et al. (2012). Pattern recognition approaches and computational systems tools for ultra performance liquid chromatography–mass spectrometry-based comprehensive metabolomic profiling and pathways analysis of biological data sets. *Analytical Chemistry* 84 (1): 428–439.

38 Zhang, A., Zhou, X., Zhao, H. et al. (2017). Metabolomics and proteomics technologies to explore the herbal preparation affecting metabolic disorders using high resolution mass spectrometry. *Molecular BioSystems* 13 (2): 320–329.

39 Wu, X.H., Sun, X.L., Zhao, C. et al. (2019). Exploring the pharmacological effects and potential targets of paeoniflorin on the endometriosis of cold coagulation and blood stasis model rats by ultra-performance liquid chromatography tandem mass spectrometry with a pattern recognition approach. *RSC Advances* 9 (36): 20796–20805.

40 Zhang, A., Sun, H., and Wang, X. (2017). Emerging role and recent applications of metabolomics biomarkers in obesity disease research. *RSC Advances* 7 (25): 14966–14973.

41 Liang, Q., Liu, H., Xie, L.X. et al. (2017). High-throughput metabolomics enables biomarker discovery in prostate cancer. *RSC Advances* 7 (5): 2587–2593.

42 Zhang, A., Sun, H., Wu, X. et al. (2012). Urine metabolomics. *Clinica Chimica Acta* 414: 65–69.

43 Liang, Q., Zhu, Y., Liu, H. et al. (2017). High-throughput lipidomics enables discovery of the mode of action of huaxian capsule impacting the metabolism of sepsis. *RSC Advances* 7 (71): 44990–44996.

44 Zhang, A.H., Ma, Z.M., Sun, H. et al. (2019). High-throughput metabolomics evaluate the efficacy of total lignans from *Acanthophanax senticosus* stem against ovariectomized osteoporosis rat. *Frontiers in Pharmacology* 10: 553.

45 Fang, H., Zhang, A.H., Sun, H. et al. (2019). High-throughput metabolomics screen coupled with multivariate statistical analysis identifies therapeutic targets in alcoholic liver disease rats using liquid chromatography–mass spectrometry. *Journal of Chromatography B* 1109: 112–120.

46 Zhang, A., Wang, H., Sun, H. et al. (2015). Metabolomics strategy reveals therapeutical assessment of limonin on nonbacterial prostatitis. *Food & Function* 6 (11): 3540–3549.

47 Zhang, Y., Liu, P., Li, Y. et al. (2017). Exploration of metabolite signatures using high-throughput mass spectrometry coupled with multivariate data analysis. *RSC Advances* 7 (11): 6780–6787.

48 Qiu, S., Zhang, A., Zhang, T. et al. (2017). Dissect new mechanistic insights for geniposide efficacy on the hepatoprotection using multiomics approach. *Oncotarget* 8 (65): 108760.

49 Zhang, A., Sun, H., Wang, P. et al. (2012). Modern analytical techniques in metabolomics analysis. *Analyst* 137 (2): 293–300.

50 Wang, X., Yang, B., Zhang, A. et al. (2012). Potential drug targets on insomnia and intervention effects of Jujuboside a through metabolic pathway analysis as revealed by UPLC/ESI-SYNAPT-HDMS coupled with pattern recognition approach. *Journal of Proteomics* 75 (4): 1411–1427.

51 Yang, B., Zhang, A., Sun, H. et al. (2012). Metabolomic study of insomnia and intervention effects of Suanzaoren decoction using ultra-performance liquid-chromatography/electrospray-ionization synapt high-definition mass spectrometry. *Journal of Pharmaceutical and Biomedical Analysis* 58: 113–124.

52 Zhang, A.H., Sun, H., Qiu, S. et al. (2013). Metabolomics in noninvasive breast cancer. *Clinica Chimica Acta* 424: 3–7.

53 Fang, H., Zhang, A., Yu, J. et al. (2016). Insight into the metabolic mechanism of scoparone on biomarkers for inhibiting Yanghuang syndrome. *Scientific Reports* 6 (1): 1–9.

54 Zhang, A., Sun, H., Dou, S. et al. (2013). Metabolomics study on the hepato-protective effect of scoparone using ultra-performance liquid chromatography/electrospray ionization quadruple time-of-flight mass spectrometry. *Analyst* 138: 353–361.

55 Wang, X., Sun, H., Zhang, A. et al. (2011). Ultra-performance liquid chromatography coupled to mass spectrometry as a sensitive and powerful technology for metabolomic studies. *Journal of Separation Science* 34 (24): 3451–3459.

56 Zhang, A., Sun, H., Han, Y. et al. (2012). Exploratory urinary metabolic biomarkers and pathways using UPLC-Q-TOF-HDMS coupled with pattern recognition approach. *Analyst* 137: 4200–4208.

57 Wang, X., Zhang, A., and Sun, H. (2012). Future perspectives of Chinese medical formulae: chinmedomics as an effector. *Omics: A Journal of Integrative Biology* 16 (7–8): 414–421.

58 Zhang, A., Sun, H., Han, Y. et al. (2013). Urinary metabolic biomarker and pathway study of hepatitis B virus infected patients based on UPLC-MS system. *PLoS One* 8 (5): e64381.

59 Wang, X., Zhang, A., Sun, H. et al. (2016). Discovery and development of innovative drug from traditional medicine by integrated chinmedomics strategies in the post-genomic era. *TrAC Trends in Analytical Chemistry* 76: 86–94.

60 Dong, H., Zhang, A., Sun, H. et al. (2012). Ingenuity pathways analysis of urine metabolomics phenotypes toxicity of Chuanwu in Wistar rats by UPLC-Q-TOF-HDMS coupled with pattern recognition methods. *Molecular BioSystems* 8 (4): 1206–1221.

61 Zhang, A., Sun, H., Yan, G. et al. (2017). Recent developments and emerging trends of mass spectrometry for herbal ingredients analysis. *TrAC Trends in Analytical Chemistry* 94: 70–76.

62 Wang, X.J., Ren, J.L., Zhang, A.H. et al. (2019). Novel applications of mass spectrometry-based metabolomics in herbal medicines and its active ingredients: current evidence. *Mass Spectrometry Reviews* 38 (4–5): 380–402.

63 Wang, X.J., Zhang, A.H., Kong, L. et al. (2019). Rapid discovery of quality-markers from Kaixin san using chinmedomics analysis approach. *Phytomedicine* 54: 371–381.

64 Chu, H., Zhang, A., Han, Y. et al. (2016). Metabolomics approach to explore the effects of Kai-Xin-san on Alzheimer's disease using UPLC/ESI-Q-TOF mass spectrometry. *Journal of Chromatography B* 1015: 50–61.

65 Wang, X., Sun, H., Zhang, A. et al. (2011). Potential role of metabolomics apporoaches in the area of traditional Chinese medicine: as pillars of the bridge between Chinese and Western medicine. *Journal of Pharmaceutical and Biomedical Analysis* 55 (5): 859–868.

66 Nan, Y., Zhou, X., Liu, Q. et al. (2016). Serum metabolomics strategy for understanding pharmacological effects of ShenQi pill acting on kidney yang deficiency syndrome. *Journal of Chromatography B* 1026: 217–226.

67 Li, X., Zhang, A., Sun, H. et al. (2017). Metabolic characterization and pathway analysis of berberine protects against prostate cancer. *Oncotarget* 8 (39): 65022.

68 Zhang, A., Sun, H., Yan, G. et al. (2013). Urinary metabolic profiling identifies a key role for glycocholic acid in human liver cancer by ultra-performance liquid-chromatography coupled with high-definition mass spectrometry. *Clinica Chimica Acta* 418: 86–90.

69 Li, X.N., Zhang, A., Wang, M. et al. (2017). Screening the active compounds of Phellodendri Amurensis cortex for treating prostate cancer by high-throughput chinmedomics. *Scientific Reports* 7 (1): 1–15.

70 Holmes, E., Loo, R.L., Stamler, J. et al. (2008). Human metabolic phenotype diversity and its association with diet and blood pressure. *Nature* 453 (7193): 396–400.

71 Ren, J.L., Zhang, A.H., Kong, L. et al. (2018). Advances in mass spectrometry-based metabolomics for investigation of metabolites. *RSC Advances* 8 (40): 22335–22350.

72 Chu, H., Zhang, A.H., Han, Y. et al. (2015). Metabolomics and its potential in drug discovery and development from TCM. *World Journal of Traditional Chinese Medicine* 1 (4): 26.

73 Song, Q., Zhang, A.H., Yan, G.L. et al. (2017). Technological advances in current metabolomics and its application in tradition Chinese medicine. *RSC Advances* 7 (84): 53516–53524.

74 Li, Y.F., Qiu, S., Gao, L.J. et al. (2018). Metabolomic estimation of the diagnosis of hepatocellular carcinoma based on ultrahigh performance liquid chromatography coupled with time-of-flight mass spectrometry. *RSC Advances* 8 (17): 9375–9382.

75 Sun, H., Zhang, A.H., Song, Q. et al. (2018). Functional metabolomics discover pentose and glucuronate interconversion pathways as promising targets for Yang Huang syndrome treatment with Yinchenhao tang. *RSC Advances* 8 (64): 36831–36839.

76 Wang, T.J., Larson, M.G., Vasan, R.S. et al. (2011). Metabolite profiles and the risk of developing diabetes. *Nature Medicine* 17 (4): 448–453.

77 Sun, H., Zhang, H.L., Zhang, A.H. et al. (2018). Network pharmacology combined with functional metabolomics discover bile acid metabolism as a promising target for mirabilite against colorectal cancer. *RSC Advances* 8 (53): 30061–30070.

78 Liang, Q., Wang, C., Li, B. et al. (2015). Metabolomics of alcoholic liver disease: a clinical discovery study. *RSC Advances* 5 (98): 80381–80387.

79 Garcia-Perez, I., Posma, J.M., Serrano-Contreras, J.I. et al. (2020). Identifying unknown metabolites using NMR-based metabolic profiling techniques. *Nature Protocols* 15 (8): 2538–2567.

80 Wang, X., Li, J., and Zhang, A.H. (2016). Urine metabolic phenotypes analysis of extrahepatic cholangiocarcinoma disease using ultra-high performance liquid chromatography-mass spectrometry. *RSC Advances* 6 (67): 63049–63057.

81 Bar, N., Korem, T., Weissbrod, O. et al. (2020). A reference map of potential determinants for the human serum metabolome. *Nature* 588 (7836): 135–140.

82 Fang, H., Zhang, A., Zhou, X. et al. (2020). High-throughput metabolomics reveals the perturbed metabolic pathways and biomarkers of Yang Huang syndrome as potential targets for evaluating the therapeutic effects and mechanism of geniposide. *Frontiers in Medicine* 14 (5): 651–663.

83 Tsoukalas, D., Fragoulakis, V., Papakonstantinou, E. et al. (2020). Prediction of autoimmune diseases by targeted Metabolomic assay of urinary organic acids. *Metabolites* 10 (12): 502.

84 Shao, Y., Ouyang, Y., Li, T. et al. (2020). Alteration of metabolic profile and potential biomarkers in the plasma of Alzheimer's disease. *Aging and Disease* 11 (6): 1459–1470.

85 Li, S., Liu, J., Zhou, J. et al. (2020). Urinary Metabolomic profiling reveals biological pathways and predictive signatures associated with childhood asthma. *Journal of Asthma and Allergy* 13: 713–724.

86 Xie, Y., Meng, W.Y., Li, R.Z. et al. (2021). Early lung cancer diagnostic biomarker discovery by machine learning methods. *Translational Oncology* 14 (1): 100907.

87 Wang, X.F., Zhang, Y.X., and Ma, H.Y. (2020). Targeted profiling of amino acid metabolome in serum by a liquid chromatography–mass spectrometry method: application to identify potential markers for diet-induced hyperlipidemia. *Analytical Methods* 12 (18): 2355–2362.

88 Brown, A.L., Sok, P., Taylor, O. et al. (2020). Cerebrospinal fluid Metabolomic profiles associated with fatigue during treatment for pediatric acute lymphoblastic leukemia. *Journal of Pain and Symptom Management* S0885-3924 (20): 30716–30718.

89 Banales, J.M., Iñarrairaegui, M., Arbelaiz, A. et al. (2019). Serum metabolites as diagnostic biomarkers for Cholangiocarcinoma, hepatocellular carcinoma, and primary Sclerosing cholangitis. *Hepatology* 70 (2): 547–562.

7

Current Status of Technical Challenges in Mass Spectrometry-Driven Metabolomics

Shi Qiu, Aihua Zhang, Hui Sun, and Xijun Wang

Heilongjiang University of Chinese Medicine, National Chinmedomics Research Center, Functional Metabolomics Laboratory, Department of Pharmaceutical Analysis, Heping Road 24, Harbin 150040, China

7.1 Introduction

In recent years, metabolomics has become increasingly popular for many applications ranging from human health to disease diagnosis [1–5]. This strategy has the ability to analyze and quantify the complete collection of metabolites present in biological samples. Different from other omics, metabolomics aims at the changes of metabolites or the changing with time in the biological systems including cells, tissues, and organs after disturbance or stimulation [6–11]. Metabolome mainly includes endogenous components with less than 1000 molecules, which participate in the biological metabolism and maintain the normal function of the organism [12–15]. At present, the technical platforms used in metabolomics, such as nuclear magnetic resonance (NMR), gas chromatography (GC)/mass spectrometry (MS), capillary electrophoresis (CE)/MS, and liquid chromatography (LC)/MS, have been widely used in the field of plant science, biomedicine, food and nutrition, drug development, and environment science for the biomarker discovery and the metabolic analysis from 10 to 100 biological samples [16–18]. Compared with NMR, MS has unique characteristics including high selectivity and high sensitivity analysis [19, 20]. Although MS analysis can obtain comprehensive data, many isobar and isomer metabolites remain unresolved, which still is a key point in the compound identification.

Metabolomics is intended to provide a comprehensive and quantitative analysis of a wide range of metabolites in biological samples, and MS as a major analytical tool in this field has great potential [21–24]. The large data acquisition from thousands of samples and the precise quantification of all the metabolites are still technically challenging, and the metabolites analysis detectable by MS is still in its infancy. Further improvement in technologies of MS for larger-scale analysis will enable more comprehensive metabolome analysis for exploring the hidden secrets of metabolism. This chapter describes the current status of technical challenges and limitations in MS-driven metabolomics analysis and discusses breakthrough innovation method required to overcome the current bottlenecks.

Mass Spectrometry-Based Metabolomics in Clinical and Herbal Medicines: Strategies, Technologies, and Applications, First Edition. Edited by Aihua Zhang and Wanying Wang.

7.2 Metabolic Fingerprinting and Profiling

Metabolomics developed in mid-1990s was a new discipline for qualitative analysis of small-molecule metabolites with a molecular weight in a single organism or cell. Metabolites are the final product of gene expression and are produced by metabolic enzymes. In fact, metabolomics research has been able to diagnose the metabolic diseases such as diabetes, obesity, and metabolic syndrome [25–31]. There are two common methods in metabolomics research. One method, called metabolic fingerprinting, uses liquid chromatography–mass spectrometry (LC–MS) to compare those respective metabolites in different samples to determine all of the metabolites [32, 33]. In essence, metabolic fingerprinting analysis involves comparing the mass spectra of metabolites in different individuals, ultimately understanding the structure of these compounds and establishing a complete set of analytical methods to identify the characteristics of these compounds [34, 35]. The other method is metabolic profiling, which means that the researchers presume a specific metabolic pathway and conduct a more in-depth study [36, 37].

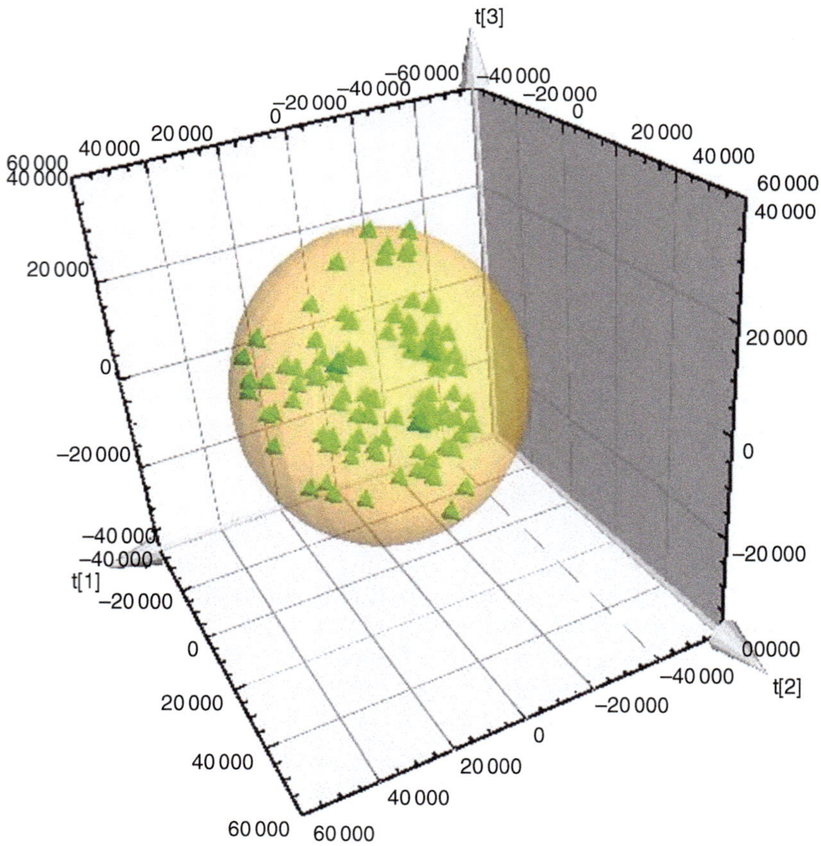

Figure 7.1 The principal component analysis for non-targeted metabolomics.

7.3 Non-targeted or Targeted Metabolomics

There are targeted and non-targeted metabolomics approaches for the qualitative analysis of metabolites [31, 38–41]. The bio-samples are first analyzed by a non-targeted metabolomics analytical method (Figure 7.1). Then, a high-quality data matrix constructed including signal intensity data of all the observed peaks in the chromatograms. The structure-related information of the observed peak signal including retention time and m/z in mass spectra can be annotated by searching the database for the fragmentation patterns. And then, the principal component analysis was used for non-targeted metabolomics (Figure 7.1); the loading ion analysis plot was used for large biological metabolites (Figure 7.2); S-plot performed by partial least squares discriminant analysis was used for non-targeted metabolomics (Figure 7.3); and the variable importance in projection was used for screening the meaningful metabolite from large biological data sets (Figure 7.4). In the case of targeted analyses, after optimizing the analytical conditions, the raw data could be obtained, and target metabolites of interest are detected using the standard compounds.

7.4 Metabolite Identification Bottleneck

Despite technological advances, currently, metabolomics studies are still limited by the difficulty of identifying all the metabolites with great chemical diversity. The biggest challenge in metabolomics research is the identification of metabolites, and

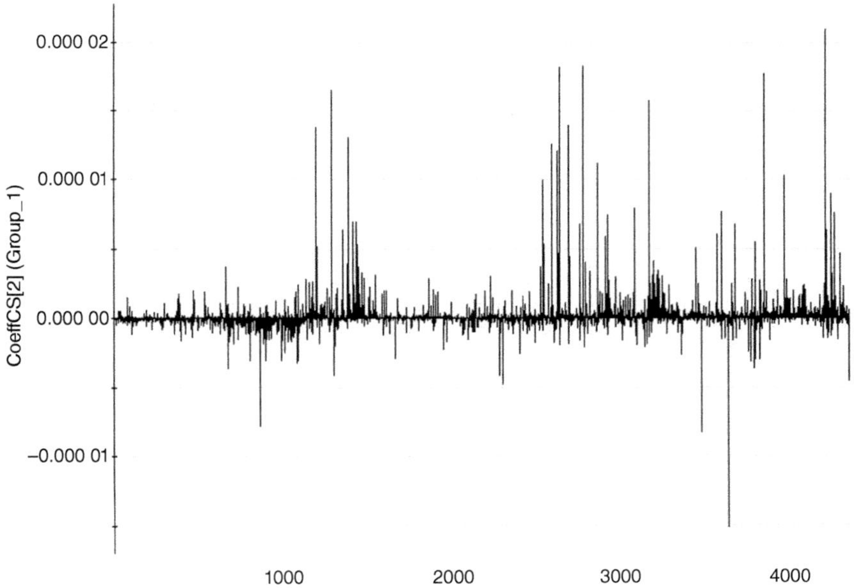

Figure 7.2 The loading ion analysis plot for large biological metabolites.

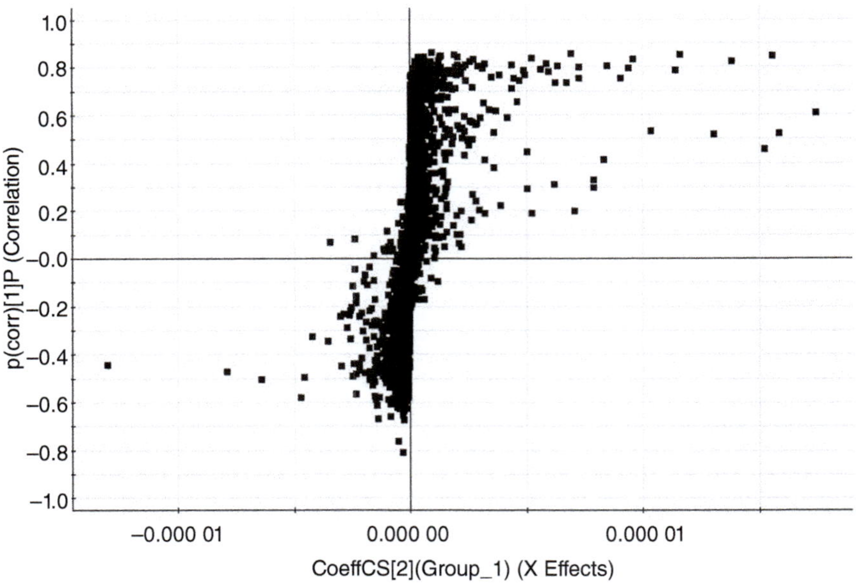

Figure 7.3 S-plot performed by partial least squares discriminant analysis for non-targeted metabolomics.

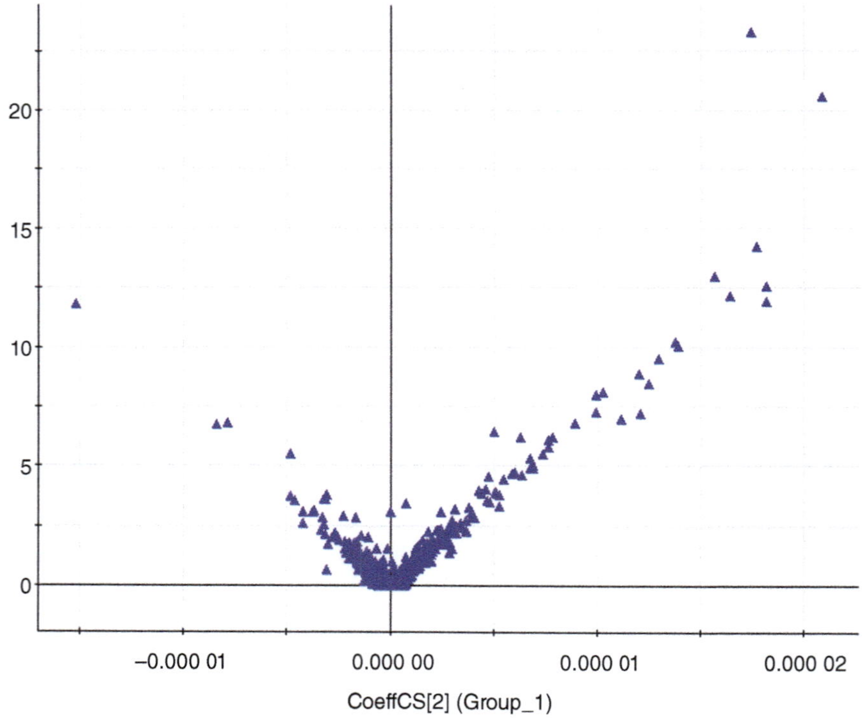

Figure 7.4 The variable importance in projection for screening the meaningful metabolite from large biological data sets.

the more challenging work is further confirming the function of all metabolites. The combination of multivariate analysis makes it easier to achieve the discovery of potential biomarkers [42–47]. The structural annotation of metabolites is essential for investigating disease mechanism and biomarker significance. Therefore, the identification of unknown structures is still the bottleneck for the development of metabolomics technology. In the identification of metabolites based on conventional metabolites research, the compounds are identified by the directly obtained MS information. There are two common ways to obtain candidate molecular structures: one is to use the molecular ion peak and MS/MS information for spectral library search and the other is that the composition of the chemical element should be determined first, and then based on the composition, database retrieval can be done. Finally, through the validation of the standard compounds, the accurate qualitative results are ultimately achieved. The elemental composition provides the basis for the molecular structure of the metabolites. However the problem is that as the relative molecular mass increases, the number of possible isomers also increases rapidly. A more convenient and reasonable way is to retrieve known chemical structures and natural product databases to remove large amounts of unreasonable or non-existent isomers. Of course, there is also a risk that may lead to the poor results for the new compounds. In recent years, the database has developed rapidly, and the database commonly used in metabolomics research has 4 types: natural compounds (Pubchem), MS database (NIST), metabolite database (HMDB [human metabolome database]), and metabolic pathway database (BioCyc) [48–54]. Success in the non-targeted approach depends on the enrichment of metabolite annotation by database searching. MS spectra-based identification of metabolites has been reported, and structural elucidation of unknown metabolites is still a difficult task, owing to lack of diversity in the measured MS data and the poor method by searching the similarity for the fragmentation patterns, retention time, and m/z in mass spectra.

7.5 Quality Control Bottleneck

Over recent years, metabolomics has been featured as the state-of-the-art technology that successfully opens the paths to understand biological mechanisms and facilitate biomarker discovery [55–61]. However, the inherent dynamic and sensitive nature of the metabolome has been challenging the accuracy of capturing the time points of interest while using biofluids such as urine and blood. The collection of samples depends on the design and form of the experiment, and the number of samples should be obtained as much as possible to reduce the effect of biological variability. Especially in metabolomics study, some influences including diet, sex, age, and genetic factors should also be concentrated. It is well known that the gender difference also can cause clinically pharmacokinetic differences and produce differences in drug efficacy of important factors [62, 63]. For children, especially newborns and infants, because of age, their respective anatomical and physiological characteristics and biochemical function have a big difference, which lead to the different metabolic capacity *in vivo* process. Thus the metabolic fingerprint analysis should be detected to cluster samples as much as possible.

Large-scale metabolome analyses usually used an experimental design employing quality control (QC) samples to correct a drift of the raw signal intensity during the whole analysis. QC samples were prepared by mixing all the samples in one metabolome analysis study [64–68]. The QC sample analyses were inserted into the start, end, and in every 5–10 actual samples in batch sequences of data acquisition. Since the QC sample theoretically includes all metabolites in the samples, a time course of signal intensity should be monitored for each metabolite. QC samples should also be analyzed to ensure experimental reliability. As for biological sample types, compared with serum, plasma, cells, and tissues samples, urine samples are widely used in metabolomics studies because it can provide a large amount of information for analysis because of its lossless availability [69, 70]. It is noteworthy that more efforts should be tried to prevent the metabolites from enzyme or oxidation process leading to its degradation or formation in the biological samples. The main methods that should be avoided this case include freezing in freezing clamping, liquid nitrogen, and acid treatment and quenching in salt containing aqueous methanol at low temperature that may lead to the data variability.

7.6 Mass Spectrometry Techniques

7.6.1 LC–MS

LC–MS (Figure 7.5) is the combination of LC and MS, where LC serves as the separation system and MS as the detection system. The advantage of chromatography is that the separation provides the most efficient choice for the separation of the mixture, but it is difficult to obtain information on the structure of the substance. MS could provide the structure information, and the sample size is also very small, but the analysis of the samples needs to be purified. Thus, it is desirable to use chromatography and MS to compensate for the shortcomings of the two instruments [71–74]. LC–MS can analyze the compounds with strong polarity, nonvolatility, and thermal instability. In addition, LC–MS has a low detection limit. By using an ion monitor, the detection capability of MS with high sensitivity can also increase more than one order of magnitude. LC–MS has the fast analysis time because the column is the narrow diameter column, shortening the analysis time and improving the separation effect. Currently, the widely used ion source is electrospray ionization (ESI) source and atmospheric pressure chemical ionization (APCI) source [75–78]. The most significant advantage is that the mass-to-charge ratio can be simplified, which can be detected by a variety of different types of quality analyzer. Besides, it can be detected in the charged state to calculate the true molecular weight of the ion and can generate highly charged and no fragmentation of the ions. Along with the development of LC–MS interface technology, MS itself is evolving and various types of MS detectors have emerged. Currently, more commonly used MS instruments include quadrupole, quadrupole ion trap, time of flight (TOF), ion cyclotron resonance, *etc.* [79–82].

Figure 7.5 Ultra-performance liquid chromatography–mass spectrometry instrumentation for metabolomics analysis.

7.6.2 GC–MS

Because LC–MS does not require inconvenient pretreatment for samples and can isolate polar, volatile, and poor thermal stability of the compound, it has a wide range of application in metabolomics researches [83, 84]. However, the first developed GC–MS technology (Figure 7.6), because of its high sensitivity, strong resolution, good reproducibility, and high-throughput advantages, can greatly expand the scope of the object by the derivative technology, which can be widely used in the analysis and detection of small-molecule metabolites such as amino acids, organic acids, fatty acids, and cholesterol with the standardized ionization conditions. Besides, in the metabolite structure identification, it has a mature and perfect standard spectrum database for query and retrieval; therefore, this technology is still one of the most common platforms for metabolomics. At present, GC–Q–MS and GC–QQQ–MS have good quantitative properties and are

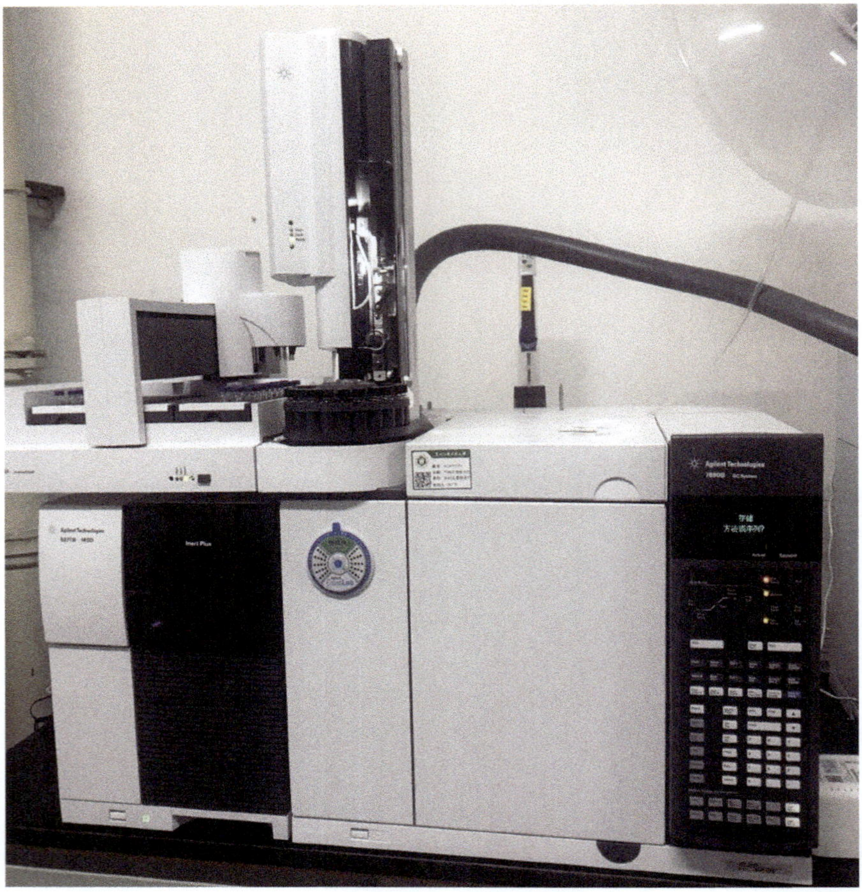

Figure 7.6 Gas chromatography–mass spectrometry instrumentation for metabolomics analysis.

commonly used in the quantitative analysis of compounds [85, 86]. In qualitative analysis, compared with other MS, TOF has a significant advantage because of its high resolution and fast screening speed. In particular, coupled with other MS, the high-resolution mass of the target ions in the primary and multistage MS can be achieved, making it ideal for structural analysis of unknown compounds from the complex samples.

7.6.3 NMR–MS

High-resolution NMR is an early analytical method used in metabolomics with high resolution, has no damage for the structure and properties of the samples, and has the ability to achieve non-selective analysis of the samples [87, 88]. However, if NMR is used individually, there is a large amount of information in the hydrogen spectrum that includes endogenous impurities when analyzing metabolites in biological samples, and it is difficult to say which signals are metabolites. Besides, the compounds detected by NMR must be a nucleus containing the nuclear magnetic moment, when it does not contain protons or protons easily exchanged with the solvent, making the signal wider and cannot be detected, NMR spectrum is difficult to provide the structural information of metabolites, such as sulfation, phosphorylation, and nitrogen oxidation of metabolites. Therefore, the latest technology is the combination of NMR and MS instruments for the structure identification of metabolites.

7.7 Prospects

MS-driven metabolomics is a novel science that discovers the multitude of low-molecular-weight metabolites in biological systems and provides insights into the pathogenic mechanisms of diseases [89, 90]. MS with high selectivity and high sensitivity and can identify the structure of metabolite has a great potential as an analytical tool [91–94]. The most important challenge of metabolomics is rooted in its advantages. The metabolic data is very powerful because the organism has many metabolites, including the relevant precursors, derivatives, and degradation products with a certain concentration, all of which are highly diverse and rapidly changing. The complexity requires sophisticated separation and detection methods. For innovations in metabolomics bioanalysis, the key breakthroughs are required not only in the separation methods but also in MS technologies. MS, because of its fast data acquisition, excellent dynamic range, and quantitative and accurate quality capabilities, is widely used in metabolomics [68, 95–99]. To help the researchers identify metabolites, the useful database of LC–MS or GC–MS should be largely developed. Some special samples, such as semen, amniotic fluid, tears, saliva and, other samples are currently lacking of more recognized standard operating procedures. How to combine metabolomics data with other omics data is also a big challenge. With the development of science and technology, the arrival of large data age and the combination of each omics will promote the development of this field.

7.8 Conclusions

Metabolomics seems to be a promising method to identify disease biomarkers and understanding the pathogenic mechanisms of diseases. It needs the advanced MS for elucidation of metabolites from larger-scale analysis. Advancement in MS technology could directly contribute to metabolomics analysis area. The development of fundamental technology and innovation method is essential for the next breakthrough in MS-driven metabolomics.

Acknowledgments

This work was supported by grants from the Key Program of Natural Science Foundation of State (Grant No. 81830110, 81861168037, 81973745, 81302905), Natural Science Foundation of Heilongjiang Province (YQ2019H030), and Heilongjiang Touyan Innovation Team Program.

References

1 Andres, D.A., Young, L., Veeranki, S. et al. (2020). Improved workflow for mass spectrometry-based metabolomics analysis of the heart. *Journal of Biological Chemistry* 295 (9): 2676–2686.

2 Zhang, A., Sun, H., and Wang, X. (2013). Power of metabolomics in biomarker discovery and mining mechanisms of obesity. *Obesity Reviews* 14 (4): 344–349.

3 Garcia-Perez, I., Posma, J.M., Serrano-Contreras, J.I. et al. (2020). Identifying unknown metabolites using NMR-based metabolic profiling techniques. *Nature Protocols* 15 (8): 2538–2567.

4 Zhang, A.H., Sun, H., Han, Y. et al. (2013). Ultraperformance liquid chromatography–mass spectrometry based comprehensive metabolomics combined with pattern recognition and network analysis methods for characterization of metabolites and metabolic pathways from biological data sets. *Analytical Chemistry* 85 (15): 7606–7612.

5 Wang, S., Blair, I.A., and Mesaros, C. (2019). Analytical methods for mass spectrometry-based metabolomics studies. *Advances in Experimental Medicine and Biology* 1140: 635–647.

6 Liang, Q., Yu, Q., Wu, H. et al. (2014). Metabolite fingerprint analysis of cervical cancer using LC-QTOF/MS and multivariate data analysis. *Analytical Methods* 6 (12): 3937–3942.

7 Bar, N., Korem, T., Weissbrod, O. et al. (2020). A reference map of potential determinants for the human serum metabolome. *Nature* 588 (7836): 135–140.

8 Zhang, A., Sun, H., Yan, G. et al. (2014). Metabolomics analysis of marker metabolites for patients with pancreatic cancer. *Medicinal Chemistry* 4 (6): 506–510.

9 Singh, K.D., Tancev, G., Decrue, F. et al. (2019). Standardization procedures for real-time breath analysis by secondary electrospray ionization high-resolution mass spectrometry. *Analytical and Bioanalytical Chemistry* 411 (19): 4883–4898.

10 Zhang, A., Sun, H., Qiu, S. et al. (2014). Metabolomics insights into pathophysiological mechanisms of nephrology. *International Urology and Nephrology* 46 (5): 1025–1030.

11 Zhang, A., Yan, G., Han, Y. et al. (2014). Metabolomics approaches and applications in prostate cancer research. *Applied Biochemistry and Biotechnology* 174 (1): 6–12.

12 Kempa, E.E., Hollywood, K.A., Smith, C.A. et al. (2019). High throughput screening of complex biological samples with mass spectrometry - from bulk measurements to single cell analysis. *Analyst* 144 (3): 872–891.

13 Liang, Q., Wang, C., Li, B. et al. (2015). Metabolic fingerprinting to understand therapeutic effects and mechanisms of silybin on acute liver damage in rat. *Pharmacognosy Magazine* 11 (43): 586.

14 Zhang, A., Sun, H., Yan, G. et al. (2015). Metabolomics for biomarker discovery: moving to the clinic. *BioMed Research International* 2015: 354671–354676.

15 Burton, C. and Ma, Y. (2019). Current trends in cancer biomarker discovery using urinary metabolomics: achievements and new challenges. *Current Medicinal Chemistry* 26 (1): 5–28.

16 Chen, Y., Zhou, Z., Yang, W. et al. (2017). Development of a data-independent targeted metabolomics method for relative quantification using liquid chromatography coupled with tandem mass spectrometry. *Analytical Chemistry* 89 (13): 6954–6962.

17 Matsuda, F. (2016). Technical challenges in mass spectrometry-based metabolomics. *Mass Spectrometry (Tokyo, Japan)* 5 (2): S0052.

18 Park, T.J., Reznick, J., Peterson, B.L. et al. (2017). Fructose-driven glycolysis supports anoxia resistance in the naked mole-rat. *Science* 356 (6335): 307–311.

19 Sun, H., Zhang, A., Yan, G. et al. (2013). Metabolomic analysis of key regulatory metabolites in hepatitis C virus–infected tree shrews. *Molecular & Cellular Proteomics* 12 (3): 710–719.

20 Qiu, S., Zhang, A.H., Guan, Y. et al. (2020). Functional metabolomics using UPLC-Q/TOF-MS combined with ingenuity pathway analysis as a promising strategy for evaluating the efficacy and discovering amino acid metabolism as a potential therapeutic mechanism-related target for geniposide against alcoholic liver disease. *RSC Advances* 10 (5): 2677–2690.

21 Pareek, V., Tian, H., Winograd, N. et al. (2020). Metabolomics and mass spectrometry imaging reveal channeled de novo purine synthesis in cells. *Science* 368 (6488): 283–290.

22 Wang, X., Zhang, A., and Sun, H. (2013). Power of metabolomics in diagnosis and biomarker discovery of hepatocellular carcinoma. *Hepatology* 57 (5): 2072–2077.

23 Guo, H., Chou, W.C., Lai, Y. et al. (2020). Multi-omics analyses of radiation survivors identify radioprotective microbes and metabolites. *Science* 370 (6516): 9097.

24 Zhang, A., Sun, H., Wang, P. et al. (2012). Recent and potential developments of biofluid analyses in metabolomics. *Journal of Proteomics* 75 (4): 1079–1088.

25 Anyfanti, P., Nikolaidou, B., and Gkaliagkousi, E. (2020). Urine metabolomic phenotyping for detection of adrenocortical carcinoma: still a long way to go. *The Lancet Diabetes & Endocrinology* 8 (11): 876–877.

26 Zhang, A., Sun, H., Yan, G. et al. (2014). Metabolomics in diagnosis and biomarker discovery of colorectal cancer. *Cancer Letters* 345 (1): 17–20.

27 Bancos, I., Taylor, A.E., Chortis, V. et al. (2020). Urine steroid metabolomics for the differential diagnosis of adrenal incidentalomas in the EURINE-ACT study: a prospective test validation study. *The Lancet Diabetes & Endocrinology* 8 (9): 773–781.

28 Zhang, H.L., Zhang, A.H., Miao, J.H. et al. (2019). Targeting regulation of tryptophan metabolism for colorectal cancer therapy: a systematic review. *RSC Advances* 9 (6): 3072–3080.

29 Liang, Q., Liu, H., Xing, H. et al. (2016). High-resolution mass spectrometry for exploring metabolic signatures of sepsis-induced acute kidney injury. *RSC Advances* 6 (36): 29863–29868.

30 van Laarhoven, A., Dian, S., Aguirre-Gamboa, R. et al. (2018). Cerebral tryptophan metabolism and outcome of tuberculous meningitis: an observational cohort study. *The Lancet Infectious Diseases* 18 (5): 526–535.

31 Zhang, A., Liu, Q., Zhao, H. et al. (2016). Phenotypic characterization of nanshi oral liquid alters metabolic signatures during disease prevention. *Scientific Reports* 6: 19333.

32 Zhang, A., Fang, H., Wang, Y. et al. (2017). Discovery and verification of the potential targets from bioactive molecules by network pharmacology-based target prediction combined with high-throughput metabolomics. *RSC Advances* 7 (81): 51069–51078.

33 Harder, J.M., Guymer, C., Wood, J. et al. (2020). Disturbed glucose and pyruvate metabolism in glaucoma with neuroprotection by pyruvate or rapamycin. *Proceedings of the National Academy of Sciences of the United States of America* 117 (52): 33619–33627.

34 Liang, Q., Liu, H., Zhang, T. et al. (2016). Discovery of serum metabolites for diagnosis of mild cognitive impairment to Alzheimer's disease progression using an optimized metabolomics method. *RSC Advances* 6: 3586–3591.

35 Zhang, A., Sun, H., and Wang, X. (2012). Saliva metabolomics opens door to biomarker discovery, disease diagnosis, and treatment. *Applied Biochemistry and Biotechnology* 168 (6): 1718–1727.

36 Vial, T., Tan, W.L., Deharo, E. et al. (2020). Mosquito metabolomics reveal that dengue virus replication requires phospholipid reconfiguration via the remodeling cycle. *Proceedings of the National Academy of Sciences of the United States of America* 117 (44): 27627–27636.

37 Liang, Q., Liu, H., Xing, H. et al. (2016). UPLC-QTOF/MS based metabolomics reveals metabolic alterations associated with severe sepsis. *RSC Advances* 6 (49): 43293–43298.

38 Zhang, A., Sun, H., Xu, H. et al. (2013). Cell metabolomics. *OMICS: A Journal of Integrative Biology* 17 (10): 495–501.

39 Ryabova, A., Cornette, R., Cherkasov, A. et al. (2020). Combined metabolome and transcriptome analysis reveals key components of complete desiccation tolerance in an anhydrobiotic insect. *Proceedings of the National Academy of Sciences of the United States of America* 117 (32): 19209–19220.

40 Zhang, A.-H., Qiu, S., Xu, H.-Y. et al. (2014). Metabolomics in diabetes. *Clinica Chimica Acta* 429: 106–110.

41 Hou, Y.C., Yu, H.C., Martin, R. et al. (2020). Precision medicine integrating whole-genome sequencing, comprehensive metabolomics, and advanced imaging. *Proceedings of the National Academy of Sciences of the United States of America* 117 (6): 3053–3062.

42 Wang, X., Wang, Q., Zhang, A. et al. (2013). Metabolomics study of intervention effects of wen-Xin-formula using ultra high-performance liquid chromatography/mass spectrometry coupled with pattern recognition approach. *Journal of Pharmaceutical and Biomedical Analysis* 74: 22–30.

43 Sovio, U., Goulding, N., McBride, N. et al. (2020). A maternal serum metabolite ratio predicts fetal growth restriction at term. *Nature Medicine* 26 (3): 348–353.

44 Liang, Q., Liu, H., Jiang, Y. et al. (2016). Novel liquid chromatography-mass spectrometry for metabolite biomarkers of acute lung injury disease. *Analytical Methods* 8 (31): 6017–6022.

45 Zhang, A.H., Sun, H., Yan, G.L. et al. (2019). Chinmedomics: a powerful approach integrating metabolomics with serum pharmacochemistry to evaluate the efficacy of traditional Chinese medicine. *Engineering* 5 (1): 60–68.

46 Zhang, A.H., Sun, H., Yan, G.L. et al. (2014). Metabolomics study of type 2 diabetes using ultra-performance LC-ESI/quadrupole-TOF high-definition MS coupled with pattern recognition methods. *Journal of Physiology and Biochemistry* 70 (1): 117–128.

47 Yachida, S., Mizutani, S., Shiroma, H. et al. (2019). Metagenomic and metabolomic analyses reveal distinct stage-specific phenotypes of the gut microbiota in colorectal cancer. *Nature Medicine* 25 (6): 968–976.

48 Zhang, A., Sun, H., Yan, G. et al. (2016). Mass spectrometry-based metabolomics: applications to biomarker and metabolic pathway research. *Biomedical Chromatography* 30 (1): 7–12.

49 Wang, X., Zhang, A., Han, Y. et al. (2012). Urine metabolomics analysis for biomarker discovery and detection of jaundice syndrome in patients with liver disease. *Molecular & Cellular Proteomics* 11 (8): 370–380.

50 Yang, Q., Zhang, A.H., Miao, J.H. et al. (2019). Metabolomics biotechnology, applications, and future trends: a systematic review. *RSC Advances* 9 (64): 37245–37257.

51 Zhang, A., Sun, H., and Wang, X. (2014). Urinary metabolic profiling of rat models revealed protective function of scoparone against alcohol induced hepatotoxicity. *Scientific Reports* 4: 6768.

52 Liu, R., Hong, J., Xu, X. et al. (2017). Gut microbiome and serum metabolome alterations in obesity and after weight-loss intervention. *Nature Medicine* 23 (7): 859–868.

53 Wang, X., Yang, B., Sun, H. et al. (2012). Pattern recognition approaches and computational systems tools for ultra performance liquid chromatography–mass spectrometry-based comprehensive metabolomic profiling and pathways analysis of biological data sets. *Analytical Chemistry* 84 (1): 428–439.

54 Wu, X.H., Sun, X.L., Zhao, C. et al. (2019). Exploring the pharmacological effects and potential targets of paeoniflorin on the endometriosis of cold coagulation and blood stasis model rats by ultra-performance liquid chromatography tandem mass spectrometry with a pattern recognition approach. *RSC Advances* 9 (36): 20796–20805.

55 Zhang, A.H., Ma, Z.M., Kong, L. et al. (2020). High-throughput lipidomics analysis to discover lipid biomarkers and profiles as potential targets for evaluating efficacy of Kai-Xin-San against APP/PS1 transgenic mice based on UPLC–Q/TOF–MS. *Biomedical Chromatography* 34 (2): e4724.

56 Zhang, A., Zhou, X., Zhao, H. et al. (2017). Metabolomics and proteomics technologies to explore the herbal preparation affecting metabolic disorders using high resolution mass spectrometry. *Molecular BioSystems* 13 (2): 320–329.

57 Zhang, A., Sun, H., and Wang, X. (2017). Emerging role and recent applications of metabolomics biomarkers in obesity disease research. *RSC Advances* 7 (25): 14966–14973.

58 Liang, Q., Zhu, Y., Liu, H. et al. (2017). High-throughput lipidomics enables discovery of the mode of action of huaxian capsule impacting the metabolism of sepsis. *RSC Advances* 7 (71): 44990–44996.

59 Chu, H., Zhang, A., Han, Y. et al. (2016). Metabolomics approach to explore the effects of Kai-Xin-San on Alzheimer's disease using UPLC/ESI-Q-TOF mass spectrometry. *Journal of Chromatography B* 1015: 50–61.

60 Liang, Q., Liu, H., Xie, L.X. et al. (2017). High-throughput metabolomics enables biomarker discovery in prostate cancer. *RSC Advances* 7 (5): 2587–2593.

61 Wang, X., Sun, H., Zhang, A. et al. (2011). Potential role of metabolomics approaches in the area of traditional Chinese medicine: as pillars of the bridge between Chinese and Western medicine. *Journal of Pharmaceutical and Biomedical Analysis* 55 (5): 859–868.

62 Zhang, A., Sun, H., Wu, X. et al. (2012). Urine metabolomics. *Clinica Chimica Acta* 414: 65–69.

63 Fang, H., Zhang, A.H., Sun, H. et al. (2019). High-throughput metabolomics screen coupled with multivariate statistical analysis identifies therapeutic targets in alcoholic liver disease rats using liquid chromatography-mass spectrometry. *Journal of Chromatography B* 1109: 112–120.

64 Zhang, A.H., Ma, Z.M., Sun, H. et al. (2019). High-throughput metabolomics evaluate the efficacy of total lignans from Acanthophanax senticosus stem against ovariectomized osteoporosis rat. *Frontiers in Pharmacology* 10: 553.

65 Wang, X.J., Ren, J.L., Zhang, A.H. et al. (2019). Novel applications of mass spectrometry-based metabolomics in herbal medicines and its active ingredients: current evidence. *Mass Spectrometry Reviews* 38 (4–5): 380–402.

66 Zhang, A., Wang, H., Sun, H. et al. (2015). Metabolomics strategy reveals therapeutical assessment of limonin on nonbacterial prostatitis. *Food & Function* 6 (11): 3540–3549.

67 Qiu, S., Zhang, A., Zhang, T. et al. (2017). Dissect new mechanistic insights for geniposide efficacy on the hepatoprotection using multiomics approach. *Oncotarget* 8 (65): 108760.

68 Zhang, Y., Liu, P., Li, Y. et al. (2017). Exploration of metabolite signatures using high-throughput mass spectrometry coupled with multivariate data analysis. *RSC Advances* 7 (11): 6780–6787.

69 Zhang, A., Sun, H., Wang, P. et al. (2012). Modern analytical techniques in metabolomics analysis. *Analyst* 137 (2): 293–300.

70 Li, Y.F., Qiu, S., Gao, L.J. et al. (2018). Metabolomic estimation of the diagnosis of hepatocellular carcinoma based on ultrahigh performance liquid chromatography coupled with time-of-flight mass spectrometry. *RSC Advances* 8 (17): 9375–9382.

71 Song, Q., Zhang, A.H., Yan, G.L. et al. (2017). Technological advances in current metabolomics and its application in tradition Chinese medicine. *RSC Advances* 7 (84): 53516–53524.

72 Chu, H., Zhang, A.H., Han, Y. et al. (2015). Metabolomics and its potential in drug discovery and development from TCM. *World Journal of Traditional Chinese Medicine* 1 (4): 26.

73 Sun, H., Zhang, A.H., Song, Q. et al. (2018). Functional metabolomics discover pentose and glucuronate interconversion pathways as promising targets for Yang Huang syndrome treatment with Yinchenhao Tang. *RSC Advances* 8 (64): 36831–36839.

74 Zhang, A., Sun, H., Dou, S. et al. (2013). Metabolomics study on the hepato-protective effect of scoparone using ultra-performance liquid chromatography/electrospray ionization quadruple time-of-flight mass spectrometry. *Analyst* 138: 353–361.

75 Wang, X., Sun, H., Zhang, A. et al. (2011). Ultra-performance liquid chromatography coupled to mass spectrometry as a sensitive and powerful technology for metabolomic studies. *Journal of Separation Science* 34 (24): 3451–3459.

76 Zhang, A., Sun, H., and Wang, X. (2012). Serum metabolomics as a novel diagnostic approach for disease: a systematic review. *Analytical and Bioanalytical Chemistry* 404 (4): 1239–1245.

77 Li, X., Zhang, A., Sun, H. et al. (2017). Metabolic characterization and pathway analysis of berberine protects against prostate cancer. *Oncotarget* 8 (39): 65022.

78 Fang, H., Zhang, A., Zhou, X. et al. (2020). High-throughput metabolomics reveals the perturbed metabolic pathways and biomarkers of Yang Huang syndrome as potential targets for evaluating the therapeutic effects and mechanism of geniposide. *Frontiers in Medicine* 14 (5): 651–663.

79 Zhang, A., Sun, H., Yan, G. et al. (2013). Urinary metabolic profiling identifies a key role for glycocholic acid in human liver cancer by ultra-performance liquid-chromatography coupled with high-definition mass spectrometry. *Clinica Chimica Acta* 418: 86–90.

80 Wang, X., Yang, B., Zhang, A. et al. (2012). Potential drug targets on insomnia and intervention effects of Jujuboside A through metabolic pathway analysis as revealed by UPLC/ESI-SYNAPT-HDMS coupled with pattern recognition approach. *Journal of Proteomics* 75 (4): 1411–1427.

81 Yang, B., Zhang, A., Sun, H. et al. (2012). Metabolomic study of insomnia and intervention effects of Suanzaoren decoction using ultra-performance liquid-chromatography/electrospray-ionization synapt high-definition mass spectrometry. *Journal of Pharmaceutical and Biomedical Analysis* 58: 113–124.

82 Fang, H., Zhang, A., Yu, J. et al. (2016). Insight into the metabolic mechanism of scoparone on biomarkers for inhibiting Yanghuang syndrome. *Scientific Reports* 6 (1): 1–9.

83 Ross, A.B., Barman, M., Hartvigsson, O. et al. (2021). Umbilical cord blood metabolome differs in relation to delivery mode, birth order and sex, maternal diet and possibly future allergy development in rural children. *PLoS One* 16 (1): e0242978.

84 Gong, Y., Lyu, W., Shi, X. et al. (2020). A serum metabolic profiling analysis during the formation of fatty liver in landes geese via GC-TOF/MS. *Frontiers in Physiology* 11: 581699.

85 Capellades, J., Junza, A., Samino, S. et al. (2021). Exploring the use of gas chromatography coupled to chemical ionization mass spectrometry (GC-CI-MS) for stable isotope labeling in metabolomics. *Analytical Chemistry* 93 (3): 1242–1248.

86 Qamar, W., Alqahtani, S., Ahamad, S.R. et al. (2020). Untargeted GC-MS investigation of serum metabolomics of coronary artery disease patients. *Saudi Journal of Biological Sciences* 27 (12): 3727–3734.

87 Zhang, A.H., Sun, H., Qiu, S. et al. (2013). NMR-based metabolomics coupled with pattern recognition methods in biomarker discovery and disease diagnosis. *Magnetic Resonance in Chemistry* 51 (9): 549–556.

88 Wang, C., He, L., Li, D.W. et al. (2017). Accurate identification of unknown and known metabolic mixture components by combining 3D NMR with fourier transform ion cyclotron resonance tandem mass spectrometry. *Journal of Proteome Research* 16 (10): 3774–3786.

89 Zhang, A., Sun, H., Han, Y. et al. (2012). Exploratory urinary metabolic biomarkers and pathways using UPLC-Q-TOF-HDMS coupled with pattern recognition approach. *Analyst* 137: 4200–4208.

90 Zhang, A.H., Sun, H., Qiu, S. et al. (2013). Metabolomics in noninvasive breast cancer. *Clinica Chimica Acta* 424: 3–7.

91 Nan, Y., Zhou, X., Liu, Q. et al. (2016). Serum metabolomics strategy for understanding pharmacological effects of ShenQi pill acting on kidney yang deficiency syndrome. *Journal of Chromatography B* 1026: 217–226.

92 Wang, X., Li, J., and Zhang, A.H. (2016). Urine metabolic phenotypes analysis of extrahepatic cholangiocarcinoma disease using ultra-high performance liquid chromatography-mass spectrometry. *RSC Advances* 6 (67): 63049–63057.

93 Zhang, A., Sun, H., Han, Y. et al. (2013). Urinary metabolic biomarker and pathway study of hepatitis B virus infected patients based on UPLC-MS system. *PLoS One* 8 (5): e64381.

94 Wang, X.J., Zhang, A.H., Kong, L. et al. (2019). Rapid discovery of quality-markers from Kaixin San using chinmedomics analysis approach. *Phytomedicine* 54: 371–381.

95 Liang, Q., Wang, C., Li, B. et al. (2015). Metabolomics of alcoholic liver disease: a clinical discovery study. *RSC Advances* 5 (98): 80381–80387.

96 Ren, J.L., Zhang, A.H., Kong, L. et al. (2018). Advances in mass spectrometry-based metabolomics for investigation of metabolites. *RSC Advances* 8 (40): 22335–22350.

97 Zhang, A.H., Sun, H., Sun, W.J. et al. (2013). Trajectory analysis of metabolomics profiling in liver injured rats using ultra-performance liquid chromatography coupled with mass spectrometry. *Analytical Methods* 5 (19): 5294–5301.

98 Vascellari, S., Melis, M., Palmas, V. et al. (2021). Clinical phenotypes of Parkinson's disease associate with distinct gut microbiota and metabolome enterotypes. *Biomolecules* 11 (2): 144.

99 Delafiori, J., Navarro, L.C., Siciliano, R.F. et al. (2021). Covid-19 automated diagnosis and risk assessment through metabolomics and machine learning. *Analytical Chemistry* https://doi.org/10.1021/acs.analchem.0c04497.

8

The Application of Metabolomics in Cancer Management

Aihua Zhang, Hui Sun, and Xijun Wang

Heilongjiang University of Chinese Medicine, National Chinmedomics Research Center, Functional Metabolomics Laboratory, Department of Pharmaceutical Analysis, Heping Road 24, Harbin, 150040, China

8.1 Introduction

Emerging high-throughput metabolomics technologies have been widely applied, aiming at the discovery of candidate biomarkers for cancer staging, prediction of recurrence and prognosis, and treatment selection [1–3]. Small-molecule metabolites play an important role in biological systems and represent attractive candidates to understand disease phenotypes [4–7]. Metabolomics allows the unparalleled opportunity to identify novel diagnostic biomarkers that could be used especially in early detection [8–13], and the identified metabolites may provide additional insight into cancer metabolism [14, 15] while also illustrating the potential of this technique. This chapter highlights examples of its application in hepatocellular carcinoma (HCC). HCC is the most frequent primary malignant liver tumor, and with high mortality, ranking fourth in the world in cancer death. It is the most common primary hepatic malignancy with a high disease burden especially in advanced stages of the disease. Current diagnostic modalities, ultrasound and α-fetoprotein, are expensive and lack sensitivity in tumor detection. Thus, biomarkers for early detection and molecular targets for treating HCC are an urgent need.

In clinic, ultrasound, magnetic resonance imaging, and computed tomography are conventional means to detect HCC. However, these methods are not effective at early detection of diseases. In fact, they are performed at an advanced stage, because there are no symptoms in the early stage of HCC. The potential biomarkers of disease in biofluids or tissue that were identified by metabolomics can enable noninvasive, objective, and reproducible assessments [16–21]. Accordingly, metabolomics is being widely used to explore disease diagnosis and treatment biomarkers with powerful analytical technique and data processing mode [22–26]. Metabolomics samples derived from the circulation or other noninvasive recoverable body fluid have an advantage than those from tissue biopsy specimens obtained via the invasive procedure. This chapter shows the medical value of metabolomics in HCC management as a case and prospects for future applications of metabolomics.

Mass Spectrometry-Based Metabolomics in Clinical and Herbal Medicines: Strategies, Technologies, and Applications, First Edition. Edited by Aihua Zhang and Wanying Wang.

8.2 Epidemiology and Diagnosis

HCC is a health problem with high incidence and mortality. The survey reported that the five-year relative survival rate of HCC was 10.1%, which was only 1/3 of average cancer survival rate. Many factors contribute to this situation, and the most dangerous one is high-risk etiological challenges [26–28]. It emphasizes that hepatitis B virus (HBV) is the primary etiological factor for HCC worldwide, particularly in high-incidence areas. Besides, environment, host inheritance, virus, and lifestyle also influence the occurrence of HCC in individual [29, 30], which are accompanied by metabolic disorders. The metabolic disorders and differences are consistently associated with an increased risk of HCC.

Early diagnosis is essential to effective treatment and optimal prognosis [31]. Auxiliary diagnostic means include immunohistochemistry, ultrasound, and serum biomarkers identification. The immunohistochemistry primarily concentrates attention on the qualitative diagnosis of primary liver cancer. Comparatively, ultrasound detection is the first choice for the surveillance of HCC, which is a noninvasive detection technology [32]. However, the ultrasound examination will be mixed with subjective factors because it is always restricted by the quality of the machine, the experience of the operator, and the actual condition of the liver *in vivo*. Although it is recognized that the serum level of α-fetoprotein (AFP) is applied to evaluate the liver cancer in endangered patients, however, currently these biomarkers lack sensitivity and specificity. Current biomarkers, such as AFP for HCC, result in too many false negatives, necessitating noninvasive approaches with improved sensitivity. Therefore, the current management of HCC with a focus on early to advanced stages that impact on patient overall survival needs to be solved. Additionally, the actual survival rate of HCC is less than reported in some bulletin as a result of incomplete surveillance methods worldwide. The more innovative technologies and novel biomarkers are urgently needed.

8.3 Prospective Biomarkers

Although some tumor biomarkers are not in the diagnostic criteria, they have potential ability to assist the diagnosis of HCC. HCC biomarkers proposed by metabolomics studies have rapidly developed, and various potential candidates have clarified for early diagnostics and therapies for the disease. Metabolomics method could help to search for candidate diagnostic and prognostic biomarkers (Figure 8.1) and to understand the mechanisms (Figure 8.2). Herein, some published studies have found biomarkers of HCC by metabolomics approach (Table 8.1). The worst prognosis is related to the late diagnosis, in which the tumor is at an advanced stage and curative treatments is not efficient in terms of increasing overall survival. Nowadays, screening and monitoring tests based on current guidelines have limited accuracy, which points to the need for the development of new biomarkers that can improve HCC detection as well as its early diagnosis.

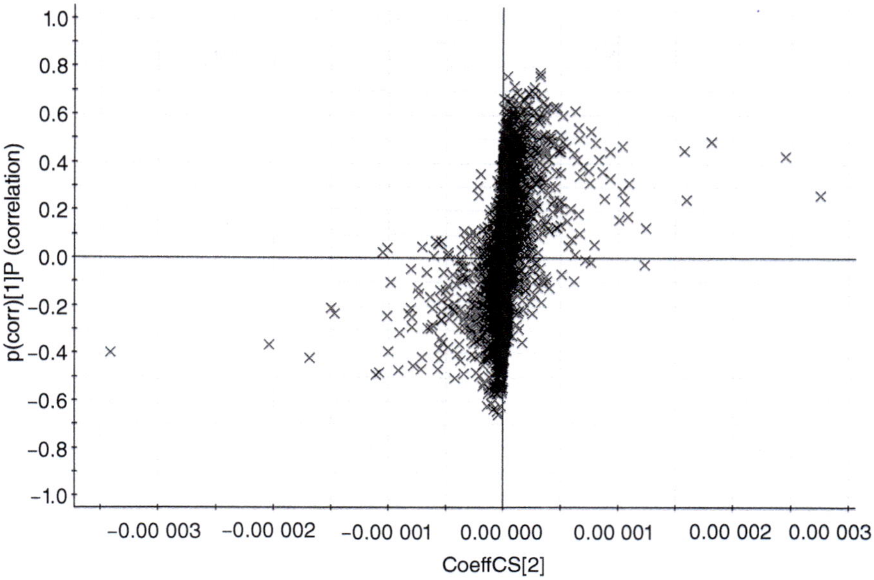

Figure 8.1 S-plot analysis for screening the candidate diagnostic and prognostic biomarkers from large biological data sets.

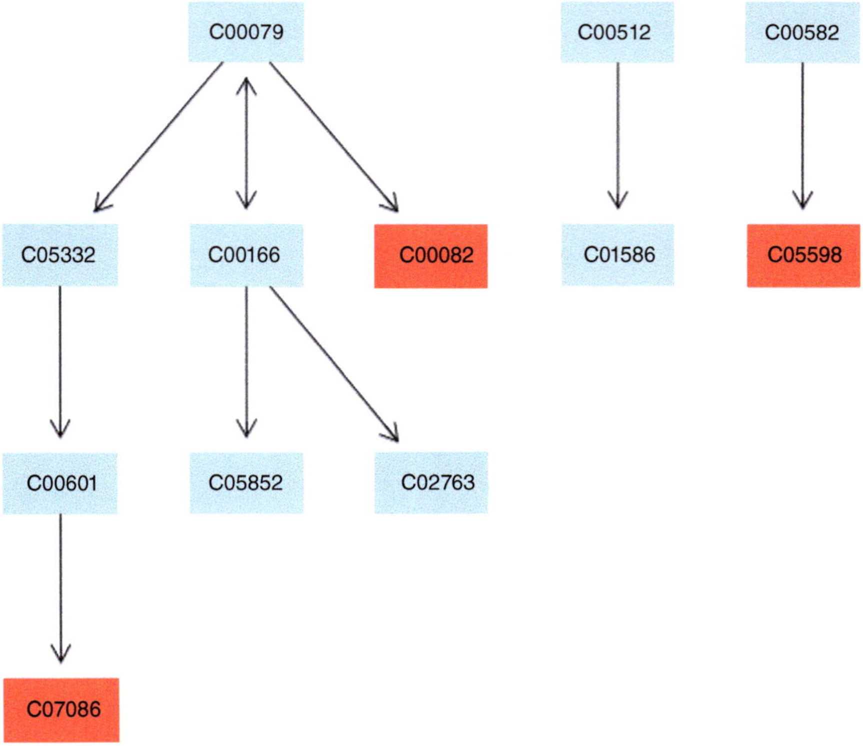

Figure 8.2 Decoding the metabolism mechanisms (phenylalanine metabolism) from metabolic networks.

Table 8.1 Summary of marker metabolites in HCC.

Specimens	Techniques	Metabolites	Potential biomarkers
Plasma	UPLC–ESI–TQMS	HCC vs. HV: glycodeoxycholate, deoxycholate-3-sulfate, bilirubin, biliverdin, fetal bile acids (↑), lysophosphocholines (↓) [33]	Glycodeoxycholate, deoxycholate-3-sulfate, bilirubin, biliverdin
Serum	NMR and LC–MS	HCC vs. HV: ascorbate, oxaloacetate, glycerol, β-hydroxybutyrate, isobutyate (↑) formate, tyrosine (↓) HCC vs. LC: ascorbate, oxaloacetate, glycerol, β-hydroxybutyrate, isobutyate formate, tyrosine (↓) [34]	Formate, tyrosine, ascorbate, oxalocetate, glycerol, β-hydroxybutyrate, isobutyate, isobutyrate
	UHPLC/Q-TOF-MS	HCC vs. HV: long-chain acylcarnitines, aromatic (↑) Medium-chain acylcarnitines, branched-chain amino (↓) [35]	Long-chain acylcarnitines, aromatic Lysophosphocholines, medium-chain acylcarnitines, branched-chain amino
	GC–MS	HCC vs. LC: L-isoleucine, L-valine, L-leucine, L-glutamic acid (↑) Critic acid, phosphatidylethanolamine, glycodeoxycholate, oleylamide, glycocholic acid (↓) [36]	Glutamic, critic acid, lactic acid, valine, isolueucine, leucine, α-tocopherol, cholesterol
	GC–TOF-MS	Glucose 6-phosphate, glutamic acid, cysteine, glycine, oxoproline, pyruvate, glutamic acid (↑), branched-chain amino acids, aromatic amino acids (↓) [37]	Asparagine, β-glutamate
Urine	UPLC–Q-TOF–HDMS	Glycocholic acid (↑) [38]	Glycocholic acid
	LC–Q-TOF-MS	Phenotypic characterization: aurocholic acid, chenodeoxycholic acid, glycocholic acid, hippuric acid [39]	Palmitic acid, α-*N*-phenylacetyl-L-glutamine, phytosphingosine, indoleacetyl glutamine, glycocholic acid

8.4 Metabolomics Screening Biomarkers

Metabolites are the endpoint of genotype functions and the biochemical phenotype in the body, and they are usually linked closely to the alteration of functions [40–44]. Metabolomics has an accurate and sensitive high-throughput technique to analyze the endogenous and exogenous substances with a large amount of information from the complex biological samples [45]. In recent years, there has become a plethora of attempts to discover biomarkers that are more reliable than AFP for the early prediction and prognosis of HCC. Metabolomics provides an opportunity, without recourse to liver biopsy, to discover biomarkers for premalignant liver disease, thereby alerting the potential of impending HCC [46].

Longitudinal change of metabolite profile and its relation to multiple risk factors was performed at the risk of developing hepatitis B-related HCC [47]. Of the 34 measured metabolites, isoleucine, glutamine, and tyrosine were consistently associated with HCC. Similar association was observed in an independent prospective analysis comprising 633 randomly selected individuals. More importantly, it suggested that an amino acid dysregulation could play a role in HBV-related HCC development and may also mediate increased HCC risks. The ultra-performance liquid chromatography–mass spectrometry (LC–MS) was used to characterize the serum metabolome of HCC patient in order to develop a new metabolic diagnostic model and to identify novel biomarkers for screening HCC [48]. The concentrations of most metabolites, including proline, were lower in patients with HCC, whereas the hydroxypurine levels were higher in these patients. Pattern recognition analysis was used to establish a diagnostic model that included proline and hydroxypurine. The area under the curve of the pattern recognition analysis model that included hydroxypurine and proline was significantly higher than the area under the curve of AFP. Proline and hydroxypurine might be novel biomarkers for HCC. Aberrant metabolisms have been hypothesized to precede the occurrence of HCC; therefore, biomarkers associated with subsequent HCC in peripheral bloods were investigated using metabolomic technologies [49]. A series of the univariate and multivariate analysis showed that phenylalanine and glutamine levels are associated with HCC.

8.5 Conclusions and Future Directions

Metabolomics plays an important role in seeking biomarkers for disease detection. Metabolite biomarkers for premalignant liver disease may help to reduce the incidence of HCC. This chapter described recent developments in metabolomics research with applications to the detection of HCC diseases, which are a common malignancy with high fatality rate. Many risk factors can lead to HCC because there are no symptoms in the early stage of this disease, which makes it difficult to diagnose accurately, and most of the patients with HCC have a poor prognosis after treatment. As a useful tool, metabolomics has been applied in disease phenotype and biomarker generation, which provides a way to demonstrate biomarkers with characteristics of sensitivity and specificity. Based on metabolomics, metabolites associated with the development of HCC are well represented to discover and identify potential markers from them. Metabolomics contributes to the diagnosis prediction and prognosis of HCC and also reveals the disease mechanism by finding related metabolic pathways. Besides, metabolomics can also be applied to dynamically study the changes of HCC metabolism in different phases at the molecular level, which is biochemistry changing disease progression.

Acknowledgments

This work was supported by grants from the Key Program of Natural Science Foundation of State (Grant No. 81830110, 81861168037, 81973745, 81302905), Natural Science Foundation of Heilongjiang Province (YQ2019H030), and Heilongjiang Touyan Innovation Team Program.

References

1 Anyfanti, P., Nikolaidou, B., and Gkaliagkousi, E. (2020). Urine metabolomic phenotyping for detection of adrenocortical carcinoma: still a long way to go. *Lancet Diabetes & Endocrinology* 8 (11): 876–877.

2 Pareek, V., Tian, H., Winograd, N. et al. (2020). Metabolomics and mass spectrometry imaging reveal channeled de novo purine synthesis in cells. *Science* 368 (6488): 283–290.

3 Qiu, S., Zhang, A.H., Guan, Y. et al. (2020). Functional metabolomics using UPLC-Q/TOF-MS combined with ingenuity pathway analysis as a promising strategy for evaluating the efficacy and discovering amino acid metabolism as a potential therapeutic mechanism-related target for geniposide against alcoholic liver disease. *RSC Advances* 10 (5): 2677–2690.

4 Yachida, S., Mizutani, S., Shiroma, H. et al. (2019). Metagenomic and metabolomic analyses reveal distinct stage-specific phenotypes of the gut microbiota in colorectal cancer. *Nature Medicine* 25 (6): 968–976.

5 Zhang, A., Sun, H., Xu, H. et al. (2013). Cell metabolomics. *OMICS: A Journal of Integrative Biology* 17 (10): 495–501.

6 Zhang, A.H., Sun, H., Yan, G.L. et al. (2014). Metabolomics study of type 2 diabetes using ultra-performance LC-ESI/quadrupole-TOF high-definition MS coupled with pattern recognition methods. *Journal of Physiology and Biochemistry* 70 (1): 117–128.

7 Zhang, A., Sun, H., Yan, G. et al. (2016). Mass spectrometry-based metabolomics: applications to biomarker and metabolic pathway research. *Biomedical Chromatography* 30 (1): 7–12.

8 Wang, X., Zhang, A., Han, Y. et al. (2012). Urine metabolomics analysis for biomarker discovery and detection of jaundice syndrome in patients with liver disease. *Molecular & Cellular Proteomics* 11 (8): 370–380.

9 Yang, Q., Zhang, A.H., Miao, J.H. et al. (2019). Metabolomics biotechnology, applications, and future trends: a systematic review. *RSC Advances* 9 (64): 37245–37257.

10 Liang, Q., Yu, Q., Wu, H. et al. (2014). Metabolite fingerprint analysis of cervical cancer using LC-QTOF/MS and multivariate data analysis. *Analytical Methods* 6 (12): 3937–3942.

11 Zhang, A., Sun, H., Qiu, S. et al. (2014). Metabolomics insights into pathophysiological mechanisms of nephrology. *International Urology and Nephrology* 46 (5): 1025–1030.

12 Zhang, A., Sun, H., Yan, G. et al. (2014). Metabolomics analysis of marker metabolites for patients with pancreatic cancer. *Medicinal Chemistry* 4 (6): 506–510.

13 Zhang, A., Yan, G., Han, Y. et al. (2014). Metabolomics approaches and applications in prostate cancer research. *Applied Biochemistry and Biotechnology* 174 (1): 6–12.

14 Burton, C. and Ma, Y. (2019). Current trends in cancer biomarker discovery using urinary metabolomics: achievements and new challenges. *Current Medicinal Chemistry* 26 (1): 5–28.

15 Wang, X., Zhang, A., and Sun, H. (2013). Power of metabolomics in diagnosis and biomarker discovery of hepatocellular carcinoma. *Hepatology* 57 (5): 2072–2077.

16 Zhang, A., Sun, H., Yan, G. et al. (2014). Metabolomics in diagnosis and biomarker discovery of colorectal cancer. *Cancer Letters* 345 (1): 17–20.

17 Zhang, H.L., Zhang, A.H., Miao, J.H. et al. (2019). Targeting regulation of tryptophan metabolism for colorectal cancer therapy: a systematic review. *RSC Advances* 9 (6): 3072–3080.

18 Wang, X., Yang, B., Sun, H. et al. (2012). Pattern recognition approaches and computational systems tools for ultra performance liquid chromatography–mass spectrometry-based comprehensive metabolomic profiling and pathways analysis of biological data sets. *Analytical Chemistry* 84 (1): 428–439.

19 Zhang, A., Sun, H., and Wang, X. (2012). Saliva metabolomics opens door to biomarker discovery, disease diagnosis, and treatment. *Applied Biochemistry and Biotechnology* 168 (6): 1718–1727.

20 Andres, D.A., Young, L., Veeranki, S. et al. (2020). Improved workflow for mass spectrometry-based metabolomics analysis of the heart. *Journal of Biological Chemistry* 295 (9): 2676–2686.

21 Zhang, A., Sun, H., and Wang, X. (2017). Emerging role and recent applications of metabolomics biomarkers in obesity disease research. *RSC Advances* 7 (25): 14966–14973.

22 Zhang, Y., Liu, P., Li, Y. et al. (2017). Exploration of metabolite signatures using high-throughput mass spectrometry coupled with multivariate data analysis. *RSC Advances* 7 (11): 6780–6787.

23 Li, Y.F., Qiu, S., Gao, L.J. et al. (2018). Metabolomic estimation of the diagnosis of hepatocellular carcinoma based on ultrahigh performance liquid chromatography coupled with time-of-flight mass spectrometry. *RSC Advances* 8 (17): 9375–9382.

24 Fang, H., Zhang, A., Yu, J. et al. (2016). Insight into the metabolic mechanism of scoparone on biomarkers for inhibiting Yanghuang syndrome. *Scientific Reports* 6 (1): 1–9.

25 Wang, S., Blair, I.A., and Mesaros, C. (2019). Analytical methods for mass spectrometry-based metabolomics studies. *Advances in Experimental Medicine and Biology* 1140: 635–647.

26 Knudsen, E.S., Gopal, P., and Singal, A.G. (2014). The changing landscape of hepatocellular carcinoma: etiology, genetics, and therapy. *American Journal of Pathology* 184 (3): 574–583.

27 Arzumanyan, A., Reis, H.M., and Feitelson, M.A. (2013). Pathogenic mechanisms in HBV- and HCV-associated hepatocellular carcinoma. *Nature Reviews Cancer* 13 (2): 123–135.

28 Blonski, W., Kotlyar, D.S., and Forde, K.A. (2010). Non-viral causes of hepatocellular carcinoma. *World Journal of Gastroenterology* 16 (29): 3603–3615.

29 Kimhofer, T., Fye, H., Taylor-Robinson, S. et al. (2015). Proteomic and metabonomic biomarkers for hepatocellular carcinoma: a comprehensive review. *British Journal of Cancer* 112 (7): 1141–1156.

30 Zhang, W., Wang, X., Huang, R. et al. (2017). Prognostic value of marital status on stage at diagnosis in hepatocellular carcinoma. *Scientific Reports* 7: 41695.

31 Hong, Y., Long, J., Li, H. et al. (2015). An analysis of immunoreactive signatures in early stage hepatocellular carcinoma. *eBioMedicine* 2 (5): 438–446.

32 Kee, K.M. and Lu, S.N. (2017). Diagnostic efficacy of ultrasound in hepatocellular carcinoma diagnosis. *Expert Review of Gastroenterology & Hepatology* 11 (4): 277–279.

33 Hong, G., Suh, K.S., Suh, S.W. et al. (2016). Alpha-fetoprotein and ^{18}F-FDG positron emission tomography predict tumor recurrence better than Milan criteria in living donor liver transplantation. *Journal of Hepatology* 64 (4): 852–859.

34 Sapisochin, G., Goldaracena, N., Laurence, J.M. et al. (2016). The extended Toronto criteria for liver transplantation in patients with hepatocellular carcinoma: a prospective validation study. *Hepatology* 64 (6): 2077–2088.

35 Toso, C., Meeberg, G., Hernandez-Alejandro, R. et al. (2015). Total tumor volume and alpha-fetoprotein for selection of transplant candidates with hepatocellular carcinoma: a prospective validation. *Hepatology* 62 (1): 158–165.

36 Sapisochin, G., Goldaracena, N., Astete, S. et al. (2015). Benefit of treating hepatocellular carcinoma recurrence after liver transplantation and analysis of prognostic factors for survival in a large Euro-American series. *Annals of Surgical Oncology* 22 (7): 2286–2294.

37 Kucukcakan, B. and Hayrulai-Musliu, Z. (2015). Challenging role of dietary aflatoxin B1 exposure and hepatitis B infection on risk of hepatocellular carcinoma. *Open Access Macedonian Journal of Medical Sciences* 3 (2): 363–369.

38 Forner, A., Llovet, J.M., and Bruix, J. (2012). Hepatocellular carcinoma. *Lancet* 379 (9822): 1245–1255.

39 Hsu, C. and Cheng, A.L. (2017). Clinical and preclinical perspectives on mechanisms of sorafenib resistance in hepatocellular carcinoma. In: *Resistance to Molecular Therapies for Hepatocellular Carcinoma*, vol. 13 (ed. A. Villanueva). Cham: Springer https://doi.org/10.1007/978-3-319-56197-4_5.

40 Zhang, A.H., Qiu, S., Xu, H.Y. et al. (2014). Metabolomics in diabetes. *Clinica Chimica Acta* 429: 106–110.

41 Zhang, A., Sun, H., Wang, P. et al. (2012). Modern analytical techniques in metabolomics analysis. *Analyst* 137 (2): 293–300.

42 Zhang, A., Zhou, X., Zhao, H. et al. (2017). Metabolomics and proteomics technologies to explore the herbal preparation affecting metabolic disorders using high resolution mass spectrometry. *Molecular BioSystems* 13 (2): 320–329.

43 Liang, Q., Liu, H., Xie, L.X. et al. (2017). High-throughput metabolomics enables biomarker discovery in prostate cancer. *RSC Advances* 7 (5): 2587–2593.

44 Zhang, A., Sun, H., Wu, X. et al. (2012). Urine metabolomics. *Clinica Chimica Acta* 414: 65–69.

45 Fang, H., Zhang, A.H., Sun, H. et al. (2019). High-throughput metabolomics screen coupled with multivariate statistical analysis identifies therapeutic targets in alcoholic liver disease rats using liquid chromatography–mass spectrometry. *Journal of Chromatography B* 1109: 112–120.

46 Toyoda, H., Kumada, T., Tada, T. et al. (2015). Tumor markers for hepatocellular carcinoma: simple and significant predictors of outcome in patients with HCC. *Liver Cancer* 4 (2): 126–136.

47 Huang, B.Y., Tsai, M.R., Hsu, J.K. et al. (2020). Longitudinal change of metabolite profile and its relation to multiple risk factors for the risk of developing hepatitis B-related hepatocellular carcinoma. *Molecular Carcinogenesis* 59 (11): 1269–1279.

48 Zhou, P.C., Sun, L.Q., Shao, L. et al. (2020). Establishment of a pattern recognition metabolomics model for the diagnosis of hepatocellular carcinoma. *World Journal of Gastroenterology* 26 (31): 4607–4623.

49 Liang, K.H., Cheng, M.L., Lo, C.J. et al. (2020). Plasma phenylalanine and glutamine concentrations correlate with subsequent hepatocellular carcinoma occurrence in liver cirrhosis patients: an exploratory study. *Scientific Reports* 10 (1): 1–7.

9

Metabolomics in Coronary Heart Disease: From Biomarker Identification to Pathomechanism Insights

Hui Sun, Aihua Zhang, Shi Qiu, and Xijun Wang

Heilongjiang University of Chinese Medicine, National Chinmedomics Research Center, Functional Metabolomics Laboratory, Department of Pharmaceutical Analysis, Heping Road 24, Harbin, 150040, China

9.1 Introduction

Over the past decade, coronary heart disease (CHD) has become a major health concern. It still accounts for about one-third of all deaths in people older than 35 years and costs for disease that approximately exceed 200 billion dollars, which have a great possibility to increase by 2030 [1, 2]. Despite the enormous progress achieved in diagnosis and medical therapy of CHD in the last decades, CHD continues to represent being the main cause of morbidity and mortality worldwide, leading to a massive healthcare cost and social burden. Due to the dynamic and complex nature of CHD, the mechanisms underlying the progression of CHD were largely unknown. Considering the etiological factors of CHD are too complex to be completely understood, there is almost no effective cure to date, suggesting that deeper pathomechanism insights are urgently needed. Metabolites are able to reflect the dynamic processes that are in disease progress (Figure 9.1), and metabolomic may therefore provide a more cost-effective and productive route to new biomarker discovery (Figure 9.2), especially in the arena for pathomechanism exploration (Figure 9.3). Changes in metabolite profiles have been used to accurately predict the disease conditions and identify the possible biomarkers and pathways [3–11]. Metabolomics offers systematic identification and quantification of all metabolic products in the body. Its applications are becoming increasingly feasible with the development of high-throughput technologies, such as mass spectrometry-based techniques [12–15]. This chapter could provide clinicians with new sets of metabolite biomarkers of CHD disease states and insights into pathomechanism.

Mass Spectrometry-Based Metabolomics in Clinical and Herbal Medicines: Strategies, Technologies, and Applications, First Edition. Edited by Aihua Zhang and Wanying Wang.

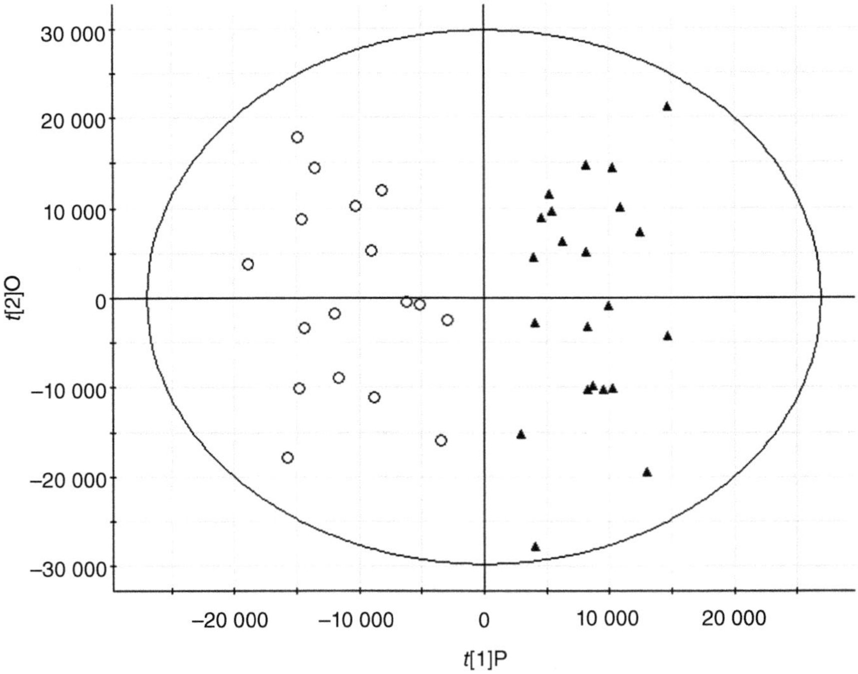

Figure 9.1 Principal component analysis of metabolomes that can reflect disease progress.

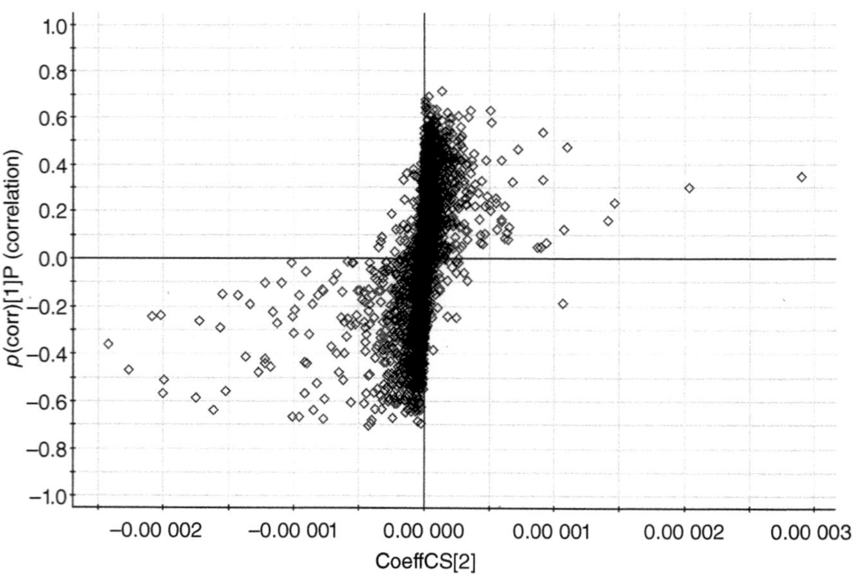

Figure 9.2 The S-plot for biomarker discovery performed by partial least squares discriminant analysis algorithm that can effectively reduce the number of variables (metabolites) in high-dimensional metabolomics data to produce robust and easy-to-interpret models.

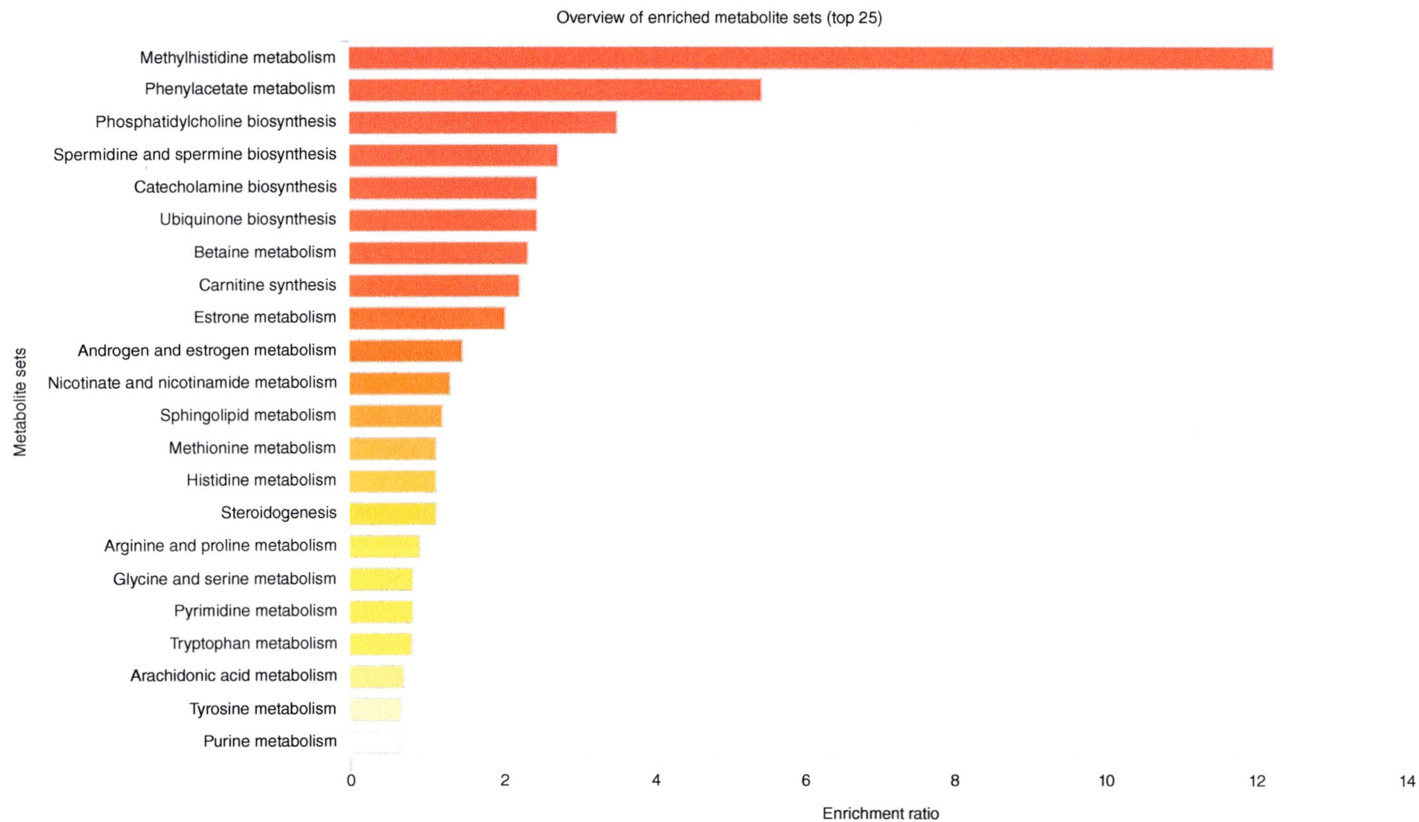

Figure 9.3 Metabolite enrichment analysis of the biologically meaningful metabolite for mechanism exploration.

9.2 Predicament

In clinical practice, the early diagnosis and the rational drug treatment of CHD are overwhelming challenges all the time. Gold standard methods for identifying CHD have helped to decrease morbidity and mortality all over the world. However, the diagnostic test suffers from a number of problems such as inconvenience, time-consuming, and high expense as well as low specificity. The additional deficiency is its lack of sensitivity and accuracy, which is closely associated with actual experience and level of doctors that lead to higher requirements for them, resulting in the late stage of disease discovery and the progress of the syndrome at the time of diagnosis. The troublesome fact for many doctors is CHD disease-related complications often appear including diabetes, organ failure, depression, and others, which not only brought difficulties to the correct diagnosis but also increased troubles for reasonable treatment project.

9.3 Metabolomics for CHD

CHD is the leading cause of death worldwide, and its pathogenesis has attracted considerable attention. Metabolomics is used as an important tool for diagnosing diseases and exploring their pathogenesis in recent years [16–18]. Metabolomics analytical system and modern techniques are being used to identify novel biomarkers for CHD [19–21]. Application of metabolomics in CHD research has provided the chance to obtain new insights into metabolic profiling change and morbidity mechanism. There is increasing evidence that metabolomics can supply vital penetrating insight into the discoveries of biomarker and the pathogenic essence of various diseases during over past decade [22].

Identifying metabolomic signatures associated with CHD may provide insight into disease pathophysiology and prevention. Metabolomic profiling was performed to select metabolites associated with incident CHD. A metabolite risk score was built to evaluate whether it can predict CHD risk beyond traditional risk factors [23]. After 30 years of follow-up, 633 incident CHD cases were observed. Thirty-two metabolites were selected and associated with CHD, and 19 of the 32 showed significant individual associations with CHD, including a sugar substitute, erythritol. Theophylline and γ-linolenic acid showed the greatest positive and negative associations with CHD, respectively. An ultra-performance liquid chromatography coupled with mass spectrometry (UPLC/MS) was used to analyze the urine metabolites of CHD patients and normal subjects [24]. A total of 14 potential biomarkers associated with CHD were identified, and among them four biomarkers were linearly correlated with the severity of CHD patients. The metabolic pathways involved were amino acid metabolism, fatty acid metabolism, and energy metabolism. Alterations in the biomarkers and metabolic pathways may be detectable in at-risk individuals before the onset of CHD.

9.4 Biomarker Dilemma

The value of biomarkers in the bio-sample for the diagnosis and prediction of CHD manifestations has been expressed in many clinical studies. The number of routinely measured biomarkers has some limitations, because CHD is regarded as a systemic disease; in many conditions pathological changes at uncontrollable rates may occur. Also, progression of CHD is often accompanied by different degrees of complications in different parts of the body, leading to a complex syndrome [25]. This dilemma could restrict theoretical results given by biomarker in the progression of CHD. The procedure of the conventional laboratory methods is complicated and it takes a long time to obtain the result. Because of the importance of early treatments and the shortcomings of conventional diagnostic methods, remarkable attention has been given to biomarkers discovery.

Metabolomics analysis may yield novel predictive biomarkers that will potentially provide value for clinical diagnosis of CHD. Developing novel biomarkers for CHD is urgently needed. In order to evaluate the potential of untargeted plasma metabolomics in biomarker discovery for characterizing CHD, the metabolites from patients newly diagnosed with CHD and controls were profiled using mass spectrometry [26]. A total of 28 differential plasma metabolites were identified, of which the concentrations of 11 increased significantly and those of 17 decreased significantly in patients with CHD. The altered metabolic pathways included the reduced phospholipid metabolism, increased monoglyceride metabolism, and abnormal fatty acid metabolism. Furthermore, binary logistic regression showed that nine metabolites could be used as potential plasma biomarkers for the diagnosis of CHD. Serum metabolite profiling between CHD patients and normal coronary artery (NCA) subjects were aimed to explore metabolic biomarkers and metabolic characteristics associated with CHD progression [27], using targeted liquid chromatography coupled with tandem mass spectrometry. The concentrations for trimethylamine oxide (TMAO), choline, creatinine, and carnitine were significantly higher in CHD patients compared with the NCA cohort. Further, the diagnostic model was constructed based on the metabolites identified, and the receiver operating characteristic (ROC) curve of the NCA subjects and CHD patients were performed. For choline and creatinine, the area under the curves (AUCs) ranged from 0.720 to 0.733. For TMAO and carnitine, the AUCs ranged from 0.568 to 0.600. It illustrates the distribution of four metabolites between CHD patients and NCA subjects.

A liquid chromatography–tandem mass spectrometry metabolomics platform was applied to measure 371 metabolites in a discovery set of postmenopausal women with validation in an independent set of postmenopausal women [28]. Eight metabolites, primarily oxidized lipids, were significantly dysregulated in cases after the adjustment for matching and CHD risk factors in both the discovery and validation data sets. These data replicate known metabolite predictors, identify novel biomarkers, and support the relationship between lipid oxidation and subsequent CHD. These studies not only yielded fundamental insights into dysregulated metabolism in CHD but also presented a combinatorial biomarker that might support the clinical diagnosis of CHD.

9.5 Lipid Metabolism

CHD refers to the abnormal lipid metabolism, which the blood lipid deposited in the originally smooth arterial intima and some atherosclerotic lipid gradually formed white plaques giving rise to the atherosclerosis lesions. The lipidomic analysis can provide a method insight into the dysregulated metabolism in early stage of CHD. Lipidomics, a branch of metabolomics, will provide insights in which lipids may play important roles in disease progression and will serve as novel markers for preventive as well as therapeutic purposes [29–35]. The high-throughput mass spectrometry approach has been used to rapidly and accurately profile hundreds of lipids in human serum, facilitating in-depth lipid phenotype for large epidemiological studies to reveal the detailed associations of individual lipids with CHD risk factors [36]. This method provided information on 360 lipids (including fatty acyls, glycerophospholipids, sphingolipids, glycerolipids, and sterol lipids). The lipids were highly correlated and exhibited a range of associations with clinical biomarkers and lifestyle factors. Change in plasma concentrations of sphingolipids was assessed in relation to change in verbal memory performance among CHD subjects undertaking a six-month cardiac rehabilitation (CR) program [37]. Plasma sphingolipid concentrations were needed to be measured using high-performance liquid chromatography and mass spectrometry. A decrease in ceramide C18 : 0 concentration was significantly associated with improvement in verbal memory performance, processing speed, visuospatial memory, and global cognition over six months of CR. Plasma ceramide C18 : 0 concentrations may be a sensitive marker of cognitive response to exercise in patients with CHD. Plasma lipid species were measured by a targeted lipidomic analysis with liquid chromatography–electrospray ionization–tandem mass spectrometry [38]. It revealed that sphingolipids, cholesteryl esters, phospholipids, and glycerolipids were associated with future cardiovascular events and cardiovascular death. The addition of seven lipid species to predict cardiovascular events increased, and the prediction of cardiovascular death needed to be improved with the incorporation of four lipid species. It demonstrated the potential of plasma lipid species as biomarkers for cardiovascular risk stratification.

9.6 Treatment of Herbal Medicine

A mass spectrometry-based metabolomics study was performed to investigate the differential metabolites to determine the effect of the classical herbal medication *Taohong Siwu decoction* on the metabolic profiles of the serum in CHD patients [39]. A total of 513 metabolites were found in the serum of CHD patients, of which 27, involved in 29 metabolic pathways, were significantly different. Among the differential metabolites, *Taohong Siwu decoction* up-regulated succinylcarnitine in 5′-methylthioadenosine and fatty acid metabolism compared with the placebo group. It indicated that 5′-methylthioadenosine, glycerophosphocholine, 8,9-DiHETrE, 3-ureidopropionate, hippurate, and indoxyl sulfate might constitute in CHD patients treated with *Taohong Siwu decoction*. Metabolic pathway analysis showed that the therapeutic effect of Xue-Fu-Zhu-Yu decoction on CHD involved in lipid metabolism, amino acid metabolism, and bile acid metabolism pathways [40]. The relevant biological pathways by assessing changes in biomarkers in response to Dan-Lou tablet (DLT), a well-known herbal prescription, intervene in CHD to reveal the potential biological mechanism of DLT [41]. In terms of metabolism, 17 biomarkers of CHD in rats were identified. Among these biomarkers, γ-linolenic acid, linoleic acid, and lysophosphatidylcholines were found to play an important role in energy metabolism and glycerophospholipid metabolism. UPLC–MS-based metabolomics were used to reveal the metabolic profiles, biomarkers, and related metabolic pathways in CHD and to evaluate the effects of *Jing Zhi Guan Xin Pian* [42]. A total of 25 biomarkers involved 9 metabolic pathways were identified associated with CHD models. Finally, targets of *Jing Zhi Guan Xin Pian* for the treatment of CHD were β-alanine metabolism and tyrosine metabolism. These studies showed that metabolomics is an effective method for exploring potential biomarkers associated with diseases and the therapeutic mechanisms of herbal medicine.

9.7 Conclusions and Future Direction

The knowledge and understanding about the pathological mechanism of CHD is still insufficient, which leads to a lot of efforts in the study of CHD for the public in much time with the expectation that the persistence brings a clinical benefit. Nowadays, clinical doctors hunt only rarely part of the metabolism information related closely to CHD. Early diagnosis and interventional therapies are essential to reduce the incidence and mortality from CHD. In practice, effective prevention and control of CHD lies on the ability to distinguish the high risk group at the asymptomatic stage of the disease or development of adverse events. Hence, there is a crucial need for more reliable diagnostic and therapeutic signature to detect CHD. Metabolomics provides more comprehensive metabolic signature as a contemporary perspective tool to solve diagnosis, prevention, and personalized treatment of disease [43–47]. Metabolomics technology is increasingly considered to be a promising tool that has the potential to identify metabolic features in an attempt to distinguish the different stages of CHD, to detect biomarkers as indicators of diagnostic process, and to measure response to treatment that may potentially be a screening tool to be used in the clinic. This chapter focuses on recent progress in metabolomics method to explore the biomarker in CHD, with emphasis on pathomechanism insights.

Acknowledgments

This work was supported by grants from the National Key Research and Development Program of China (2018YFC1706103), Key Program of Natural Science Foundation of State (Grant No. 81830110, 81861168037, 81973745, 81302905), and Heilongjiang Touyan Innovation Team Program.

References

1 Mozaffarian, D., Benjamin, E.J., Go, A.S. et al. (2015). Heart disease and stroke statistics – 2015 update: a report from the American Heart Association. *Circulation* 131 (4): e29–e322.

2 Writing Group Members, Mozaffarian, D., Benjamin, E.J. et al. (2016). Executive summary: heart disease and stroke statistics – 2016 update: a report from the American Heart Association. *Circulation* 133 (4): 447–454.

3 Wang, X., Zhang, A., and Sun, H. (2013). Power of metabolomics in diagnosis and biomarker discovery of hepatocellular carcinoma. *Hepatology* 57 (5): 2072–2077.

4 Sun, H., Zhang, A., Yan, G. et al. (2013). Metabolomic analysis of key regulatory metabolites in hepatitis C virus-infected tree shrews. *Molecular & Cellular Proteomics* 12 (3): 710–719.

5 Zhang, A., Sun, H., and Wang, X. (2013). Power of metabolomics in biomarker discovery and mining mechanisms of obesity. *Obesity Reviews* 14 (4): 344–349.

6 Zhang, A.H., Sun, H., Han, Y. et al. (2013). Ultraperformance liquid chromatography–mass spectrometry based comprehensive metabolomics combined with pattern recognition and network analysis methods for characterization of metabolites and metabolic pathways from biological data sets. *Analytical Chemistry* 85 (15): 7606–7612.

7 Qiu, S., Zhang, A.H., Guan, Y. et al. (2020). Functional metabolomics using UPLC-Q/TOF-MS combined with ingenuity pathway analysis as a promising strategy for evaluating the efficacy and discovering amino acid metabolism as a potential therapeutic mechanism-related target for geniposide against alcoholic liver disease. *RSC Advances* 10 (5): 2677–2690.

8 Zhang, A., Sun, H., Qiu, S. et al. (2014). Metabolomics insights into pathophysiological mechanisms of nephrology. *International Urology and Nephrology* 46 (5): 1025–1030.

9 Zhang, A., Sun, H., Yan, G. et al. (2014). Metabolomics in diagnosis and biomarker discovery of colorectal cancer. *Cancer Letters* 345 (1): 17–20.

10 Yang, Q., Zhang, A.H., Miao, J.H. et al. (2019). Metabolomics biotechnology, applications, and future trends: a systematic review. *RSC Advances* 9 (64): 37245–37257.

11 Zhang, A., Fang, H., Wang, Y. et al. (2017). Discovery and verification of the potential targets from bioactive molecules by network pharmacology-based target prediction combined with high-throughput metabolomics. *RSC Advances* 7 (81): 51069–51078.

12 Zhang, A., Yan, G., Han, Y. et al. (2014). Metabolomics approaches and applications in prostate cancer research. *Applied Biochemistry and Biotechnology* 174 (1): 6–12.

13 Zhang, A., Sun, H., Yan, G. et al. (2015). Metabolomics for biomarker discovery: moving to the clinic. *BioMed Research International* 2015: 354671–354676.

14 Wang, X., Zhang, A., Han, Y. et al. (2012). Urine metabolomics analysis for biomarker discovery and detection of jaundice syndrome in patients with liver disease. *Molecular & Cellular Proteomics* 11 (8): 370–380.

15 Zhang, A., Sun, H., Yan, G. et al. (2016). Mass spectrometry-based metabolomics: applications to biomarker and metabolic pathway research. *Biomedical Chromatography* 30 (1): 7–12.

16 Liang, Q., Liu, H., Zhang, T. et al. (2016). Discovery of serum metabolites for diagnosis of mild cognitive impairment to Alzheimer's disease progression using an optimized metabolomics method. *RSC Advances* 6: 3586–3591.

17 Zhang, A., Sun, H., and Wang, X. (2012). Saliva metabolomics opens door to biomarker discovery, disease diagnosis, and treatment. *Applied Biochemistry and Biotechnology* 168 (6): 1718–1727.

18 Liang, Q., Liu, H., Xing, H. et al. (2016). High-resolution mass spectrometry for exploring metabolic signatures of sepsis-induced acute kidney injury. *RSC Advances* 6 (36): 29863–29868.

19 Cohain, A.T., Barrington, W.T., Jordan, D.M. et al. (2021). An integrative multiomic network model links lipid metabolism to glucose regulation in coronary artery disease. *Nature Communications* 12 (1): 547.

20 Dugani, S.B., Moorthy, M.V., Li, C. et al. (2021). Association of lipid, inflammatory, and metabolic biomarkers with age at onset for incident coronary heart disease in women. *JAMA Cardiology* 6 (4): 437–447.

21 Qamar, W., Alqahtani, S., Ahamad, S.R. et al. (2020). Untargeted GC–MS investigation of serum metabolomics of coronary artery disease patients. *Saudi Journal of Biological Sciences* 27 (12): 3727–3734.

22 Cheng, P., Chen, Z.Q., and Wang, D.S. (2015). Metabonomics research on coronary heart disease patients of phlegm turbidity syndrome and qi deficiency syndrome. *Zhongguo Zhong Xi Yi Jie He Za Zhi* 35 (2): 193–197.

23 Wang, Z., Zhu, C., Nambi, V. et al. (2019). Metabolomic pattern predicts incident coronary heart disease. *Arteriosclerosis, Thrombosis, and Vascular Biology* 39 (7): 1475–1482.

24 Huang, M., Zhao, H., Gao, S. et al. (2019). Identification of coronary heart disease biomarkers with different severities of coronary stenosis in human urine using non-targeted metabolomics based on UPLC-Q-TOF/MS. *Clinica Chimica Acta* 497: 95–103.

25 Chen, Q., Zhang, Y., Ding, D. et al. (2016). Metabolic syndrome and its individual components with mortality among patients with coronary heart disease. *International Journal of Cardiology* 224: 8–14.

26 Fu, H., Zhu, K., Zhou, D. et al. (2019). Identification and validation of plasma metabolomics reveal potential biomarkers for coronary heart disease. *International Heart Journal* 60 (6): 1387–1397.

27 Zhong, Z., Liu, J., Zhang, Q. et al. (2019). Targeted metabolomic analysis of plasma metabolites in patients with coronary heart disease in southern China. *Medicine* 98 (7): e14309.

28 Paynter, N.P., Balasubramanian, R., Giulianini, F. et al. (2018). Metabolic predictors of incident coronary heart disease in women. *Circulation* 137 (8): 841–853.

29 Liang, Q., Zhu, Y., Liu, H. et al. (2017). High-throughput lipidomics enables discovery of the mode of action of huaxian capsule impacting the metabolism of sepsis. *RSC Advances* 7 (71): 44990–44996.

30 Zhang, A.H., Ma, Z.M., Kong, L. et al. (2020). High-throughput lipidomics analysis to discover lipid biomarkers and profiles as potential targets for evaluating efficacy of Kai-Xin-san against APP/PS1 transgenic mice based on UPLC–Q/TOF–MS. *Biomedical Chromatography* 34 (2): e4724.

31 Qiu, S., Zhang, A., Zhang, T. et al. (2017). Dissect new mechanistic insights for geniposide efficacy on the hepatoprotection using multiomics approach. *Oncotarget* 8 (65): 108760.

32 Gao, H.L., Zhang, A.H., Yu, J.B. et al. (2018, 1092). High-throughput lipidomics characterize key lipid molecules as potential therapeutic targets of Kaixinsan protects against Alzheimer's disease in APP/PS1 transgenic mice. *Journal of Chromatography B: Analytical Technologies in the Biomedical and Life Sciences*: 286–295.

33 Liang, Q., Liu, H., Jiang, Y. et al. (2016). Discovering lipid phenotypic changes of sepsis-induced lung injury using high-throughput lipidomic analysis. *RSC Advances* 6: 38233–38237.

34 Zhang, H., Zhang, A., Zhou, X. et al. (2018). High-throughput lipidomics reveal mirabilite regulating lipid metabolism as anticancer therapeutics. *RSC Advances* 8 (62): 35600–35610.

35 Liang, Q., Wang, C., Li, B. et al. (2015). Lipidomics analysis based on liquid chromatography mass spectrometry for hepatocellular carcinoma and intrahepatic cholangiocarcinoma. *RSC Advances* 5 (78): 63711–63718.

36 Harshfield, E.L., Koulman, A., Ziemek, D. et al. (2019). An unbiased lipid phenotyping approach to study the genetic determinants of lipids and their association with coronary heart disease risk factors. *Journal of Proteome Research* 18 (6): 2397–2410.

37 Saleem, M., Herrmann, N., Dinoff, A. et al. (2017). A Lipidomics approach to assess the association between plasma sphingolipids and verbal memory performance in coronary artery disease patients undertaking cardiac rehabilitation: a C18:0 signature for cognitive response to exercise. *Journal of Alzheimer's disease* 60 (3): 829–841.

38 Alshehry, Z.H., Mundra, P.A., Barlow, C.K. et al. (2016). Plasma Lipidomic profiles improve on traditional risk factors for the prediction of cardiovascular events in type 2 diabetes mellitus. *Circulation* 134 (21): 1637–1650.

39 Tao, T., He, T., Mao, H. et al. (2020). Non-targeted metabolomic profiling of coronary heart disease patients with Taohong Siwu decoction treatment. *Frontiers in Pharmacology* 11: 651.

40 Yi, M., Li, Q., Zhao, Y. et al. (2019). Metabolomics study on the therapeutic effect of traditional Chinese medicine Xue-Fu-Zhu-Yu decoction in coronary heart disease based on LC-Q-TOF/MS and GC-MS analysis. *Drug Metabolism and Pharmacokinetics* 34 (5): 340–349.

41 Li, Z., Yang, L., Liu, Y. et al. (2019). Anti-inflammatory and antioxidative effects of Dan-Lou tablets in the treatment of coronary heart disease revealed by metabolomics integrated with molecular mechanism studies. *Journal of Ethnopharmacology* 240: 111911.

42 Sun, H., Li, X.N., Zhang, A.H. et al. (2019). Exploring potential biomarkers of coronary heart disease treated by Jing Zhi Guan Xin Pian using high-throughput metabolomics. *RSC Advances* 9 (20): 11420–11432.

43 Liu, C., Zong, W.J., Zhang, A.H. et al. (2018). Lipidomic characterisation discovery for coronary heart disease diagnosis based on high-throughput ultra-performance liquid chromatography and mass spectrometry. *RSC Advances* 8 (2): 647–654.

44 Liang, Q., Liu, H., Zhang, T. et al. (2016). Untargeted lipidomics study of coronary artery disease by FUPLC-Q-TOF-MS. *Analytical Methods* 8 (6): 1229–1234.

45 Zhang, A., Sun, H., Han, Y. et al. (2012). Exploratory urinary metabolic biomarkers and pathways using UPLC-Q-TOF-HDMS coupled with pattern recognition approach. *Analyst* 137 (18): 4200–4208.

46 Zhang, Y., Liu, P., Li, Y. et al. (2017). Exploration of metabolite signatures using high-throughput mass spectrometry coupled with multivariate data analysis. *RSC Advances* 7 (11): 6780–6787.

47 Bar, N., Korem, T., Weissbrod, O. et al. (2020). A reference map of potential determinants for the human serum metabolome. *Nature* 588 (7836): 135–140.

10

Metabolomics Applications in Neurological Disease

Shi Qiu, Ying Han, Aihua Zhang, Hui Sun, and Xijun Wang

Heilongjiang University of Chinese Medicine, National Chinmedomics Research Center, Functional Metabolomics Laboratory, Department of Pharmaceutical Analysis, Heping Road 24, Harbin 150040, China

10.1 Introduction

Neurodegenerative disorders are one of the greatest global challenges for health care in the twenty-first century. Global incidence of central nervous system (CNS) such as Parkinson's disease (PD) and Alzheimer's disease (AD) is rapidly increasing; however their diagnosis at the early stage is still challenging. Therefore, the reliable biomarkers to diagnose them and monitor the development of therapeutics are more critical than ever. Metabolomics enables the assessment of a broad range of endogenous and exogenous metabolites and has a great impact on discovering biomarkers, diagnosing diseases, and identifying the perturbed pathways [1, 2]. The small-molecule metabolites including carbohydrates, vitamins, lipids, amino acids, peptides, thiols, nucleic acids, and organic acids, as primary indicators, represent attractive candidates to understand disease phenotypes [3–5]. Many diseases result in characteristic changes in the biochemical profiles prior to the development of clinical symptoms. The early detection of disease biomarkers will result in suffering reduction, efficient treatments, and lower mortality rates.

Recent progress in metabolomics resulted in an increasing number of metabolomics applications in neurological research [6–10]. The identification of specific metabolites illustrates the potential of metabolomics to address problems in CNS diseases. Metabolites biomarkers are small molecules derived from the metabolic process found in different biological media, such as cerebrospinal fluid (CSF), urine, and plasma [11–15]. CSF is the obvious choice because it provides a rich pool of biochemical markers in the brain, reflecting the biochemical changes of the CNS. The metabolites present in the CSF represent actual metabolism of CNS and can be helpful in the determination of markers for neurological disorders. Although the detection of CNS is based on several neuropsychological tests and imaging and biological analyses, none of the biomarkers allows a clear understanding of the pathophysiological mechanisms. Metabolomics could help to discover therapeutic targets, to identify disease mechanisms, and to monitor the therapeutic response.

Mass Spectrometry-Based Metabolomics in Clinical and Herbal Medicines: Strategies, Technologies, and Applications, First Edition. Edited by Aihua Zhang and Wanying Wang.

10.2 CSF Sampling

Metabolomics provides key information on the biomarkers to understand neurological pathology. Great attention has been directed to practical considerations and limitations of several major aspects of metabolomic analysis of human CSF. CSF analysis is used in biomarker discovery studies for various neurodegenerative CNS disorders. CSF samples can be accurately screened by metabolomics but requires reliable collection and storage methods. Gray et al. had collected CSF in polypropylene tubes by lumbar puncture, and then CSF samples were centrifuged, frozen, and stored at −80 °C [16]. Herman et al. also collected CSF samples in a polypropylene tubes by a lumbar puncture, and then after the samples were centrifuged and supernatant transferred, they were gently mixed, aliquoted, and stored at −80 °C [17]. Rosenling et al. recommended that the CSF samples should be centrifuged immediately after collection, and then freeze the supernatant in liquid nitrogen and avoid freeze and thaw cycles [18].

10.3 Application of Metabolomics in CNS

The advances in metabolomics contribute to the understanding of the pathophysiology and diagnosis and have opened new avenues toward biomarker discovery, which is the identification of signaling molecules associated with early detection of disease [19–23]. Metabolic profiling of biofluids is emerging as an important area with a promising number of applications in clinical medicine, including early diagnosis of numerous diseases [24–27]. CSF metabolic profiling has been shown to be a promising method for the early detection of response to therapy in wider clinical settings [28, 29]. The analysis of CSF is used in biomarker discovery studies for various CNS disorders. Metabolic profiling of CSF exhibits a large number of quantifiable metabolites, and hence CSF metabolomic methodology is potentially a useful tool for the discovery of biomarkers in the clinic.

The metabolic profile across manifest Huntington's disease (HD) from early to advanced clinical staging was explored by metabolomic analysis [30]. In CSF, HD was associated with significant increases in saturated long-chain free fatty acids, arginine, NAD^+, diacylglycerides, triacylglycerides, and sphingomyelins. It suggested that defects in the urea cycle, glycine, and serine metabolism may be associated with the progression of HD pathology and merit further study for possible therapeutic relevance. CSF metabolomics has been used to identify 19 brain-related phenotype associations [31]. To investigate metabolic alterations in AD, a large-scale metabolome phenotypic study was conducted from a multi-center clinical cohort, consisting of individuals clinically diagnosed with AD, mild cognitive impairment (MCI), and age-matched controls [32]. It revealed significantly lower metabolite concentrations of tryptophan pathway metabolites in the AD. A significant increase in the kynurenine/tryptophan ratio suggests that this may be a result of a shift to the kynurenine metabolic route. Modulation of the pathways could help improve serotonin bioavailability and signaling in AD patients.

Amino acids acting as neuromodulators and neurotransmitters play many important roles in the CNS. The study was conducted to evaluate if serum amino acid levels change along the continuum from no cognitive impairment to moderate dementia and to identify putative biomarkers for early diagnosis of neurodegenerative diseases [33]. The most exciting result is the significantly elevated concentration of arginine in patients with both stages of dementia as compared with MCI individuals. It suggested the implication of changed arginine metabolism in the pathogenesis of CNS diseases. Ultra-performance liquid chromatography–mass spectrometry-based untargeted metabolomics were used to measure the concentrations of plasma metabolites in a cohort of subjects with AD and cognitively normal controls (Ctrl) [34]. It showed marked reductions in levels of polyunsaturated fatty acids, acyl-carnitines, degradation products of tryptophan, and elevated levels of bile acids. The concentrations of these metabolites were significantly correlated with disease severity. It also suggested that altered bile acid profiles in AD and MCI might indicate early risk for the development of AD. These findings may allow for the development of new approaches for diagnosis of AD and may provide novel insights into AD pathogenesis.

MCI is regarded as a transition phase between normal aging and AD. Targeted metabolomics analysis of serum samples was performed for identification of biomarkers that can distinguish AD at an early stage from MCI [35]. Receiver operating characteristic curve analysis of a combination of three lipid metabolites (SM (OH) C24 : 1, SM C24 : 0, and PC ae C44 : 3) showed an acceptable discrimination between the early-stage AD and MCI patients. Urinary metabolomics was performed for delineating the metabolic changes and potential early biomarkers in transgenic mice [36]. A total of 24 differentially regulated metabolites were identified. Metabolic pathway analysis found that glyoxylate and dicarboxylate metabolism, pentose and glucuronate interconversions, starch and sucrose metabolism, and citrate cycle were disturbed. It revealed that levels of endogenous metabolites changed prior to the emergence of learning and cognitive impairment, which may be associated with abnormal nitric oxide production pathways and metabolic disorders of monoaminergic neurotransmitters.

10.4 Future Direction

Metabolomics technologies and their application offer new opportunities for biomarker discovery in complex disorders. It enables map perturbations of early biochemical changes in disease and hence exerts an opportunity to develop predictive biomarkers, and it could also provide valuable insights about mechanisms of disease. This chapter reveals the current status and potential applications of CSF as a source of biomarkers, with emphasis on metabolites, for early detection and diagnosis of diseases. Despite the extensive resources concentrated on this area, there are very few biomarkers currently available that qualify and are satisfactorily validated for mental disorders, and there is still a major lack of biomarkers for typifying neurodegenerative disorders. Continuing improvements in analytical technology along with an increased interest in performing comprehensive and quantitative metabolic profiling is leading to increase and develop centralized metabolite reference resources for CSF biofluids [37–40].

Finding early disease markers using noninvasive and widely available methods is essential to develop a successful therapy for CNS. There are, still, limitations to predict the occurrence and prognosis of CNS. Biomarkers are molecules that can change in different conditions, a feature that makes them potential tools to improve the diagnosis of disease, establish a prognosis, and monitor treatments. They can be identified using various strategies, targeted or untargeted experiments. CSF metabolomics provides information on the metabolite phenotype of the human and is therefore appropriate to study the status of the global system [34, 41]. It is expected that future studies using these strategies will provide more accurate and less expensive biomarkers and explore whether a combination of different biomarkers can improve diagnostic accuracy. Upon further validation, such markers will offer great benefits for the public health in the long term. With advances in methods and technology, many candidates will be discovered, leading to early diagnosis, detection, monitoring, and efficient treatment of CNS. The validated method allows sensitive and specific measurement of metabolites in CSF and may be applied for clinical diagnoses.

10.5 Conclusions

Most clinical chemistry tests available today rely on old technologies, and these tests are neither sensitive nor specific for any particular disease. The clinical diagnosis of major CNS remain unsatisfactory based on current clinical criteria. Emerging metabolomics provides a powerful platform for discovering novel biomarkers and biochemical pathways to improve diagnostic, prognostication, and therapy. It can provide potential advantages that classical diagnostic approaches do not, based on following discovery of a suite of clinically relevant biomarkers for diseases. The great asset of this platform allows for the assessment of global metabolic profiles in biofluids and biomarker discovery to distinguish diseased status information, with the promise to enhance clinical diagnostics.

Identifying multiple novel biomarkers for CNS has been greatly enhanced with recent advances in metabolomics that are more accurate than routine clinical practice. CSF is known to be a rich source of small-molecule biomarkers for neurological and neurodegenerative diseases and is in close contact with diseased areas in neurological disorders, and it could potentially be used for disease diagnosis. Metabolomics will drive CSF analyses, facilitate and improve the development of diseases treatments, and offer great benefits for the public health in the long term. Thus, this chapter showed various aspects of CSF metabolomics and discusses their significance in the post-genomic era, especially highlighting the potential roles of the endogenous small-molecule metabolites in this emerging field.

Acknowledgments

This work was supported by grants from the Key Program of Natural Science Foundation of State (Grant No. 81973745, 81302905), Natural Science Foundation of Heilongjiang Province (YQ2019H030), and the University Nursing Program for Young Scholars with Creative Talents in Heilongjiang Province (UNPYSCT-2015118).

References

1 Garcia-Perez, I., Posma, J.M., Serrano-Contreras, J.I. et al. (2020). Identifying unknown metabolites using NMR-based metabolic profiling techniques. *Nature Protocols* 15 (8): 2538–2567.

2 Zhang, A., Sun, H., Han, Y. et al. (2012). Exploratory urinary metabolic biomarkers and pathways using UPLC-Q-TOF-HDMS coupled with pattern recognition approach. *Analyst* 137 (18): 4200–4208.

3 Li, Y.F., Qiu, S., Gao, L.J. et al. (2018). Metabolomic estimation of the diagnosis of hepatocellular carcinoma based on ultrahigh performance liquid chromatography coupled with time-of-flight mass spectrometry. *RSC Advances* 8 (17): 9375–9382.

4 Zhang, A., Sun, H., and Wang, X. (2017). Emerging role and recent applications of metabolomics biomarkers in obesity disease research. *RSC Advances* 7 (25): 14966–14973.

5 Zhang, A., Sun, H., Yan, G. et al. (2014). Metabolomics analysis of marker metabolites for patients with pancreatic cancer. *Medicinal Chemistry* 4 (6): 506–510.

6 Wang, X., Sun, H., Zhang, A. et al. (2011). Ultra-performance liquid chromatography coupled to mass spectrometry as a sensitive and powerful technology for metabolomic studies. *Journal of Separation Science* 34 (24): 3451–3459.

7 Zhang, A., Sun, H., Yan, G. et al. (2013). Urinary metabolic profiling identifies a key role for glycocholic acid in human liver cancer by ultra-performance liquid-chromatography coupled with high-definition mass spectrometry. *Clinica Chimica Acta* 418: 86–90.

8 Zhang, A., Sun, H., and Wang, X. (2012). Serum metabolomics as a novel diagnostic approach for disease: a systematic review. *Analytical and Bioanalytical Chemistry* 404 (4): 1239–1245.

9 Yachida, S., Mizutani, S., Shiroma, H. et al. (2019). Metagenomic and metabolomic analyses reveal distinct stage-specific phenotypes of the gut microbiota in colorectal cancer. *Nature Medicine* 25 (6): 968–976.

10 Zhang, A., Sun, H., and Wang, X. (2014). Urinary metabolic profiling of rat models revealed protective function of scoparone against alcohol induced hepatotoxicity. *Scientific Reports* 4: 6768.

11 Liang, Q., Yu, Q., Wu, H. et al. (2014). Metabolite fingerprint analysis of cervical cancer using LC-QTOF/MS and multivariate data analysis. *Analytical Methods* 6 (12): 3937–3942.

12 Zhang, A.H., Sun, H., Yan, G.L. et al. (2014). Metabolomics study of type 2 diabetes using ultra-performance LC-ESI/quadrupole-TOF high-definition MS coupled with pattern recognition methods. *Journal of Physiology and Biochemistry* 70 (1): 117–128.

13 Zhang, H.L., Zhang, A.H., Miao, J.H. et al. (2019). Targeting regulation of tryptophan metabolism for colorectal cancer therapy: a systematic review. *RSC Advances* 9 (6): 3072–3080.

14 Wang, X., Wang, Q., Zhang, A. et al. (2013). Metabolomics study of intervention effects of Wen-Xin-formula using ultra high-performance liquid chromatography/mass spectrometry coupled with pattern recognition approach. *Journal of Pharmaceutical and Biomedical Analysis* 74: 22–30.

15 Zhang, A., Sun, H., Wang, P. et al. (2012). Recent and potential developments of biofluid analyses in metabolomics. *Journal of Proteomics* 75 (4): 1079–1088.

16 Gray, E., Larkin, J.R., Claridge, T.D. et al. (2015). The longitudinal cerebrospinal fluid metabolomic profile of amyotrophic lateral sclerosis. *Amyotrophic Lateral Sclerosis and Frontotemporal Degeneration* 16 (7–8): 456–463.

17 Herman, S., Khoonsari, P.E., Tolf, A. et al. (2018). Integration of magnetic resonance imaging and protein and metabolite CSF measurements to enable early diagnosis of secondary progressive multiple sclerosis. *Theranostics* 8 (16): 4477–4490.

18 Rosenling, T., Slim, C.L., Christin, C. et al. (2009). The effect of preanalytical factors on stability of the proteome and selected metabolites in cerebrospinal fluid (CSF). *Journal of Proteome Research* 8 (12): 5511–5522.

19 Zhang, A., Liu, Q., Zhao, H. et al. (2016). Phenotypic characterization of nanshi oral liquid alters metabolic signatures during disease prevention. *Scientific Reports* 6: 19333.

20 Wang, X., Yang, B., Sun, H. et al. (2012). Pattern recognition approaches and computational systems tools for ultra performance liquid chromatography–mass spectrometry-based comprehensive metabolomic profiling and pathways analysis of biological data sets. *Analytical Chemistry* 84 (1): 428–439.

21 Chu, H., Zhang, A., Han, Y. et al. (2016). Metabolomics approach to explore the effects of Kai-Xin-San on Alzheimer's disease using UPLC/ESI-Q-TOF mass spectrometry. *Journal of Chromatography B* 1015: 50–61.

22 Zhang, A., Zhou, X., Zhao, H. et al. (2017). Metabolomics and proteomics technologies to explore the herbal preparation affecting metabolic disorders using high resolution mass spectrometry. *Molecular BioSystems* 13 (2): 320–329.

23 Liang, Q., Liu, H., Xie, L.X. et al. (2017). High-throughput metabolomics enables biomarker discovery in prostate cancer. *RSC Advances* 7 (5): 2587–2593.

24 Zhang, A., Sun, H., Wang, P. et al. (2012). Modern analytical techniques in metabolomics analysis. *Analyst* 137 (2): 293–300.

25 Zhang, A., Sun, H., Wu, X. et al. (2012). Urine metabolomics. *Clinica Chimica Acta* 414: 65–69.

26 Zhang, Y., Liu, P., Li, Y. et al. (2017). Exploration of metabolite signatures using high-throughput mass spectrometry coupled with multivariate data analysis. *RSC Advances* 7 (11): 6780–6787.

27 Zhang, A., Wang, H., Sun, H. et al. (2015, 2015). Metabolomics strategy reveals therapeutical assessment of limonin on nonbacterial prostatitis. *Food & Function* 6 (11): 3540–3549.

28 Koch, M., Acharjee, A., Ament, Z. et al. (2021). Machine learning-driven Metabolomic evaluation of cerebrospinal fluid: insights into poor outcomes after aneurysmal subarachnoid hemorrhage. *Neurosurgery*, https://doi.org/10.1093/neuros/nyaa557 88 (5): 1003–1011.

29 Zhang, A.H., Sun, H., and Wang, X.J. (2013). Recent advances in metabolomics in neurological disease, and future perspectives. *Analytical and Bioanalytical Chemistry* 405 (25): 8143–8150.

30 McGarry, A., Gaughan, J., Hackmyer, C. et al. (2020). Cross-sectional analysis of plasma and CSF metabolomic markers in Huntington's disease for participants of varying functional disability: a pilot study. *Scientific Reports* 10 (1): 20490.

31 Panyard, D.J., Kim, K.M., Darst, B.F. et al. (2021). Cerebrospinal fluid metabolomics identifies 19 brain-related phenotype associations. *Communications Biology* 4 (1): 63.

32 Whiley, L., Chappell, K.E., D'Hondt, E. et al. (2021). Metabolic phenotyping reveals a reduction in the bioavailability of serotonin and kynurenine pathway metabolites in both the urine and serum of individuals living with Alzheimer's disease. *Alzheimer's Research & Therapy* 13 (1): 20.

33 Socha, E., Kośliński, P., Koba, M. et al. (2021). Serum amino acid profiles in patients with mild cognitive impairment and in patients with mild dementia or moderate dementia. *Amino Acids* 53 (1): 97–109.

34 Shao, Y., Ouyang, Y., Li, T. et al. (2020). Alteration of metabolic profile and potential biomarkers in the plasma of Alzheimer's disease. *Aging and Disease* 11 (6): 1459–1470.

35 Weng, W.C., Huang, W.Y., Tang, H.Y. et al. (2019). The differences of serum metabolites between patients with early-stage Alzheimer's disease and mild cognitive impairment. *Frontiers in Neurology* 10: 1223.

36 Yu, J., Kong, L., Zhang, A. et al. (2017). High-throughput metabolomics for discovering potential metabolite biomarkers and metabolic mechanism from the APPswe/PS1dE9 transgenic model of Alzheimer's disease. *Journal of Proteome Research* 16 (9): 3219–3228.

37 Song, Q., Zhang, A.H., Yan, G.L. et al. (2017). Technological advances in current metabolomics and its application in tradition Chinese medicine. *RSC Advances* 7 (84): 53516–53524.

38 Zhang, A.H., Ma, Z.M., Sun, H. et al. (2019). High-throughput metabolomics evaluate the efficacy of total lignans from *Acanthophanax senticosus* stem against ovariectomized osteoporosis rat. *Frontiers in Pharmacology* 10: 553.

39 Fang, H., Zhang, A.H., Sun, H. et al. (2019). High-throughput metabolomics screen coupled with multivariate statistical analysis identifies therapeutic targets in alcoholic liver disease rats using liquid chromatography–mass spectrometry. *Journal of Chromatography B* 1109: 112–120.

40 Liang, Q., Liu, H., Jiang, Y. et al. (2016). Novel liquid chromatography–mass spectrometry for metabolite biomarkers of acute lung injury disease. *Analytical Methods* 8 (31): 6017–6022.

41 Hajjar, I., Liu, C., Jones, D.P., and Uppal, K. (2020). Untargeted metabolomics reveal dysregulations in sugar, methionine, and tyrosine pathways in the prodromal state of AD. *Alzheimer's & Dementia* 12 (1): e12064.

11

Metabolomics Toward Precision Medicine

Aihua Zhang, Shi Qiu, Hui Sun, Guangli Yan, and Xijun Wang

Heilongjiang University of Chinese Medicine, National Chinmedomics Research Center, Sino-America Chinmedomics Technology Collaboration Center, Functional Metabolomics Laboratory, Heping Road 24, Harbin 150040, China

11.1 Introduction

Precision medicine promises to get the right patient, the right test, the right diagnosis, the right treatment, and in the right amount. It is a biomedical strategy of health care to improve patient-specific and individualized diagnoses, medical decisions, medications, therapies, and prognoses and to make the life quality better. Aiming to generate knowledge applicable to the whole range of health and disease, researchers are increasingly attempting to customize treatments in the basis of a person's genetics or environment and based on molecular, genotypic, and phenotypic information, thereby facilitating a personalized medicine decision regarding diagnostic and therapeutic approaches. Each patient's disease can be analyzed according to genomics and systems biology data to establish the patient's disease process at the molecular level and select right treatments [1–3].

The patient's response and treatments could closely monitor disease with direct or surrogate biomarkers. The term biomarker implies that a molecular change has the capacity to identify increased risk, presence of disease, or change in prognosis or outcome in response to a particular therapy. Predictive biomarkers suggest a course of therapeutic action; diagnostic biomarkers facilitate obtaining an accurate diagnosis; prognostic biomarkers provide data on risk of disease progression, thereby helping to determine which patients need additional treatment. Thanks to genomic analysis, predictive and prognostic biomarkers have been identified for many diseases that can guide which treatment to be prescribed in the course of diseases. However, major challenges remain. Many patients do not respond well to treatments and suffer adverse drug reactions, with no validated biomarkers to enable effective drug selection for individual patients, necessitating a more tailored treatment approach. Better biomarkers are needed to assist with disease detection and to help guide treatment, particularly for common acquired conditions without a strong genetic predisposition. The lack of precise biomarkers presents unique challenges for the management of diseases. Currently, unresolved issues in the precision medicine

Mass Spectrometry-Based Metabolomics in Clinical and Herbal Medicines: Strategies, Technologies, and Applications, First Edition. Edited by Aihua Zhang and Wanying Wang.

will be addressed by the development of new biomarker testing. Overall advances in precision medicine will likely improve the benefit to cost ratio for biomarker testing, suggesting that biomarker testing will expand as a routine component of disease care [4, 5]. There is an urgent need to identify and validate disease-specific and mechanism-based biomarkers to monitor and develop precision medicine.

Metabolomics offers a powerful, complex, and robust data sets as complement role obtained by genomics and other omics methodologies that bring new insights into disease pathophysiology and mechanisms of underlying differences in drug responses and beneficial responses to drug treatment by metabolic phenotype studies [6–15]. This chapter provides an overview on the essential role of precision medicine, increasingly more frequent use of metabolomics to generate knowledge from clinical research. Especially, we focus on discussing the application and new trends as well as serious challenges of metabolomics in precision medicine.

11.2 Precision Medicine: A New Era of Medicine

Precision medicine put forward based on personalized medicine developed with the rapid progress of genome sequencing technology and the cross-application of biological information and large data science [16]. In essence, it aimed at accurately finding pathogenesis, therapeutic target, disease stage, and sub-classification by using omics such as genome, proteome, and metabolomics as well as other medical cutting-edge technology for biomarker analysis, identification, validation, and application in large sample population and specific disease types sample so as to realize personalized and accurate therapy to improve the effectiveness of disease prevention and treatment [17–19]. Precise medicine can help clinicians set out tailor-made treatment option based on the individual characteristics of each patient to avoid pointless and general drug design that are not effective to everyone [20]. In the twenty-first century, there are many different definition of medicine such as the well-known 4P [21, 22]. Throughout the changes of the medical model, doctors in the past used to make judgments from the experience according to patient symptoms whether fever, abdominal pain, disordered function, and others, which was called empirical medicine [23]. Now diagnosis and treatment project from doctors in clinic is based on the physical examination, interrogation, imaging, and laboratory examination, which is named evidence-based medicine under condition that there must be enough evidence for further diagnosis and treatment [24, 25]. Modern medicine is from the "4P medicine" to "5P medicine" with "precision" [26]. Precision medicine, a great change and prominent features in medical model, means that doctors get the required data from computer and digital platform to guide medical practice in the future, which provides a new way for diagnosis and monitoring of human health status with the perfect combination of clinical pathology and micro-molecular phenotype and then put forward precise diagnosis, prediction, and treatment strategies on the basis of individual characteristics [27, 28].

To find more and better treatment for cancer is the short-term goal of precision medicine [29]. In recent years, the development of specific molecular-driven drugs, antibody applications, targeted therapy, and new immunization methods have been increasing and achieved remarkable results [30]. In order to further understand cancer and develop new treatments, it is necessary to exploit new tools for molecular diagnostics, to analyze more tumor genomes, and to build more reliable preclinical models and database platforms for the purpose of trying to solve the difficult problem, which is beyond the understanding of precision medicine such as drug resistance and tumor genetic heterogeneity, and to improve inadequate assessment of drug efficiency and tumor recurrence as well as the limitation knowledge of drug combination [31, 32]. The long-term goal is to achieve health management and disease prevention. From the late twentieth century, significant changes have taken place in human disease spectrum, which the main disease affecting human health transforms from infectious diseases into cardiovascular and cerebrovascular diseases, malignancies, diabetes, and other non-communicable diseases [33]. However, there are not yet efficient solutions of preventing or treating the great bulk of chronic diseases. Precise medicine provides a valuable way to achieve personalized treatment of multiple diseases. It can be seen that the current international precision medicine is facing many challenges [34–36]. More support from large data sharing, new technologies, and new academic field are needed to achieve.

11.3 Precision Biomarkers

The limited treatment options and late diagnosis contribute to the high mortality of the disease. There is an urgent need of precision medicine for patients to prolong patients' survival and to improve their quality of life. A major goal of modern medicine is to increase patient specificity so that the right treatment is administered to the right patient at the right time in the right dose. As the part of translational medicine, precision medicine is a further step to understand medicine and help health care. The precision medicine as a new emerging area and therapeutic strategy has occurred and will bring more new therapeutic strategies, drug discovery, and development. Precision medicine can be carried out by integrating omics technologies, disease-specific biomarkers, clinical bioinformatics, and efficacy therapies [37, 38]. Traditional clinical decision making is based on reacting to a relatively limited number of phenotypes that are determined by history, physical examination, and conventional laboratory tests. Precision medicine depends on highly detailed profiling of the patient's genetic, morphologic, and metabolic makeup. Precision medicine is a novel emerging field that plays an essential role in identifying new sub-classifications of diseases and in providing guidance in treatment that is based on multi-omics approaches including the use of genomics, transcriptomics, proteomics, and metabolomics data to unravel the disease complexity, to predict prognostic biomarkers, and to identify new targets (Figure 11.1). Translation of these studies into clinically valuable biomarkers for predictive, diagnostic, and prognostic biomarkers is critical to the development of precision medicine. The field of biomarkers has at least two goals: personalized therapy for a particular patient

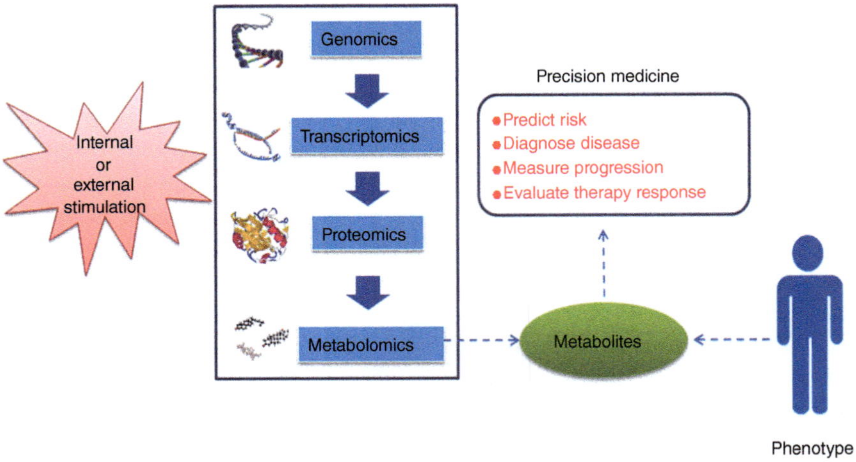

Figure 11.1 The relationship between metabolomics and precision medicine and the prospective benefits of clinical treatment by the application of metabolite biomarker.

and discovery of genetic and molecular abnormalities relevant to diseases, which lead to identification of optimal targets. Further experimental studies are ongoing to detect novel biomarkers for diagnosis, subtype, treatment response prediction, or clinical outcome. We should aim for precision medicine to optimize diagnosis and treatment choices and to monitor and improve clinical outcomes in diseases.

11.4 Metabolomics as a Powerful Tool for Potential Biomarkers

Modern recent advancement in technique has given the researchers and the clinicians the opportunity to help the clinical evaluation and to improve their ability to distinguish various types of diseases by using emerging methodologies [39–45]. A biomarker has the appraisal characteristic from an objective point, which measures and depicts normal or abnormal physiological condition, diagnoses pathogenic processes promptly, or responses to drug therapeutic intervention. An "ideal" biomarker is characterized with high sensitivity and specificity; capacity to predict disease prognosis, progression, and therapeutic effects from molecular mechanisms and pathology perspective; easy to operate; and noninvasiveness, simple, accurate, and cost-effective. "Omics" technologies have shown greater contribution in this pursuit. The change of the organism from health state to disease state is the overall result of the metabolic function fluctuation in organ, which metabolites species and concentration change constantly. The metabolome is close to representative of the phenotype, which is the end-product and most stable among genome, transcriptome, and proteome levels. It is one of the most important goals for metabolomics research to discover specific and sensitive biomarkers that are capable of unequivocally detecting disease and seem to be an effective way to comprehend the phenotypic changes associated with disease.

An individual's metabolic profile has been referred to as his or her metabotype. As such, metabolomic profiles could provide insights about the mechanism of action and response to treatment. Metabolomics is capable of the parallel assessment of a broad range variation of metabolites *in vivo* and *in vitro*, which has been reported to have a significant impact on physiology research, diagnosing diseases, biomarkers discovery, and seeking disturbed pathways related with disease or treatment [46–48]. Complete process framework of metabolomics processes covers sample collection, pretreatment, compound separation, data analysis, and metabolic variation interpretation. Combination of different metabolomics technique approaches could gain all-sided and sufficient metabolites information in intricate biological samples [49, 50]. To retain the overall information of metabolite in the sample as much as possible for accurate analysis of the metabolic components in biological systems is the inevitable trend of metabolomics analysis. At the end of the metabolomics study stage, bioinformatics platforms are used to analyze and interpret the data. The metabolic fingerprints obtained by bioanalysis are converted to multidimensional complex data and then new data disposed by modern chemometrics as well as biostatistics means [51–55]. According to the object and purpose of the study, metabolite profiling can be performed in a targeted or untargeted manner [56–60]. The combination of non-targeted or targeted metabolomics with stable isotope technologies provides concentrations and labeling patterns of metabolites.

11.5 Practical Application of Metabolomics

Currently, the lack of accurate diagnostic approaches and unknown molecular basis of pathophysiology will result in patients being unable to receive a precise and timely treatment. Therefore, it is urgent to identify early diagnostic biomarkers and develop therapeutic strategies. Fortunately, metabolomics offers promising tools to discover novel indicators and further understanding of pathological mechanisms and has been proposed and heralded as the key to advancing precision medicine.

Biomarker discovery and implementation are at the forefront of the precision medicine movement. Many of the metabolites are derived from or directly reflective of metabolism [61]. Metabolomics using ultra-performance liquid chromatography (UPLC)–tandem mass spectrometry (MS) was used to identify biochemical features of the supragingival biofilm associated with early childhood caries (ECC) prevalence and severity [62]. There were 503 metabolites identified, and 16 metabolites were significantly associated with ECC, including fucose and N-acetylneuraminate with higher ECC prevalence, as well as catechin and epicatechin with lower. Catechin, epicatechin, imidazole propionate, fucose, 9,10-DiHOME, and *N*-acetylneuraminate were the top metabolites in terms of ECC classification importance. A non-targeted metabolomics approach investigated the serum metabolite profile that associates with non-alcoholic fatty liver disease with the related metabolic risk factors [63]. It found that serum levels of the non-alcoholic steatohepatitis-related aromatic amino acids and the branched-chain amino acids demonstrate divergent associations with serum lipids. Nontargeted metabolomics studies revealed that indoxyl sulfate and *N*-acetylglutamate in urine may be the potential biomarkers for early diagnosis and prevention of hyperuricemia [64].

Concentrations of 15 bile acids (BAs) in plasma were measured using UPLC–MS to discover the specific BAs as biomarkers for cholangiocarcinoma (CCA) diagnosis [65]. Binary logistic regression analysis was used to build a diagnostic model for CCA. The specific changes in plasma concentrations of BAs may serve as diagnostic biomarkers for distinguishing CCA from benign biliary diseases and healthy controls, with higher performance than carbohydrate antigen 19–9. Twenty-four-hour rhythms were evaluated in vitals and putatively identified plasma metabolites individually [66]. Intensive care unit (ICU) patients' 24-hour rhythms show significant desynchronization compared with healthy controls. Clustering of plasma metabolic profiles suggests that metabolomics could be used to track individual patients' clinical courses longitudinally for patient care. Metabolic signatures of myocardial and vascular health in young adulthood specify the pathways of metabolic dysfunction relevant to cardiovascular disease (CVD) [67]. Efforts to include precision measures of precise metabolic risk phenotypes for CVD at its earliest stage are warranted. Metabolomics was used to explore the evolution from cognitively normal to Alzheimer's disease (AD) [68]. A panel of six metabolites that predicted subjective memory decline, mild cognitive impairment and AD with receiver operating characteristic values from 0.88 to 0.99 and sensitivity values from 88% to 95%, respectively. It showed that blood biomarkers had diagnostic relevance and indicated the preclinical symptoms of AD.

11.6 Precision Medicine for Herbal Medicine

Sir William Osler (1849–1919) recognized that "Variability is the law of life, and as no two faces are the same, so no two bodies are alike, and no two individuals react and behave alike under the abnormal conditions we know as disease." During recent years, personalized medicine has attracted significant enthusiasm and gained a visible public momentum. It represents a new model that seeks to identify unique characteristics within each patient that can serve as a basis for personalized treatment that will be tailored to the future therapies. Life science research has in the past very much focused on diseases and how to reestablish human health after illness. Regardless, medicines are needed not only for treatment of diseases but also for prevention of diseases and promotion of quality of life of individuals. The question "who should be treated and what to be used" is central to the practice of personalized medicine.

Patient treatment is now based on a more holistic and integrated approach to manage the specific disease and promote the well-being of the patients. Herbal medicine (HM), regarded as intellectual experience of humans, has been accumulating since time immemorial for the prevention and treatment of diseases along with skillful application of natural medicines. With the development of life science, systems biology attempts to elucidate the complex interaction between genes, proteins, and metabolites to provide a mechanistic understanding of cellular function and how this function is affected by disease processes or drug efficacy. Such concepts are transforming current diagnostic and therapeutic approaches to medicine and, together with new technologies, will enable a personalized medicine that will lead to precision medicine.

Precision medicine is a biomedical strategy of health care to get the right patient, the right test, the right diagnosis, the right treatment, and in the right amount. Each patient's disease can be analyzed according to genomics and systems biology data to establish the patient's disease process at the molecular level and select right treatments. Precision medicine currently focuses largely on treatment that can reduce morbidity. Disease prevention and treatment are important for the entire population. The question "what and how to be used" of drugs is central to the practice of precision medicine. Moreover, in addition to the diagnosis, treatment is more important, so Obama's philosophy of precision medicine only reflects one aspect of precision medicine, not all. Even if the diagnosis is accurate, and there are no effective drugs, we cannot achieve the goals of precision medicine. Phytomedicine is a part of healthcare systems around the world. HM is one of the oldest phytomedicine systems of health care. Recent data suggest that at least 11 146 plant species, 1581 animal species, and 80 minerals and substances, ranging from precious stones to mineralize fossils, are used as HM medicines [69]. HM treatment with therapeutic strategies in holistic fashion is based on an overall symptoms and signs of differential syndrome and concerns the disease from various dynamic functional aspects as outlined from a nonlinear point of view [70–72]. It focuses on the patient rather than the disease, enhancing the quality of life. HM often uses formulae that are comprised by several types of crude herbs tailored to an individual's condition based on subjective diagnostic methods. It now provides perhaps the best opportunity to innovate and improve clinical success rates in drug development. HM advocates combinatory therapeutic strategies by formulae, in which one represents the key component, and others serve as adjuvant ones to assist the effects or facilitate the delivery of the key components [73]. For example, researchers found that a combination of multiple compounds could be more effective than using a single compound [74, 75].

The era of personalized precision medicine is about to emerge. Disease treatment through personalized precision medicine will become more important as the obvious avenue of research in life sciences. As the part of translational medicine, precision medicine is a further step to understand HM and will significantly improve the health of the general public and cut down healthcare costs. Precision medicine as a new emerging area and therapeutic strategy has occurred and will bring more new therapeutic strategies, drug discovery, and development. Potentially, more people in the world could benefit from precision medicine to optimize treatment choices and learn from HM. Future personalized health care is expected to benefit from the integration of precision medicine into the HM, such as exploring new targets and mechanisms, discovering lead compounds, and making innovative drug design. Scientists, governments, pharmaceutical companies, and patients should work closely together to ensure the success of this transformation. With economic development and rising living standards, people's concepts of health and way of life have undergone great changes. Disease prevention through personalized precision medicine will become more important as the obvious avenue of research in life sciences. Personalized HM will catalyze fundamental changes in the future of healthcare systems and appears to be following a general societal trend.

11.7 Conclusions and Future Perspectives

The main purpose of PM is to obtain the best response, which is to improve health care and to lower costs. The precision medicine as a new emerging area and therapeutic strategy has occurred and will bring more new therapeutic strategies, drug discovery, and development. We should aim for precision medicine to optimize diagnosis and treatment choices and to monitor and improve clinical outcomes in diseases. Precision medicine can be carried out by metabolomics technologies, metabolic biomarkers, and efficacy therapies [76–79]. Translation of these studies into clinically valuable biomarkers for predictive, diagnostic, and prognostic biomarkers is critical to the development of precision medicine [80–82]. Metabolomics, as a molecular technology in the era of precision medicine, possesses the strongest correlation between the health care and clinical features involving the integrated characterization of metabolite metabolism in biological systems. This chapter provided the basic information and the vital role of precision medicine as core of personalized therapy in new medicine era, and then we focus on discussing potential effect and new trends of metabolomics on precision medicine. Finally, we put forward the future research directions and challenges toward the holistic systems-based strategy of precision medicine with the help of metabolomics for turning into the new age of the medical success.

Metabolomics, as a relatively young discipline compared with other omics fields, is changing biomedical research that can effectively monitor the human body from the healthy state to the pathological changes in the whole process, while metabolic phenotype can also fully reflect the individual differences. Metabolomics technology is a weapon in the implementation of precision medical strategy allowing deeper interpretation of a disease scenario in precision medicine. Over the past decade, important advances in metabolomics such as instruments, methods, substantial informatics resources, and software tools and protocols for data have occurred than ever before. As it is uniquely able to perform the multitude and detailed phenotypic measurements, it realized the benefits that metabolomics can bring to precision medicine. Understanding patient's metabolomics will significantly lead to determining predisposition, diagnostic, prognostic, and predictive biomarkers and paths, ultimately providing optimal and personalized care for chronic and acute diseases. The recent application of metabolomics to clinical medicine and HM has demonstrated their potential role in better understanding of diseases' pathophysiology and therapeutic effect. More research is needed to validate the recent metabolomics findings in order to integrate multidimensional data for future translation into precision medicine for disease. Functional metabolomics facilitated a better option to decipher therapeutic effects of HM from the characterization of functional compounds, target identification, and annotation of pharmacological mechanisms and eventually to improve the performance of new drug discovery. It will also need to focus on how to convert a discovery into a device, a drug, or a clinical test, and biomarkers will need to be validated and moved to the clinic for the development of precision medicine.

Acknowledgments

This work was supported by grants from the Key Program of Natural Science Foundation of State (Grant No. 81973745, 81302905), Natural Science Foundation of Heilongjiang Province (YQ2019H030), and the University Nursing Program for Young Scholars with Creative Talents in Heilongjiang Province (UNPYSCT-2015118).

References

1 Prasuhn, J., Davis, R.L., and Kumar, K.R. (2021). Targeting mitochondrial impairment in Parkinson's disease: challenges and opportunities. *Frontiers in Cell and Development Biology* 8: 615461.
2 Yang, A., Kantor, B., and Chiba-Falek, O. (2021). *APOE*: the new frontier in the development of a therapeutic target towards precision medicine in late-onset Alzheimer's. *International Journal of Molecular Sciences* 22 (3): 1244.
3 Schechter, M., Leibowitz, G., and Mosenzon, O. (2020). Paving the way to precision medicine for diabetic kidney disease: the PRIORITY trial. *Annals of Translational Medicine* 8 (24): 1698.
4 Ashbrook, D.G., Arends, D., Prins, P. et al. (2021). A platform for experimental precision medicine: the extended BXD mouse family. *Cell Systems* S2405-4712 (20): 30503–30502.
5 Rasouly, H.M., Aggarwal, V., Bier, L. et al. (2021). Cases in precision medicine: genetic testing to predict future risk for disease in a healthy patient. *Ann Intern Med.* 2021 Apr; 174 (4): 540–547.
6 Sun, H., Zhang, A., Yan, G. et al. (2013). Metabolomic analysis of key regulatory metabolites in hepatitis C virus-infected tree shrews. *Molecular & Cellular Proteomics* 12 (3): 710–719.
7 Zhang, A., Sun, H., and Wang, X. (2013). Power of metabolomics in biomarker discovery and mining mechanisms of obesity. *Obesity Reviews* 14 (4): 344–349.
8 Wang, X., Zhang, A., and Sun, H. (2013). Power of metabolomics in diagnosis and biomarker discovery of hepatocellular carcinoma. *Hepatology* 57 (5): 2072–2077.
9 Zhang, A.H., Sun, H., Han, Y. et al. (2013). Ultraperformance liquid chromatography–mass spectrometry based comprehensive metabolomics combined with pattern recognition and network analysis methods for characterization of metabolites and metabolic pathways from biological data sets. *Analytical Chemistry* 85 (15): 7606–7612.

10 Qiu, S., Zhang, A.H., Guan, Y. et al. (2020). Functional metabolomics using UPLC-Q/TOF-MS combined with ingenuity pathway analysis as a promising strategy for evaluating the efficacy and discovering amino acid metabolism as a potential therapeutic mechanism-related target for geniposide against alcoholic liver disease. *RSC Advances* 10 (5): 2677–2690.
11 Zhang, A., Sun, H., Qiu, S. et al. (2014). Metabolomics insights into pathophysiological mechanisms of nephrology. *International Urology and Nephrology* 46 (5): 1025–1030.
12 Yang, Q., Zhang, A.H., Miao, J.H. et al. (2019). Metabolomics biotechnology, applications, and future trends: a systematic review. *RSC Advances* 9 (64): 37245–37257.
13 Zhang, A., Fang, H., Wang, Y. et al. (2017). Discovery and verification of the potential targets from bioactive molecules by network pharmacology-based target prediction combined with high-throughput metabolomics. *RSC Advances* 7 (81): 51069–51078.
14 Wang, X., Zhang, A., Han, Y. et al. (2012). Urine metabolomics analysis for biomarker discovery and detection of jaundice syndrome in patients with liver disease. *Molecular and Cellular Proteomics* 11 (8): 370–380.
15 Zhang, A., Sun, H., Yan, G. et al. (2014). Metabolomics in diagnosis and biomarker discovery of colorectal cancer. *Cancer Letters* 345 (1): 17–20.
16 Daneshjou, R., Wang, Y., Bromberg, Y. et al. (2017). Working toward precision medicine: predicting phenotypes from exomes in the critical assessment of genome interpretation (CAGI) challenges. *Human Mutation* 38 (9): 1182–1192.
17 Barnes, J.W. and Dweik, R.A. (2017). Pulmonary hypertension and precision medicine through the "omics" looking glass. *American Journal of Respiratory and Critical Care Medicine* 195 (12): 1558–1560.
18 Parikh, V.N. and Ashley, E.A. (2017). Next-generation sequencing in cardiovascular disease: present clinical applications and the horizon of precision medicine. *Circulation* 135 (5): 406–409.
19 Lu, Z.R. and Minko, T. (2017). Molecular imaging for precision medicine. *Advanced Drug Delivery Reviews* 113: 1–2.
20 Ding, X.X., Zhu, Q.G., Zhang, S.M. et al. (2017). Precision medicine for hepatocellular carcinoma: driver mutations and targeted therapy. *Oncotarget* 8 (33): 55715–55730.
21 Maier, M., Takano, T., and Sapir-Pichhadze, R. (2017). Changing paradigms in the management of rejection in kidney transplantation: evolving from protocol-based care to the era of P4 medicine. *Canadian Journal of Kidney Health and Disease* 4: 2054358116688227.
22 Younesi, E. and Hofmann-Apitius, M. (2013). From integrative disease modeling to predictive, preventive, personalized and participatory (P4) medicine. *EPMA Journal* 4 (1): 23.
23 AlRawi, S.N., Khidir, A., Maha, S. et al. (2017). Traditional Arabic & Islamic medicine: validation and empirical assessment of a conceptual model in Qatar. *BMC Complementary and Alternative Medicine* 17: 157.
24 Nejašmić, D., Miošić, I., Vrdoljak, D. et al. (2017). Awareness and use of evidence-based medicine information among patients in Croatia: a nation-wide cross-sectional study. *Croatian Medical Journal* 58 (4): 300–309.

25 Cirocchi, R., Afshar, S., Di Saverio, S. et al. (2017). A historical review of surgery for peritonitis secondary to acute colonic diverticulitis: from Lockhart-mummery to evidence-based medicine. *World Journal of Emergency Surgery* 12: 14.

26 Ozdemir, V., Fisher, E., Dove, E.S. et al. (2012). End of the beginning and public health pharmacogenomics: knowledge in 'Mode 2' and P5 medicine. *Current Pharmacogenomics and Personalized Medicine* 10 (1): 1–6.

27 Liu, X. and Meltzer, S.J. (2017). Gastric cancer in the era of precision medicine. *Cellular and Molecular Gastroenterology and Hepatology* 3 (3): 348–358.

28 Foran, D.J., Chen, W., Chu, H. et al. (2017). Roadmap to a comprehensive clinical data warehouse for precision medicine applications in oncology. *Cancer Informatics* 16: 1–10. 1176935117694349.

29 Deng, X. and Nakamura, Y. (2017). Cancer precision medicine: from cancer screening to drug selection and personalized immunotherapy. *Trends in Pharmacological Sciences* 38 (1): 15–24.

30 Voest, E.E. and Bernards, R. (2016). DNA-guided precision medicine for cancer: a case of irrational exuberance? *Cancer Discovery* 6 (2): 130–132.

31 Bar-Zeev, M., Livney, Y.D., and Assaraf, Y.G. (2017). Targeted nanomedicine for cancer therapeutics: towards precision medicine overcoming drug resistance. *Drug Resistance Updates* 31: 15–30.

32 Yi, S. and Sahni, N. (2017). Regulome networks and mutational landscape in liver cancer: an informative path to precision medicine. *Hepatology* 66 (1): 280–282.

33 Nakatani, H. (2016). Global strategies for the prevention and control of infectious diseases and non-communicable diseases. *Journal of Epidemiology* 26 (4): 171–178.

34 Duffy, D.J. (2016). Problems, challenges and promises: perspectives on precision medicine. *Briefings in Bioinformatics* 17 (3): 494–504.

35 Knudsen, E.S., O'Reilly, E.M., Brody, J.R. et al. (2016). Genetic diversity of pancreatic ductal adenocarcinoma and opportunities for precision medicine. *Gastroenterology* 150 (1): 48–63.

36 Saad, E.D., Paoletti, X., Burzykowski, T. et al. (2017). Precision medicine needs randomized clinical trials. *Nature Reviews Clinical Oncology* 14 (5): 317–323.

37 Mesleh, A.G., Abdulla, S.A., and El-Agnaf, O. (2021). Paving the way toward personalized medicine: current advances and challenges in multi-OMICS approach in autism Spectrum disorder for biomarkers discovery and patient stratification. *J Pers Med.* 2021 Jan 13; 11 (1): 41.

38 El-Achkar, T.M., Eadon, M.T., Menon, R. et al. (2021). A multimodal and integrated approach to interrogate human kidney biopsies with rigor and reproducibility: guidelines from the kidney precision medicine project. *Physiol Genomics.* 2021 Jan 1; 53 (1): 1–11.

39 Liang, Q., Liu, H., Xing, H. et al. (2016). High-resolution mass spectrometry for exploring metabolic signatures of sepsis-induced acute kidney injury. *RSC Advances* 6 (36): 29863–29868.

40 Zhang, A., Yan, G., Han, Y. et al. (2014). Metabolomics approaches and applications in prostate cancer research. *Applied Biochemistry and Biotechnology* 174 (1): 6–12.

41 Zhang, A., Sun, H., Yan, G. et al. (2016). Mass spectrometry-based metabolomics: applications to biomarker and metabolic pathway research. *Biomedical Chromatography* 30 (1): 7–12.
42 Zhang, A., Sun, H., and Wang, X. (2017). Emerging role and recent applications of metabolomics biomarkers in obesity disease research. *RSC Advances* 7 (25): 14966–14973.
43 Zhang, A., Sun, H., Yan, G. et al. (2015). Metabolomics for biomarker discovery: moving to the clinic. *BioMed Research International*: 2015.
44 Liang, Q., Liu, H., Zhang, T. et al. (2016). Discovery of serum metabolites for diagnosis of mild cognitive impairment to Alzheimer's disease progression using an optimized metabolomics method. *RSC Advances* 6: 3586–3591.
45 Zhang, A., Sun, H., Yan, G. et al. (2014). Metabolomics analysis of marker metabolites for patients with pancreatic cancer. *Medicinal Chemistry* 4 (6): 506–510.
46 Wang, X., Sun, H., Zhang, A. et al. (2011). Ultra-performance liquid chromatography coupled to mass spectrometry as a sensitive and powerful technology for metabolomic studies. *Journal of Separation Science* 34 (24): 3451–3459.
47 Zhang, A., Sun, H., and Wang, X. (2012). Saliva metabolomics opens door to biomarker discovery, disease diagnosis, and treatment. *Applied Biochemistry and Biotechnology* 168 (6): 1718–1727.
48 Zhang, A., Sun, H., and Wang, X. (2012). Serum metabolomics as a novel diagnostic approach for disease: a systematic review. *Analytical and Bioanalytical Chemistry* 404 (4): 1239–1245.
49 Wang, X., Wang, Q., Zhang, A. et al. (2013). Metabolomics study of intervention effects of Wen–Xin-formula using ultra high-performance liquid chromatography/mass spectrometry coupled with pattern recognition approach. *Journal of Pharmaceutical and Biomedical Analysis* 74: 22–30.
50 Zhang, A., Sun, H., Yan, G. et al. (2013). Urinary metabolic profiling identifies a key role for glycocholic acid in human liver cancer by ultra-performance liquid-chromatography coupled with high-definition mass. *Clinica Chimica Acta* 418: 86–90.
51 Zhang, A., Sun, H., Han, Y. et al. (2012). Exploratory urinary metabolic biomarkers and pathways using UPLC–Q–TOF–HDMS coupled with pattern recognition approach. *Analyst* 137 (18): 4200–4208.
52 Li, Y.F., Qiu, S., Gao, L.J. et al. (2018). Metabolomic estimation of the diagnosis of hepatocellular carcinoma based on ultrahigh performance liquid chromatography coupled with time-of-flight mass spectrometry. *RSC Advances* 8 (17): 9375–9382.
53 Zhang, A., Sun, H., and Wang, X. (2014). Urinary metabolic profiling of rat models revealed protective function of scoparone against alcohol induced hepatotoxicity. *Scientific Reports* 4: 6768.
54 Liang, Q., Yu, Q., Wu, H. et al. (2014). Metabolite fingerprint analysis of cervical cancer using LC–QTOF/MS and multivariate data analysis. *Analytical Methods* 6 (12): 3937–3942.

55 Zhang, A.H., Sun, H., Yan, G.L. et al. (2014). Metabolomics study of type 2 diabetes using ultra-performance LC–ESI/quadrupole-TOF high-definition MS coupled with pattern recognition methods. *Journal of Physiology and Biochemistry* 70 (1): 117–128.

56 Wang, X., Yang, B., Sun, H. et al. (2012). Pattern recognition approaches and computational systems tools for ultra performance liquid chromatography–mass spectrometry-based comprehensive metabolomic profiling and pathways analysis of biological data sets. *Analytical Chemistry* 84 (1): 428–439.

57 Zhang, A., Sun, H., Wang, P. et al. (2012). Modern analytical techniques in metabolomics analysis. *Analyst* 137 (2): 293–300.

58 Zhang, A., Sun, H., Wang, P. et al. (2012). Recent and potential developments of biofluid analyses in metabolomics. *Journal of Proteomics* 75 (4): 1079–1088.

59 Chu, H., Zhang, A., Han, Y. et al. (2016). Metabolomics approach to explore the effects of Kai–Xin–San on Alzheimer's disease using UPLC/ESI–Q–TOF mass spectrometry. *Journal of Chromatography B* 1015: 50–61.

60 Zhang, A., Zhou, X., Zhao, H. et al. (2017). Metabolomics and proteomics technologies to explore the herbal preparation affecting metabolic disorders using high resolution mass spectrometry. *Molecular BioSystems* 13 (2): 320–329.

61 McCann, M.R., George De la Rosa, M.V., Rosania, G.R. et al. (2021). L-Carnitine and acylcarnitines: mitochondrial biomarkers for precision medicine. *Metabolites* 11 (1): 51.

62 Heimisdottir, L.H., Lin, B.M., Cho, H. et al. (2021). Metabolomics insights in early childhood caries. *J Dent Res.* 2021 Jun; 100 (6): 615–622.

63 de Mello, V.D., Sehgal, R., Männistö, V. et al. (2020). Serum aromatic and branched-chain amino acids associated with NASH demonstrate divergent associations with serum lipids. *Liver Int.* 2021 Apr; 41 (4): 754–763.

64 Wang, L.M., Wang, P., Teka, T. et al. (2020). 1H NMR and UHPLC/Q-Orbitrap–MS-based metabolomics combined with 16S rRNA gut microbiota analysis revealed the potential regulation mechanism of nuciferine in hyperuricemia rats. *Journal of Agricultural and Food Chemistry* 68 (47): 14059–14070.

65 Zhang, X., Yang, Z., Shi, Z. et al. (2021). Analysis of bile acid profile in plasma to differentiate cholangiocarcinoma from benign biliary diseases and healthy controls. *Journal of Steroid Biochemistry and Molecular Biology* 105775: 205.

66 Lusczek, E.R., Parsons, L.S., Elder, J. et al. (2020). Metabolomics pilot study identifies desynchronization of 24-H rhythms and distinct intra-patient variability patterns in critical illness: a preliminary report. *Frontiers in Neurology* 11: 533915.

67 Murthy, V.L., Reis, J.P., Pico, A.R. et al. (2020). Comprehensive metabolic phenotyping refines cardiovascular risk in young adults. *Circulation* 142 (22): 2110–2127.

68 Di Costanzo, A., Paris, D., Melck, D. et al. (2020). Blood biomarkers indicate that the preclinical stages of Alzheimer's disease present overlapping molecular features. *Scientific Reports* 10 (1): 15612.

69 Liu, X., Wu, W.Y., Jiang, B.H. et al. (2013). Pharmacological tools for the development of traditional Chinese medicine. *Trends in Pharmacological Sciences* 34 (11): 620–628.

70 Wang, X., Zhang, A., and Sun, H. (2012). Future perspectives of Chinese medical formulae: chinmedomics as an effector. *Omics: A Journal of Integrative Biology* 16 (7–8): 414–421.

71 Wang, X.J., Ren, J.L., Zhang, A.H. et al. (2019). Novel applications of mass spectrometry-based metabolomics in herbal medicines and its active ingredients: current evidence. *Mass Spectrometry Reviews* 38 (4–5): 380–402.

72 Zhang, A., Sun, H., Yan, G. et al. (2017). Recent developments and emerging trends of mass spectrometry for herbal ingredients analysis. *TrAC Trends in Analytical Chemistry* 94: 70–76.

73 Sun, H., Zhang, A., and Wang, X. (2012). Potential role of metabolomic approaches for Chinese medicine syndromes and herbal medicine. *Phytotherapy Research* 26 (10): 1466–1471.

74 Zhang, A., Sun, H., Wang, X. et al. (2012). Simultaneous *in vivo* RP–HPLC–DAD quantification of multiple-component and drug–drug interaction by pharmacokinetics, using 6,7-dimethylesculetin, geniposide and rhein as examples. *Biomedical Chromatography* 26 (7): 844–850.

75 Wang, X., Zhang, A., Wang, P. et al. (2013). Metabolomics coupled with proteomics advancing drug discovery toward more agile development of targeted combination therapies. *Molecular and Cellular Proteomics* 12 (5): 1226–1238.

76 Kedzierski, P. and Torbicki, A. (2019). Precision medicine: the future of diagnostic approach to pulmonary hypertension? *Anatolian Journal of Cardiology* 22 (4): 168–171.

77 Olivier, M., Asmis, R., Hawkins, G.A. et al. (2019). The need for multi-omics biomarker signatures in precision medicine. *International Journal of Molecular Sciences* 20 (19): 4781.

78 Wang, Q., Peng, W.X., Wang, L., and Ye, L. (2019). Toward multiomics-based next-generation diagnostics for precision medicine. *Personalized Medicine* 16 (2): 157–170.

79 Ahmed, Z. (2020). Practicing precision medicine with intelligently integrative clinical and multi-omics data analysis. *Human Genomics* 14 (1): 35.

80 Nassar, S.F., Raddassi, K., Ubhi, B. et al. (2020). Precision medicine: steps along the road to combat human cancer. *Cell* 9 (9): 2056.

81 Assadi, M., Jokar, N., Ghasemi, M. et al. (2020). Precision medicine approach in prostate cancer. *Current Pharmaceutical Design* 26 (31): 3783–3798.

82 Klonoff, D.C., Florez, J.C., German, M., and Fleming, A. (2020). The need for precision medicine to be applied to diabetes. *Journal of Diabetes Science and Technology* 14 (6): 1122–1128.

12

Mass Spectrometry-Based Metabolomics Toward Biological Function Analysis

Shi Qiu[1], Wanying Wang[2], Aihua Zhang[1], and Xijun Wang[1]

[1] *Heilongjiang University of Chinese Medicine, National Chinmedomics Research Center, Functional Metabolomics Laboratory, Department of Pharmaceutical Analysis, Heping Road 24, Harbin, 150040, China*
[2] *University of Science and Technology, State Key Laboratory of Quality Research in Chinese Medicine, Avenida Wai Long, Taipa, China*

12.1 Introduction

Entering the twenty-first century, the major challenges for medical research are doing research into disease etiology for better understanding of the pathological mechanism, especially clarifying the impact of genetic and environmental factors on the health state and vigorously promoting the development of individual treatment [1–3]. With the increasing incidence and mortality of various diseases, the late diagnosis and the ineffective treatment decisions are primarily relying on the physiological histology, limited serological markers, and incomplete clinical presentation. It is essential to detect the disease at an early stage for the purpose of successful treatment [4, 5]. As novel "omics" technology, metabolomics offers systematic identification and quantification of all metabolic products such as amino acids, fatty acids, peptides, vitamins, nucleic acids, organic acids, and carbohydrates. It could provide clinicians with new sets of diagnostic biomarkers for disease states and treatment at an individualized level. Metabolic profiling can be performed using mass spectrometry (MS) technique, utilizing a variety of biological samples. MS possesses prominent advantages, such as low sample detection requirement, high sensitivity, and wide linear range, and the metabolites with large difference in concentration can also achieve good detection results; thus it has become an increasingly important role of technology platform for metabolomics research [6–11]. In practice, MS metabolomics technologies are now conveniently offering us the opportunity to measure metabolites toward biological function analysis [12–15]. MS-based metabolomics could determine the sensitivity, specificity, and potential value of the metabolite biomarkers, providing biological function analysis of diseases. This chapter could improve the understanding of MS-based metabolomics for discovering small-molecule metabolism and reveal the metabolic phenotype toward biochemical functions and associated mechanisms.

Mass Spectrometry-Based Metabolomics in Clinical and Herbal Medicines: Strategies, Technologies, and Applications, First Edition. Edited by Aihua Zhang and Wanying Wang.

12.2 Current Trend and Challenge of Disease Phenotype

Nowadays, the human is being tortured by the number of unprecedented diseases, which probably will continue into the future. From the late twentieth century, significant changes have taken place in human disease spectrum, which the main disease affecting human health transforms from infectious diseases into cardiovascular and cerebrovascular diseases, diabetes, malignancies, etc. [16]. Meanwhile, the concept of human health has undergone a major change from the "no disease is healthy" into "a healthy state is mind and social adaptability." In the biological–psychological–social medical pattern, infectious diseases replaced by a growing concern about the dangers of chronic diseases cannot constitute a threat. Clinical feature has shifted from the acute or epidemic into chronic or sudden [17]. The transition from short-term to the long-term, systematic, and from high mortality to high morbidity has been submitted in treatment [18, 19]. However, there are not yet efficient solutions of preventing or treating the vast bulk of chronic diseases. The reasons for this situation may include measurement being performed too late or too short in the course of the disease, inadequate treatment targets on account of a lacking of pathogenic knowledge, and insufficient knowledge on the specific clinical features of the diseases. The burden of non-communicable diseases grows with each passing day in the whole world, leading to the need for monitoring system to track trends and assess interventions [20]. In the face of complex disease, we not only need to understand the medical treatment cause from the macro level but also to understand the nature of the disease from the microscopic level to the initial detection, for timely and accurate treatment of disease. Using adequate clinical and biomarker tools is a beneficial way that may offer key information to prediction, occurrence and development, and therapeutic targets in patients [21, 22].

12.3 Advances in Mass Spectrometry Analytic Technique

Choosing the correct appropriate research techniques is meaningful to successful advancement of metabolites research. MS, as a dominating tool utilized for metabolites analysis, plays a number of important roles in the phases of metabolites identification from the novel chemical structures to biomarker discovery [23–26]. It has been sought to make manifest the chemical structure of metabolites as an efficient technique for measuring the peaks and relative intensities, exact mass, molecular formula, and the relative error. MS identifies metabolite species in the light of their mass/charge ratio (m/z). Metabolite components are often infused directly into the mass spectrometer and then carried out to separate. GC–MS, as a highly sensitive, repeatability, and cost-efficient form of MS analysis method, is time-consuming in sample preparation, resulting to degradation during the operation and metabolite identification. In addition, capillary electrophoresis–MS (CE–MS) is also the main emerging spectroscopic technique utilized in metabolomics analysis. LC/MS is a dominant analytical platform in metabolomics given the ability to produce large numbers of ions and because of its high sensitivity and resolution, thus enabling large-scale coverage of metabolome [27, 28]. Chromatographic separation refers to two processes including dissolving the sample into solvent as mobile phase and passing them through a stationary phase of specific interaction chemistries that allow a portion of the dissolved metabolite to remain in the stationary phase and a portion to pass at a different flow rate. Subsequently, the analyzed compounds are subjected to ionization for forming mass separation fragments. The analytical integration of these tandems can result to the ability to detect and analyze thousands of metabolites with a low molar range.

12.4 Advantages of Mass Spectrometry-Based Metabolomics

Metabolomics is capable of allowing the concurrent and relative quantification of countless diverse kinds of metabolites within a treated specimen by MS methodology. A crucial and difficult assignment in medical research is to detect and diagnose the illness as soon as possible before the onset of symptoms. For purpose of realizing the goal, MS-based metabolomics provides novel chances for biomarker discovery in various kinds of complicated diseases, and many opportunities to fully understand pathological mechanisms may lead to appropriate treatment for patients beyond traditional technologies [29–35]. MS-based metabolomics allows the parallel detection of a wide range of metabolites levels and has been manifested to carry huge implications on the study of biomarkers, disease diagnosis, treatment evaluation, and the investigation of disturbed pathways resulting from biological mechanisms [36–41].

12.5 Mass Spectrometry-Based Metabolomics for Biological Function

Metabolomics has emerged as a cutting-edge method to qualify and quantify the metabolites in different biological matrixes, and it has the extraordinary capacity to interrogate the biological significance that underlies metabolic modification and modulation [42]. MS, as a robust platform for metabolomics analysis, has increased in popularity due to its excellent sensitivity, throughput, and versatility. It provided us the metabolic phenotype data for revealing the biochemical functions and associated mechanisms. MS-based metabolomics has many advantages [43]. Firstly, the functional roles of metabolites and related metabolic enzymes are focused; secondly, the experiments *in vivo* and *in vitro* are conducted to validate the findings, therefore increasing the reliability of metabolomics study and producing the new knowledge; thirdly, MS-based metabolomics can be used by biologists to investigate the functions of metabolites. Due to the tremendous advantages of functional metabolomics, it will have a bright future. The general workflow of mass spectrometry-based metabolomics toward biological function analysis mainly includes the mass spectrometry instrumentation, fingerprint chromatogram,

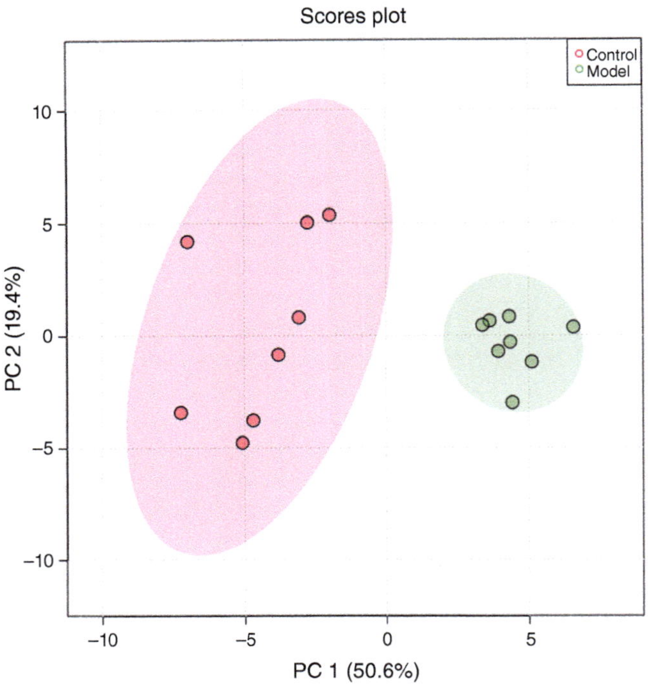

Figure 12.1 The principal component pattern recognition of mass spectrometry-based metabolomics.

principal component pattern recognition (Figure 12.1), partial least squares pattern recognition (Figure 12.2), K-means feature clustering (Figure 12.3), the correlation network analysis (Figure 12.4), the reconstruction of metabolic pathways, the metabolites biological function analysis (Figure 12.5), etc. Metabolic pathway network analysis (Figure 12.6) of the identified potential biomarkers could be performed with MetaboAnalyst 4.0 (www.metaboanalyst.ca) and Kyoto Encyclopedia of Genes and Genomes (KEGG; http://www.kegg.jp). The biological interaction network of compound–target–metabolite can be visualized using Cytoscape software (Cytoscape Consortium, CA, USA).

An MS-based approach was subsequently performed for targeted analysis of the metabolic marker *N*-acetylneuraminic acid [44]. The functional metabolomics strategy was proposed to investigate the role of Neu5Ac in the progression of coronary artery diseases (CAD). A total of 36 differential metabolites were identified, 35 of which were confirmed with reference compounds. Elevation of Neu5Ac was observed in plasma during CAD progression. Mechanistically, Neu5Ac was able to trigger myocardial injury *in vitro* and *in vivo* by the activation of the protein kinase signaling pathway. Silencing neuraminidase-1, the enzyme that regulates Neu5Ac generation, ameliorated oxygen–glucose deprivation-induced injury in cardiomyocytes. Functional metabolomics identified a key role for Neu5Ac in acute

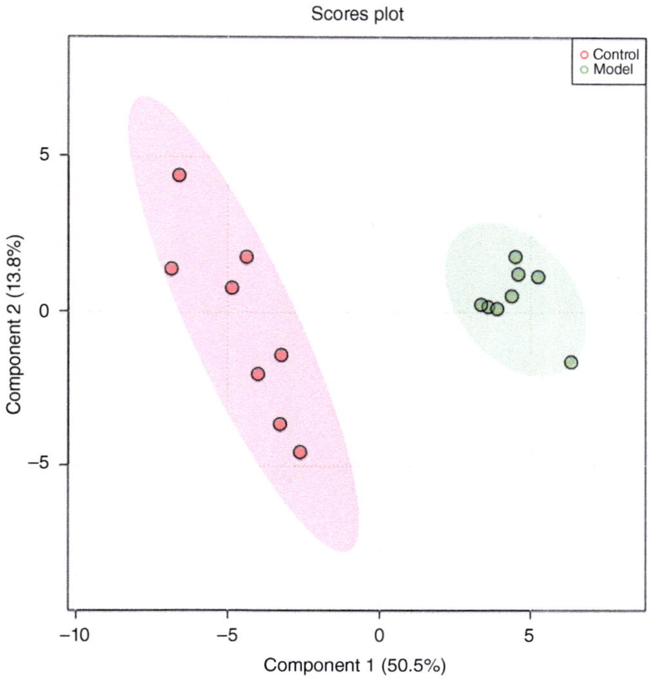

Figure 12.2 The partial least squares pattern recognition of mass spectrometry-based metabolomics.

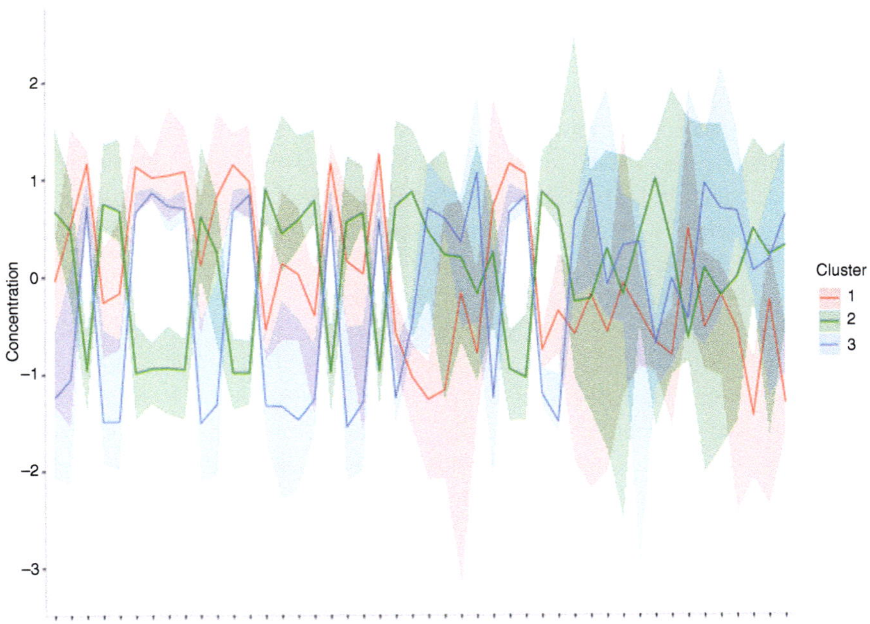

Figure 12.3 The K-means feature clustering of mass spectrometry-based metabolomics.

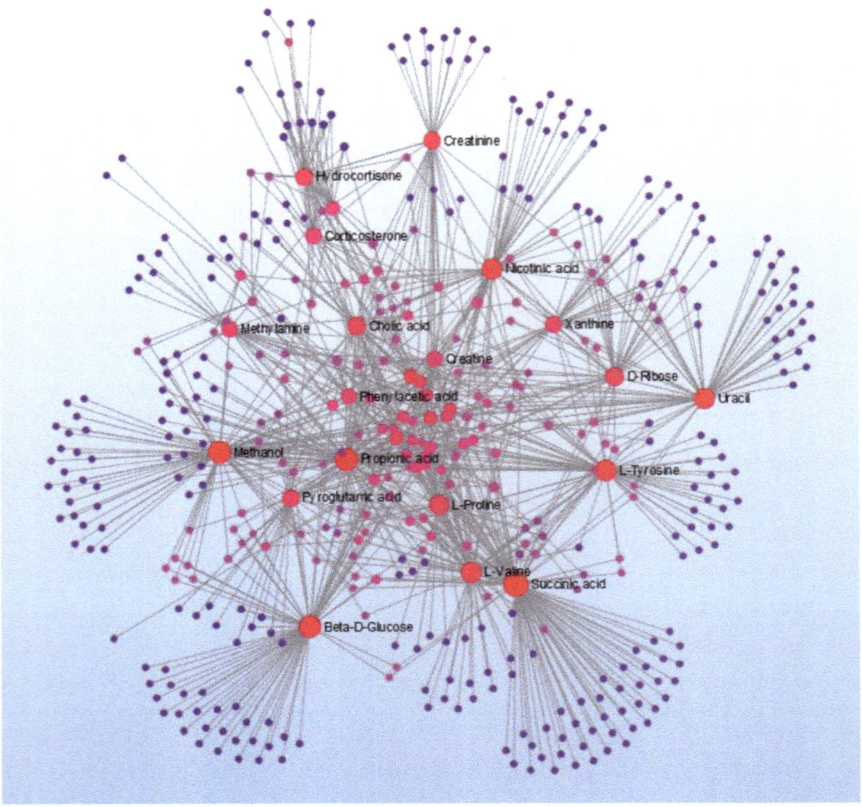

Figure 12.4 The correlation network analysis of mass spectrometry-based metabolomics.

myocardial infarction, and targeting neuraminidase-1 may represent an unrecognized therapeutic intervention for CAD. MS-based metabolomics is a widely used approach in biomedical research and offer deeply insight into the therapeutic mechanisms of rhein [45]. It suggested chronic glomerulonephritis treatment with rhein, which may be mediated via interference with metabolic pathway such as amino acid metabolism, arachidonic acid metabolism, and tricarboxylic acid (TCA) cycle to regulate inflammation, oxidation response, and immune regulation.

Plasma metabolomics were used to investigate and identify novel biomarkers and pathophysiological pathways of incident atrial fibrillation [46]. A total of 200 endogenous metabolites in plasma/serum by non-targeted ultra-performance liquid chromatography coupled to time-of-flight MS in three independent population-based samples were identified, of which 9-decenoylcarnitine associates with incident atrial fibrillation. A non-targeted metabolomics analysis disclosed 1 novel replicated biomarker, 9-decenoylcarnitine, for atrial fibrillation. High-throughput mass spectrometry metabolomics was performed to analyze the cell metabolic

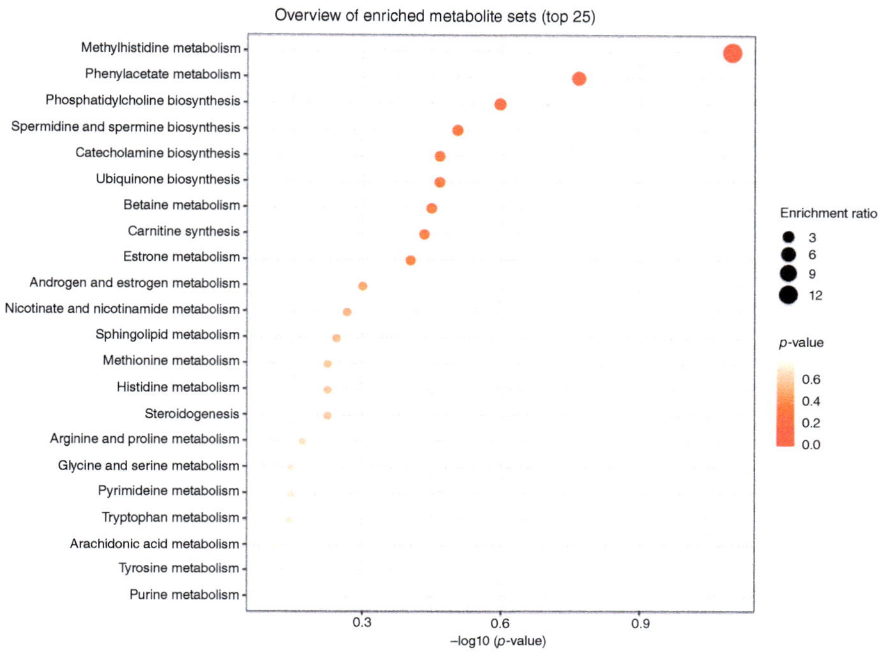

Figure 12.5 The reconstruction of metabolic pathways of metabolites enrichment analysis of mass spectrometry-based metabolomics.

profiles and pathway of obacunone on prostate cancer [47]. Obacunone effectually control proliferation and promote apoptosis in 22RV1 prostate cancer cells, which were related to 21 metabolites in prostate cancer cell and 9 vital metabolic pathways that were identified from large amounts of data.

Coronavirus disease 2019 continues to spread rapidly with high mortality. Metabolomics profiling of coronavirus disease 2019 patients were carried out to identify potential diagnostic/prognostic biomarkers and to understand the underlying pathologic pathways [48]. Arginine/kynurenine ratio could accurately identify coronavirus disease 2019 status, whereas creatinine/arginine ratio could predict coronavirus disease 2019-associated death. The administration of tryptophan, arginine, and sarcosine may be considered as potential adjunctive therapies. Metabolome-based method determined the metabolic signatures of chronic hepatitis B disease progression and developed accurate model for disease diagnosis, thereby giving a scientific basis for treatment [49]. Perturbations in glutamine and glutamate metabolism, ammonia detoxification, methionine metabolism, TCA cycle, and dysregulation of branched-chain amino acids are the key factors involved in the disease progression. Fluctuations increasing in glutamine, methionine, aspartate, and glutamate are the main fingerprints of chronic hepatitis B disease progression.

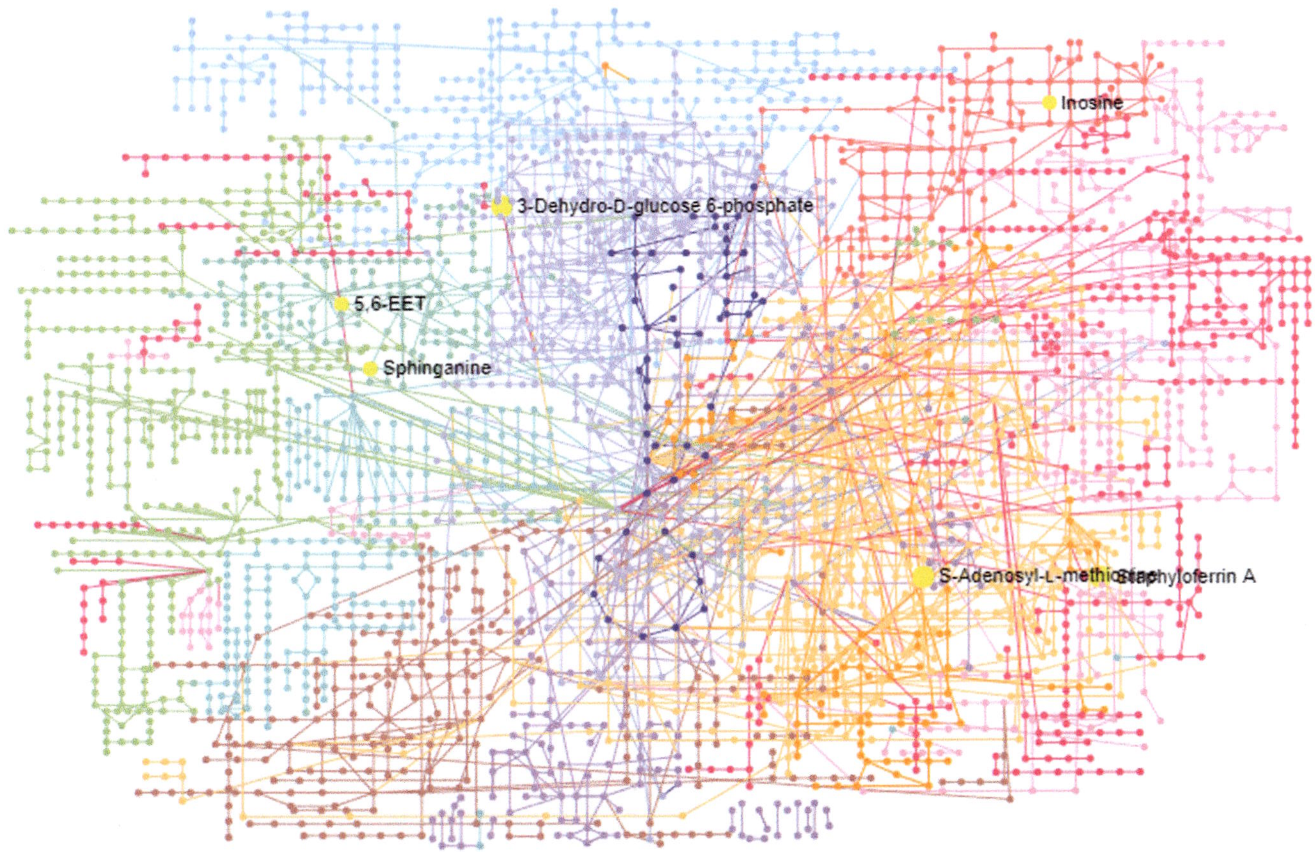

Figure 12.6 Metabolic network analysis of the metabolites biomarkers that performed with MetaboAnalyst online tool (www.metaboanalyst.ca).

12.6 Limitations of MS-Based Metabolomics

In clinical applications, MS-metabolomics facing the greatest challenge may be attributed to the advantages, which data of metabolomics is strong because the organism possesses many metabolites including related precursors, derivatives, and degradation products, and they are highly diverse and rapidly changing. The complexity leads to the need for sophisticated separation and detection methods that forms a major bottleneck to the growth of the metabolomics field. Following consideration of the extraordinary complexity of biological mixtures, the vast majority of MS analytical methods refer to pre-separation using LC, GC, or CE. In routine clinical practice, sample collection from patients and analysis methods should take into account permitting accurate quantification of metabolites, rapid measurements, and low cost, but these requirements could be hard to achieve for physicians and patients. In addition, MS is not in a position to measure all kinds of compounds at the same time, which itself is so sensitive that we often get countless substances in clinic.

High-throughput MS metabolomics could allow the detection of hundreds to thousands of metabolites and produces information-rich data sets on known and unknown metabolites. One of the challenges is the annotation and identification of metabolites from the complex data sets, since the number of authentic standards is lower than the number of mass spectral features [50]. Metabolites such as decomposition products converted from enzymes or with microbial activity remain unknown, and hence difficult to differentiate. Recently, confusion between researchers and clinical investigators is that the results obtained by different independent groups could not be compared and correlated. Building up much of the infrastructure for measuring metabolomics analysis of metabolites is actively performed by hospital laboratories around the world. The candidate biomarkers in the experimental research will be validated in the real clinical practice, or whether the identified biomarker in the disease could predict the disease state with the variability of disease progress. Many biomarker have been explored in a small number of clinical settings [51–54], but they have not been validated in practice, which is worth pondering, owning to potential confounding factor in the study.

12.7 Future Perspectives

For the development of MS-metabolomics, people are actively taking measures from the following aspects: firstly, the development of high-throughput and precision equipment for the identification of unknown metabolites; secondly, the development of standard data sets for available to query; thirdly, to strengthen the data integration system of metabolomics and other omics; and fourthly, building up the bridge of basic research and clinical application. With the growing understanding of metabolomics, many biomarkers are being discovered for better treatment of the disease [55–60]. However, none of the biomarkers has fulfilled the research community for biomarker selection to tackle with all aspects of the disorder. We still should depend on conventional risk factors, in line with clinical parameters, laboratory, and imaging examinations to estimate diseases. The emerging candidate biomarkers own a promising future, but numerous and extensive clinical studies have an essential role in validation for disease. Developing standard operating procedures for biomarkers in disease diagnostic routine must be improved to realize effective prevention and treatment before its clinical expression. Technological advances in MS-metabolomics provide the possibility of offering more data in a time-efficient and clinically effective manner, affording more comprehensive and in-depth candidate characterization that has been demonstrated to empower clinical discovery efforts.

12.8 Conclusions

With the development of MS techniques, the ability of measurement in biological samples offers a robust platform for the low-molecular weight metabolites that are closely associated with human disease. High-definition MS-metabolomics has accomplished to get a large of biological data on metabolic profiling toward biological function analysis in a comprehensive manner. It could determine the potential value of the metabolite biomarkers and will be used in exploring biomarkers related to judge the disease incidence, the disease severity, and the rationality of treatment, providing an effective method for biological function analysis of diseases.

Acknowledgments

This work was supported by grants from the Key Program of Natural Science Foundation of State (Grant No. 81830110, 81861168037, 81973745, 81302905) and Natural Science Foundation of Heilongjiang Province (YQ2019H030).

References

1 Lernmark, A. (2016). Environmental factors in the etiology of type 1 diabetes, celiac disease, and narcolepsy. *Pediatric Diabetes* 17 (22): 65–72.
2 Tosto, G. and Reitz, C. (2016). Genomics of Alzheimer's disease: value of high-throughput genomic technologies to dissect its etiology. *Molecular and Cellular Probes* 30 (6): 397–403.
3 Marcinkiewicz-Siemion, M., Ciborowski, M., Kretowski, A. et al. (2016). Metabolomics – a wide-open door to personalized treatment in chronic heart failure? *International Journal of Cardiology* 219: 156–163.
4 Schneider, M., Levant, B., Reichel, M. et al. (2016). Lipids in psychiatric disorders and preventive medicine. *Neuroscience and Biobehavioral Reviews* 76 (Pt B): 336–362.

5 Gualtero, D.F., Suarez Castillo, A. et al. (2016). Biomarkers in saliva for the detection of oral squamous cell carcinoma and their potential use for early diagnosis: a systematic review. *Acta Odontologica Scandinavica* 74 (3): 170–177.

6 Zhang, A., Wang, H., Sun, H. et al. (2015). Metabolomics strategy reveals therapeutical assessment of limonin on nonbacterial prostatitis. *Food & Function* 6 (11): 3540–3549.

7 Li, Y.F., Qiu, S., Gao, L.J. et al. (2018). Metabolomic estimation of the diagnosis of hepatocellular carcinoma based on ultrahigh performance liquid chromatography coupled with time-of-flight mass spectrometry. *RSC Advances* 8 (17): 9375–9382.

8 Zhang, A., Sun, H., and Wang, X. (2012). Serum metabolomics as a novel diagnostic approach for disease: a systematic review. *Analytical and Bioanalytical Chemistry* 404 (4): 1239–1245.

9 Qiu, S., Zhang, A., Zhang, T. et al. (2017). Dissect new mechanistic insights for geniposide efficacy on the hepatoprotection using multiomics approach. *Oncotarget* 8 (65): 108760.

10 Zhang, Y., Liu, P., Li, Y. et al. (2017). Exploration of metabolite signatures using high-throughput mass spectrometry coupled with multivariate data analysis. *RSC Advances* 7 (11): 6780–6787.

11 Zhang, A., Sun, H., Wang, P. et al. (2012). Modern analytical techniques in metabolomics analysis. *Analyst* 137 (2): 293–300.

12 Zhang, A.-H., Qiu, S., Xu, H.-Y. et al. (2014). Metabolomics in diabetes. *Clinica Chimica Acta* 429: 106–110.

13 Liang, Q., Liu, H., Xing, H. et al. (2016). UPLC-QTOF/MS based metabolomics reveals metabolic alterations associated with severe sepsis. *RSC Advances* 6 (49): 43293–43298.

14 Qiu, S., Zhang, A.H., Guan, Y. et al. (2020). Functional metabolomics using UPLC-Q/TOF-MS combined with ingenuity pathway analysis as a promising strategy for evaluating the efficacy and discovering amino acid metabolism as a potential therapeutic mechanism-related target for geniposide against alcoholic liver disease. *RSC Advances* 10 (5): 2677–2690.

15 Sun, H., Zhang, A., Yan, G. et al. (2013). Metabolomic analysis of key regulatory metabolites in hepatitis C virus–infected tree shrews. *Molecular & Cellular Proteomics* 12 (3): 710–719.

16 Nakatani, H. (2016). Global strategies for the prevention and control of infectious diseases and non-communicable diseases. *Journal of Epidemiology* 26 (4): 171–178.

17 Zhang, L. and Zuo, L. (2016). Current burden of end-stage kidney disease and its future trend in China. *Clinical Nephrology* 86 (13): 27–28.

18 Guo, Y. (2015). Progress in monoclonal antibody-based immunotherapy for cancer treatment. *Sheng Wu Gong Cheng Xue Bao* 31 (6): 857–870.

19 Yokoi, Y. (2011). Global vascular intervention for peripheral vascular disease: endovascular integrated approach and future trend. *Nihon Rinsho* 69 (2): 295–300.

20 Kroll, M., Phalkey, R.K., and Kraas, F. (2015). Challenges to the surveillance of non-communicable diseases – a review of selected approaches. *BMC Public Health* 15: 1243.

21 McGrath, S., Christidis, D., Perera, M. et al. (2016). Prostate cancer biomarkers: are we hitting the mark? *Prostate International* 4 (4): 130–135.

22 Gao, G. and Smith, D.I. (2015). Mate-pair sequencing as a powerful clinical tool for the characterization of cancers with a DNA viral etiology. *Viruses* 7 (8): 4507–4528.

23 Zhang, A., Liu, Q., Zhao, H. et al. (2016). Phenotypic characterization of nanshi oral liquid alters metabolic signatures during disease prevention. *Scientific Reports* 6: 19333.

24 Zhang, A., Fang, H., Wang, Y. et al. (2017). Discovery and verification of the potential targets from bioactive molecules by network pharmacology-based target prediction combined with high-throughput metabolomics. *RSC Advances* 7 (81): 51069–51078.

25 Liang, Q., Liu, H., Jiang, Y. et al. (2016). Novel liquid chromatography-mass spectrometry for metabolite biomarkers of acute lung injury disease. *Analytical Methods* 8 (31): 6017–6022.

26 Zhang, A.H., Sun, H., Yan, G.L. et al. (2019). Chinmedomics: a powerful approach integrating metabolomics with serum pharmacochemistry to evaluate the efficacy of traditional Chinese medicine. *Engineering* 5 (1): 60–68.

27 Zhang, A.H., Sun, H., Yan, G.L. et al. (2014). Metabolomics study of type 2 diabetes using ultra-performance LC-ESI/quadrupole-TOF high-definition MS coupled with pattern recognition methods. *Journal of Physiology and Biochemistry* 70 (1): 117–128.

28 Zhang, A., Sun, H., and Wang, X. (2014). Urinary metabolic profiling of rat models revealed protective function of scoparone against alcohol induced hepatotoxicity. *Scientific Reports* 4: 6768.

29 Zhang, A.H., Sun, H., Han, Y. et al. (2013). Ultraperformance liquid chromatography–mass spectrometry based comprehensive metabolomics combined with pattern recognition and network analysis methods for characterization of metabolites and metabolic pathways from biological data sets. *Analytical Chemistry* 85 (15): 7606–7612.

30 Zhang, A., Sun, H., and Wang, X. (2013). Power of metabolomics in biomarker discovery and mining mechanisms of obesity. *Obesity Reviews* 14 (4): 344–349.

31 Wang, S., Blair, I.A., and Mesaros, C. (2019). Analytical methods for mass spectrometry-based metabolomics studies. *Advances in Experimental Medicine and Biology* 1140: 635–647.

32 Liang, Q., Yu, Q., Wu, H. et al. (2014). Metabolite fingerprint analysis of cervical cancer using LC-QTOF/MS and multivariate data analysis. *Analytical Methods* 6 (12): 3937–3942.

33 Zhang, A., Sun, H., Yan, G. et al. (2014). Metabolomics analysis of marker metabolites for patients with pancreatic cancer. *Medicinal Chemistry* 4 (6): 506–510.

34 Bar, N., Korem, T., Weissbrod, O. et al. (2020). A reference map of potential determinants for the human serum metabolome. *Nature* 588 (7836): 135–140.

35 Zhang, A., Sun, H., Qiu, S. et al. (2014). Metabolomics insights into pathophysiological mechanisms of nephrology. *International Urology and Nephrology* 46 (5): 1025–1030.

36 Zhang, A., Yan, G., Han, Y. et al. (2014). Metabolomics approaches and applications in prostate cancer research. *Applied Biochemistry and Biotechnology* 174 (1): 6–12.

37 Liang, Q., Wang, C., Li, B. et al. (2015). Metabolic fingerprinting to understand therapeutic effects and mechanisms of silybin on acute liver damage in rat. *Pharmacognosy Magazine* 11 (43): 586.

38 Zhang, A., Sun, H., Wang, P. et al. (2012). Recent and potential developments of biofluid analyses in metabolomics. *Journal of Proteomics* 75 (4): 1079–1088.

39 Bancos, I., Taylor, A.E., Chortis, V. et al. (2020). Urine steroid metabolomics for the differential diagnosis of adrenal incidentalomas in the EURINE-ACT study: a prospective test validation study. *Lancet Diabetes & Endocrinology* 8 (9): 773–781.

40 Zhang, A., Sun, H., Yan, G. et al. (2014). Metabolomics in diagnosis and biomarker discovery of colorectal cancer. *Cancer Letters* 345 (1): 17–20.

41 Zhang, A., Sun, H., Yan, G. et al. (2016). Mass spectrometry-based metabolomics: applications to biomarker and metabolic pathway research. *Biomedical Chromatography* 30 (1): 7–12.

42 Hu, L., Liu, J., Zhang, W. et al. (2020). Functional metabolomics decipher biochemical functions and associated mechanisms underlie small-molecule metabolism. *Mass Spectrometry Reviews* 39 (5–6): 417–433.

43 Yan, M. and Xu, G. (2018). Current and future perspectives of functional metabolomics in disease studies – a review. *Analytica Chimica Acta* 1037: 41–54.

44 Zhang, L., Wei, T.T., Li, Y. et al. (2018). Functional metabolomics characterizes a key role for *N*-acetylneuraminic acid in coronary artery diseases. *Circulation* 137 (13): 1374–1390.

45 Yu, W., Yang, W., Zhao, M.Y., and Meng, X.L. (2020). Functional metabolomics analysis elucidating the metabolic biomarker and key pathway change associated with the chronic glomerulonephritis and revealing action mechanism of Rhein. *Frontiers in Pharmacology* 11: 554783.

46 Lind, L., Salihovic, S., Sundström, J. et al. (2021). Multicohort metabolomics analysis discloses 9-decenoylcarnitine to be associated with incident atrial fibrillation. *Journal of the American Heart Association* 10 (2): 1–25.

47 Xie, J., Zhang, A.H., Qiu, S. et al. (2019). Identification of the perturbed metabolic pathways associating with prostate cancer cells and anticancer affects of obacunone. *Journal of Proteomics* 206: 103447.

48 Fraser, D.D., Slessarev, M., Martin, C.M. et al. (2020). Metabolomics profiling of critically ill coronavirus disease 2019 patients: identification of diagnostic and prognostic biomarkers. *Critical Care Explorations* 2 (10): e0272.

49 Nguyen, H.T.T., Wimmer, R., Le, V.Q., and Krarup, H.B. (2021). Metabolic fingerprint of progression of chronic hepatitis B: changes in the metabolome and novel diagnostic possibilities. *Metabolomics* 17 (2): 16.

50 Naake, T., Gaquerel, E., and Fernie, A.R. (2020). Annotation of specialized metabolites from high-throughput and high-resolution mass spectrometry metabolomics. *Methods in Molecular Biology* 2104: 209–225.

51 Zhang, A., Sun, H., and Wang, X. (2017). Emerging role and recent applications of metabolomics biomarkers in obesity disease research. *RSC Advances* 7 (25): 14966–14973.

52 Yang, Q., Zhang, A.H., Miao, J.H. et al. (2019). Metabolomics biotechnology, applications, and future trends: a systematic review. *RSC Advances* 9 (64): 37245–37257.

53 Wang, X., Zhang, A., Han, Y. et al. (2012). Urine metabolomics analysis for biomarker discovery and detection of jaundice syndrome in patients with liver disease. *Molecular & Cellular Proteomics* 11 (8): 370–380.

54 Wang, X., Yang, B., Sun, H. et al. (2012). Pattern recognition approaches and computational systems tools for ultra performance liquid chromatography–mass spectrometry-based comprehensive metabolomic profiling and pathways analysis of biological data sets. *Analytical Chemistry* 84 (1): 428–439.

55 Guo, H., Chou, W.C., Lai, Y. et al. (2020). Multi-omics analyses of radiation survivors identify radioprotective microbes and metabolites. *Science* 370 (6516): 9097.

56 Liang, Q., Liu, H., Xie, L.X. et al. (2017). High-throughput metabolomics enables biomarker discovery in prostate cancer. *RSC Advances* 7 (5): 2587–2593.

57 Zhang, A., Sun, H., Wu, X. et al. (2012). Urine metabolomics. *Clinica Chimica Acta* 414: 65–69.

58 Fang, H., Zhang, A.H., Sun, H. et al. (2019). High-throughput metabolomics screen coupled with multivariate statistical analysis identifies therapeutic targets in alcoholic liver disease rats using liquid chromatography-mass spectrometry. *Journal of Chromatography B* 1109: 112–120.

59 Zhang, A.H., Ma, Z.M., Sun, H. et al. (2019). High-throughput metabolomics evaluate the efficacy of total lignans from Acanthophanax senticosus stem against ovariectomized osteoporosis rat. *Frontiers in Pharmacology* 10: 553.

60 Ashrafian, H., Sounderajah, V., Glen, R. et al. (2020). Metabolomics – the stethoscope for the 21st century. *Medical Principles and Practice* https://doi.org/10.1159/000513545.

13

Mass Spectrometry-Driven Active Ingredients Discovery from Herbal Medicine

Wanying Wang[1], *Aihua Zhang*[2], *Shi Qiu*[2], *and Xijun Wang*[2]

[1] *Macau University of Science and Technology, State Key Laboratory of Quality Research in Chinese Medicine, Heping Road 24, Avenida Wai Long, Taipa, Macau, 150040, China*
[2] *Heilongjiang University of Chinese Medicine, National Chinmedomics Research Center, Functional Metabolomics Laboratory, Department of Pharmaceutical Analysis, Heping Road 24, Harbin, 150040, China*

13.1 Introduction

Herbal medicine (HM) is derived from natural plants, seeds, berries, roots, leaves, bark, and flowers that are utilized to treat diseases. These products are complicated mixtures of organic chemicals that may come from any raw or processed part of a plant. HM is adopted by every system of alternative medicine existing around the world. Natural products have been recognized as an important reservoir of bioactive lead compounds that may become drugs [1]. The drugs that are available now, including artemisinin, camptothecin, paclitaxel, and silibinin, were directly derived from HM. Artemisinin, WHO-recommended and used in therapies against malaria, is an HM product isolated from *Artemisia annua* (*Asteraceae*) [2]. The herbal medicinal plant has specific groups in various plants. Especially in China, there are many kinds of HM recorded in pharmaceutical works since ancient times, which make a contribution to medical practice and applied to the prevention and treatment of diseases [3–8]. Modern pharmacological studies show that HMs have actions on various diseases, such as cardiovascular diseases, diabetes mellitus, hypertension, hyperlipidemia, and psychiatric disorders, and some of HMs can be used in pregnancy [9–13]. Nonetheless, adverse reaction of HMs is reported endlessly, and hepatotoxicity is common; besides that, gastrointestinal toxicity, carcinogenic activity, and thyroid damage are also usually observed [14–16]. Then the problem that prompts us cautiously is how to make sure the pharmacological components of HMs can avoid the danger of toxic ingredients. To solve this problem, it is essential to understand the quality control, material basis, drug screening, and pharmacodynamic mechanism.

Compared with the chemical drugs, herbal preparation (HP) that contains multiple plant extracts has the characteristics of being multi-component, multi-target, and multi-pathway [17–20]. However, the compounds from HP are complex enough, and not all components are effective, and a series of biochemical reaction process, such as oxidation, reduction, and hydrolysis, will happen *in vivo*. Although

Mass Spectrometry-Based Metabolomics in Clinical and Herbal Medicines: Strategies, Technologies, and Applications, First Edition. Edited by Aihua Zhang and Wanying Wang.

HP has proven its value in the therapy, the process of screening and identification of active compounds from HP is time-consuming and strenuous [21–25]. Thus, there is a need for a rapid method to screen the active compounds from HP. The commonly used bio-guided screening approaches, including the separation, extraction, purification, and identification of the structure, have been used for many years; however these processes are time-consuming and strenuous. Currently, the characteristic of highly efficient mass spectrometry (MS) has been widely used to identify the structure and molecular weight, which makes MS suitable for discovering active compounds from HMs [26–31]. It can provide the appraisal of the identification of HMs and screen the active compounds *in vivo* [32–36]. It is important to explain the efficacy of material base and mechanism of action in HP [37–40]. MS technology, with a rapid analysis speed and high sensitivity, can be used to aid identification, understand the mechanism of action, and evaluate efficacy, when coupled with appropriate compound libraries. In this chapter, we summarized the discovery and screening of active compounds from HMs using MS based on the distinct analytical techniques. Finally, examples are provided to show how this technology has been applied to enhance the functional ingredients discovery process from HMs. The applications of MS approaches in the discovery of lead compounds were also reviewed, and future perspectives of this remarkable technique were proposed.

13.2 An Overview of Herbal Medicine

HM has a long history of clinical use in disease treatment and often uses a variety of herbs in different combinations to strengthen the synergistic effects. Moreover, some studies have showed that many herbal formulation (composed of multiple herbs) can promote a cure via multi-target effects [4, 41]. To restore the dynamic balance among body systems and achieve the best curative effect, herbalists chose a combination of several herbs based on distinct disease presentations. The combinational formulation usage of HM plays a very important role in the treatment of disease. Nowadays, HM is popularly used for treatment of various diseases, such as hepatitis, cardiovascular disease, diabetes, and cancer. Due to antibacterial, antifungal, anticancer, antiviral, and anti-inflammatory activities and other pharmacological activities, HMs have received increased attention for clinical efficacy in health care. HMs or herbal formula also applied in the therapies of mental illnesses as a result of less unfavorable effect in mind of patients. A number of herbs or herbal formulations have demonstrated significant therapeutic effects against the disease, especially during its early stages [42–44]. However, the scientific evidence to support the use of HM for disease treatment remains scarce and is often inconclusive. The underlying causes are multiple, but primarily due to the complex chemicals and pharmacological properties of HM. Due to the compositional complexity, it is a challenge to identify the chemical constituents from HM. The conventional separation method fails to reveal the efficacy of a single herb or herbal formulation. A better understanding of the key components and the underlying mechanisms responsible for the effects is required.

13.3 Mass Spectrometry to Identify the Chemical Constituents

Numerous tools have been used to identify the structure and assess action mechanism of natural products. The proper selection of the detection techniques is fundamental to successful advancement of drug candidates. The principal tools used for HM analysis are nuclear magnetic resonance spectroscopy (NMR) and MS, which play a key role in the discovery phases for natural products, ranging from the characterization of chemical structures to mechanism analysis [45, 46]. There are many analytical technologies that have been used for chemical identification and separation techniques coupled with NMR and MS [47–52]. Particularly, MS provides an efficient tool for revealing the peaks and relative intensities, exact mass, molecular formula, and relative error therein. The advantages of MS for identifying chemical constituents of HM include its high-throughput power, which is essential for compound screening, and high sensitivity, compared with other techniques. Especially, it is ideal for quantitative determination of known compounds in the selected reaction monitoring mode and providing high sensitivity and mass accuracy for the structural elucidation of unknown compounds. Recently, UNIFI™ platform has been developed for MS-based data to provide more information in the natural product discovery process [53, 54]. Its workspace, illustrated in Figure 13.1, provides a table of the compound and MS and MS/MS data for a selected compound. Within this user interface, the parent compound structure is imported and modified manually to match the assigned formula, and it can clearly see the organizational characteristics.

13.4 Mass Spectrometry for Constituent Analysis *In Vivo*

As we all know, the compounds from HM and HP have the essential effects through *in vivo* processes. The blood samples are analyzed by MS in gradient parameters and methods, providing an array of information about the structure of *in vivo* compounds. MS instruments offer software packages, such as MetWorks (Thermo Electron), MetaboLynx (Waters), Metabolite Pilot (AB Sciex), etc., to facilitate the constituents and metabolite identification process. According to the MS data, the prototype components can be obtained through the retention time, MS data, and MS/MS fragment. The metabolites can be identified by MetaboLynx software according to the data processing of the plasma samples. Selective reaction monitoring (SRM) and multi-reaction monitoring (MRM) are chosen to analyze the parent ions with the corresponding conditions of ionization and energy. Matching the fragment ions with the mol files of these compounds, the structure of the *in vivo* compounds can be determined. In addition, MS^E, a new data collection mode, can provide more fragment ions to make the identification of the structure easier by scanning with the crossing high and low collision [55, 56], and the fragment ions can undergo target screening based on the chemical structure types in the high and low collision MS data, which can also accomplish the identification of the structure.

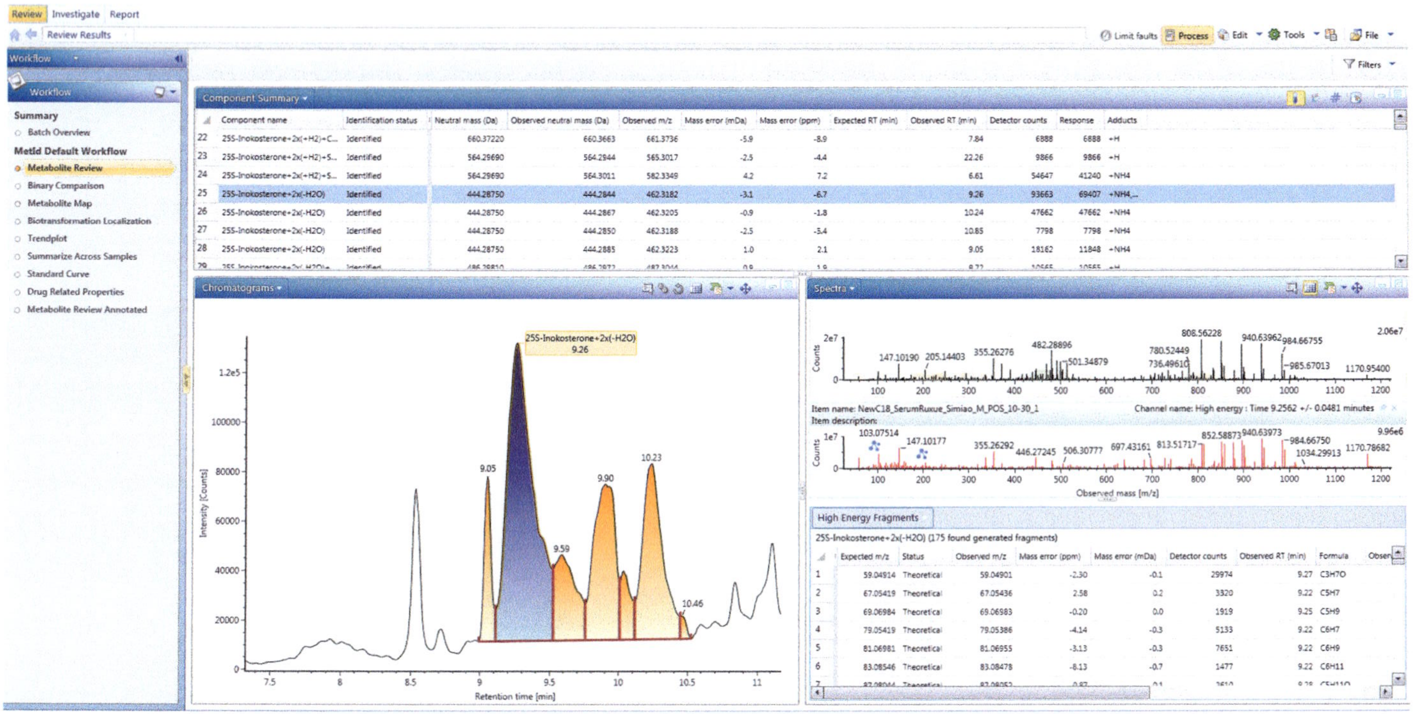

Figure 13.1 An example of the UNIFI library platform viewer.

Combination of the assay *in vivo* and *in vitro* could not only reflect the effects of the prototype compounds from HM but also elucidate the interaction between the corresponding metabolism and endogenous substance.

13.5 Mass Spectrometry-Based Chinmedomics Strategy

Evaluation of HM efficacy is an essential requirement for the discovery of the effective constituents. Therefore, to develop an innovative method for the screening of active compounds, chinmedomics strategy has been developed based on the MS [57, 58]. Chinmedomics strategy has integrated serum pharmacochemistry with metabolomics technology and is capable of evaluating the HM efficacy and discovering biomarkers as well as holds promise for improving the effective substances discovery process from HM [43, 59–64]. It is part of the outstanding achievements in the field of HM research, which aims to improve the understanding of formulation and seeks to elucidate the therapeutic properties with integrated traditional and modern biological techniques. Kai-Xin San (KXS) is a classic prescription for the treatment of memory decline applied for Alzheimer's disease (AD). Chinmedomics was applied for assessing the efficacy and discovering the effective ingredients [65]. KXS could regulate 20 biomarkers of AD. The effective material basis of KXS was ginsenoside Rf, ginsenoside F1, 20-*O*-glucopyranosyl ginsenoside Rf, dehydropachymic acid, and E-3,4,5-trimethoxycinnamic acid. It demonstrated that KXS significantly improved cognitive function of AD, repaired the damage caused by Aβ, regulated amino acid metabolism and lipid metabolism, and determined the effective material basis of KXS treating AD. Chinmedomics was also used to evaluate the efficacy of Sijunzi decoction (SJZD) against spleen qi deficiency syndrome (SQDS) to discover the effective potential ingredients [66]. A total of 56 compounds in SJZD were characterized *in vitro*, and 23 compounds were discovered *in vivo*. A total of 58 biomarkers were related to SQDS, and SJZD can regulate the metabolic profile to the health status by adjusting a large proportion of biomarkers to normal level. A total of 10 constituents were absorbed as effective ingredients that were associated with overall efficacy.

13.6 Application

MS screening strategy was used to analyze the alkaloids in Eomecon chionantha Hance (ECH), a traditional folk herb, which has analgesic and anti-inflammatory properties [67]. The structures of 95 alkaloids in ECH were identified, and six types of benzylisoquinoline alkaloids were identified in ECH. The identification of the chemical composition in Paris polyphylla var. yunnanensis (Franch) Hand Mazz is helpful to discover the active ingredients for further quality control [68]. A total of 222 compounds were characterized by comparing their main fragmentation and pathways, and 54 steroidals and 8 phenylpropanoids were discovered and identified for the first time. A high-performance liquid chromatography tandem mass spectrometry (HPLC–MS/MS) method was developed for analyzing 14 antidepressants and 13 antipsychotics in human plasma for routine clinical therapeutic drug monitoring [69]. The method was precise with acceptable intra- and inter-day precision and accuracy of 85.51–114.77%. This dynamic method is efficient and serves as a reference for plasma concentrations of supra-alert level samples.

Ultra-high-performance liquid chromatography combined with mass spectrometry (UPLC/MS) was used for identifying the polymeric phenolic acids and flavonoids in complex chemical constituents of Clerodendranthi Spicati Herba (CSH) [70]. A total of 118 compounds, including 33 flavonoids and 85 polymeric phenolic acids, were identified in CSH by comparing their main fragmentation pathways. These findings showed that this approach was possible and rational for the discovery and identification of natural products. American ginseng, the dried roots of Pawajc quinquefolium L. (Araliaceae), is a valuable HM that is associated with numerous active ingredients. The *in vitro* and *in vivo* constituents of American ginseng using UPLC–MS were investigated [71]. Correlation analysis of 41 metabolites with 14 serum components showed that pseudoginsenoside F11 and ginsenoside Rd are highly correlated with the therapeutic effects. These compounds can be regarded as potential quality-markers for American ginseng. A rapid and reliable LC–MS method was developed for the quantitation of major components in Folium Artemisiae Argyi (mugwort) [72]. A total of 5 phenolic acids and 17 flavonoids were separated and simultaneously determined. Gray relational analysis and partial least square regression were used to evaluate the relevance between chemicals and bioactivities, and the results indicated that chlorogenic acid, isochlorogenic acid, eriodictyol, jaceosidin, and eupatilin had the key contribution to antioxidant activity and possess potential value for the activity-oriented quality control of mugwort.

13.7 Conclusions and Future Perspectives

This chapter described how the conjunction of MS with related techniques and methods could provide us with useful structural information aiding the discovery of new lead compounds from HM. HM has made a major contribution to the drug discovery process with regard to identifying natural products. Currently, more attention has been focused on drug discovery from natural compounds of HM. Despite the rapid advancement of modern analytical techniques, drug discovery is still a difficult and lengthy process. The components from HP are complex and the mechanism is intricacy. It is essential to develop an efficient and high-throughput method integrated with a MS platform for early screening of effective ingredients from HM. Establishing a feasible method to screen the active compounds is critical to realize the modernization of HM. Recent advances in MS applications offer the capability of providing more data in a time-efficient manner. The improvements in MS-based strategies have radically changed this situation. MS, a technology with great potential development, has been utilized in HM and gets many successful achievements. Fortunately, MS can provide us with useful structural information for effective ingredients discovery and has been recognized as a sensitive and high-throughput technology for advancing effective ingredients discovery from HM.

Acknowledgments

This work was supported by grants from the National Key Research and Development Program of China (2018YFC1706103), Key Program of Natural Science Foundation of State (Grant No. 81830110, 81861168037, 81973745, 81302905), Natural Science Foundation of Heilongjiang Province (YQ2019H030), and Heilongjiang Touyan Innovation Team Program.

References

1 Liu, X., Wu, W.Y., Jiang, B.H. et al. (2013). Pharmacological tools for the development of traditional Chinese medicine. *Trends in Pharmacological Sciences* 34 (11): 620–628.
2 Tu, Y. (2011). The discovery of artemisinin (qinghaosu) and gifts from Chinese medicine. *Nature Medicine* 17 (10): 1217–1220.
3 Zhang, A., Sun, H., Wang, X. et al. (2012). Simultaneous in vivo RP-HPLC-DAD quantification of multiple-component and drug-drug interaction by pharmacokinetics, using 6,7-dimethylesculetin, geniposide and rhein as examples. *Biomedical Chromatography* 26 (7): 844–850.
4 Sun, H., Zhang, A., and Wang, X. (2012). Potential role of metabolomic approaches for Chinese medicine syndromes and herbal medicine. *Phytotherapy Research* 26 (10): 1466–1471.
5 Wang, X., Yang, B., Sun, H. et al. (2012). Pattern recognition approaches and computational systems tools for ultra performance liquid chromatography-mass spectrometry-based comprehensive metabolomic profiling and pathways analysis of biological data sets. *Analytical Chemistry* 84 (1): 428–439.
6 Zhang, A., Sun, H., and Wang, X. (2018). Mass spectrometry-driven drug discovery for development of herbal medicine. *Mass Spectrometry Reviews* 37 (3): 307–320.
7 Sun, H., Dong, W., Zhang, A. et al. (2012). Pharmacokinetics study of multiple components absorbed in rat plasma after oral administration of Stemonae radix using ultra-performance liquid-chromatography/mass spectrometry with automated MetaboLynx software analysis. *Journal of Separation Science* 35 (24): 3477–3485.
8 Sun, H., Dong, W., Zhang, A. et al. (2012). Ultra-performance liquid-chromatography with tandem mass spectrometry performing pharmacokinetic and biodistribution studies of croomine, neotuberostemonine and tuberostemonine alkaloids absorbed in the rat plasma after oral administration of Stemonae radix. *Fitoterapia* 83 (8): 1699–1705.

9 Yin, Q., Sun, H., Zhang, A. et al. (2012). Pharmacokinetics and tissue distribution study of scoparone in rats by ultraperformance liquid-chromatography with tandem high-definition mass spectrometry. *Fitoterapia* 83 (4): 795–800.

10 Dong, W., Wang, P., Meng, X. et al. (2012). Ultra-performance liquid chromatography-high-definition mass spectrometry analysis of constituents in the root of Radix Stemonae and those absorbed in blood after oral administration of the extract of the crude drug. *Phytochemical Analysis* 23 (6): 657–667.

11 Zhang, A., Sun, H., Yuan, Y. et al. (2011). An in vivo analysis of the therapeutic and synergistic properties of Chinese medicinal formula Yin-Chen-Hao-Tang based on its active constituents. *Fitoterapia* 82 (8): 1160–1168.

12 Wang, X., Sun, H., Zhang, A. et al. (2011). Pharmacokinetics screening for multi-components absorbed in the rat plasma after oral administration traditional Chinese medicine formula Yin-Chen-Hao-Tang by ultra performance liquid chromatography-electrospray ionization/quadrupole-time-of-flight mass spectrometry combined with pattern recognition methods. *The Analyst* 136 (23): 5068–5076.

13 Yang, B., Dong, W., Zhang, A. et al. (2011). Ultra-performance liquid chromatography coupled with electrospray ionization/quadrupole-time-of-flight mass spectrometry for rapid analysis of constituents of Suanzaoren decoction. *Journal of Separation Science* 34 (22): 3208–3215.

14 Wang, X., Wang, H., Zhang, A. et al. (2012). Metabolomics study on the toxicity of aconite root and its processed products using ultraperformance liquid-chromatography/electrospray-ionization synapt high-definition mass spectrometry coupled with pattern recognition approach and ingenuity pathways analysis. *Journal of Proteome Research* 11 (2): 1284–1301.

15 Dong, H., Zhang, A., Sun, H. et al. (2012). Ingenuity pathways analysis of urine metabolomics phenotypes toxicity of Chuanwu in Wistar rats by UPLC-Q-TOF-HDMS coupled with pattern recognition methods. *Molecular BioSystems* 8 (4): 1206–1221.

16 Sun, H., Ni, B., Zhang, A. et al. (2012). Metabolomics study on Fuzi and its processed products using ultra-performance liquid-chromatography/electrospray-ionization synapt high-definition mass spectrometry coupled with pattern recognition analysis. *The Analyst* 137 (1): 170–185.

17 Yang, D.Z., Sun, G., Zhang, A. et al. (2015). Screening and analyzing the potential bioactive components from rhubarb, using a multivariate data processing approach and ultra-high performance liquid chromatography coupled with time-of-flight mass spectrometry. *Analytical Methods* 7 (2): 650–661.

18 Zhang, A., Zou, D., Yan, G. et al. (2014). Identification and characterization of the chemical constituents of Simiao Wan by ultra high performance liquid chromatography with mass spectrometry coupled to an automated multiple data processing method. *Journal of Separation Science* 37 (14): 1742–1747.

19 Liu, J.H., Sun, H., Zhang, A.H. et al. (2014). Serum pharmacochemistry combined with multiple data processing approach to screen the bioactive components and their metabolites in Mutan Cortex by ultra-performance liquid chromatography tandem mass spectrometry. *Biomedical Chromatography* 28 (4): 500–510.

20 Xue, C., Zhang, A., Sun, H. et al. (2014). An improved ultra-performance liquid chromatography-electrospray ionization/quadrupole-time-of-flight high-definition mass spectrometry method for determining ingredients of herbal Fructus corni in blood samples. *Pharmacognosy Magazine* 10 (40): 422–429.

21 Cao, H., Zhang, A., Zhang, F.M. et al. (2014). Ultra-performance liquid chromatography tandem mass spectrometry combined with automated MetaboLynx analysis approach to screen the bioactive components and their metabolites in Wen-Xin-Formula. *Biomedical Chromatography* 28 (12): 1774–1781.

22 Wang, X., Zhang, A., Yan, G. et al. (2014). UHPLC-MS for the analytical characterization of traditional Chinese medicines. *TrAC Trends in Analytical Chemistry* 63: 180–187.

23 Wang, P., Yin, Q.W., Zhang, A.H. et al. (2014). Preliminary identification of the absorbed bioactive components and metabolites in rat plasma after oral administration of Shaoyao-Gancao decoction by ultra-performance liquid chromatography with electrospray ionization tandem mass spectrometry. *Pharmacognosy Magazine* 10 (40): 497–502.

24 Qiu, S., Sun, H., Zhang, A.H. et al. (2014). Natural alkaloids: basic aspects, biological roles, and future perspectives. *Chinese Journal of Natural Medicines* 12 (6): 401–406.

25 Zou, D., Zhang, A.H., Yan, G.L. et al. (2014). UPLC-MS coupled with a dynamic multiple data processing method for the comprehensive detection of the chemical constituents of the herbal formula San-Miao-Wan. *Analytical Methods* 6 (9): 2848–2854.

26 Zhou, Y., Zhang, A., Sun, H. et al. (2014). Plant-derived natural products as leads to antitumor drugs. *Plant Science Today* 1 (2): 46–63.

27 Zhang, A.H., Sun, H., Han, Y. et al. (2013). Ultraperformance liquid chromatography-mass spectrometry based comprehensive metabolomics combined with pattern recognition and network analysis methods for characterization of metabolites and metabolic pathways from biological data sets. *Analytical Chemistry* 85 (15): 7606–7612.

28 Zhang, A.H., Wang, P., Sun, H. et al. (2013). High-throughput ultra-performance liquid chromatography-mass spectrometry characterization of metabolites guided by a bioinformatics program. *Molecular BioSystems* 9 (9): 2259–2265.

29 Zhang, A., Sun, H., Qiu, S. et al. (2013). Advancing drug discovery and development from active constituents of yinchenhao tang, a famous traditional chinese medicine formula. *Evidence-based Complementary and Alternative Medicine: Ecam* 2013: 257909.

30 Yan, G.L., Zhang, A.H., Sun, H. et al. (2013). An effective method for determining the ingredients of Shuanghuanglian formula in blood samples using high-resolution LC-MS coupled with background subtraction and a multiple data processing approach. *Journal of Separation Science* 36 (19): 3191–3199.

31 Wang, H., Yan, G., Zhang, A. et al. (2013). Rapid discovery and global characterization of chemical constituents and rats metabolites of Phellodendri

amurensis cortex by ultra-performance liquid chromatography-electrospray ionization/quadrupole-time-of-flight mass spectrometry coupled with pattern recognition approach. *The Analyst* 138 (11): 3303–3312.
32 Liu, C., Zhang, A., Han, Y. et al. (2015). Ultra-high performance liquid chromatography coupled with time-of-flight mass spectrometry screening and analysis of potential bioactive compounds from traditional chinese medicine Kai-Xin-San, using a multivariate data processing approach and the MetaboLynx tool. *RSC Advances* 5 (1): 85–92.
33 Sun, H., Liu, C., Zhang, A.H. et al. (2015). Rapid discovery and global characterization of multiple constituents from Kai-Xin-San using an integrated MS E data acquisition mode strategy based on ultra-performance liquid chromatography coupled to electrospray ionization/quadrupole-time-of-flight mass spectrometry. *Analytical Methods* 7 (1): 279–286.
34 Yan, G., Zou, D., Zhang, A. et al. (2015). UPLC-Q-TOF-MS/MS fingerprinting for rapid identification of the chemical constituents of Ermiao Wan. *Analytical Methods* 7 (3): 846–862.
35 Han, Y., Wu, F., Zhang, A. et al. (2015). Characterization of multiple constituents in rat plasma after oral administration of Shengmai San using ultra-performance liquid chromatography coupled with electrospray ionization/quadrupole-time-of-flight high-definition mass spectrometry. *Analytical Methods* 7 (3): 830–837.
36 Zhang, Y., Zhang, A., Zhang, Y. et al. (2016). Application of ultra-performance liquid chromatography with time-of-flight mass spectrometry for the rapid analysis of constituents and metabolites from the extracts of Acanthopanax senticosus harms leaf. *Pharmacognosy Magazine* 12 (46): 145–152.
37 Sun, H., Wang, H., Zhang, A. et al. (2016). Chemical discrimination of cortex Phellodendri amurensis and cortex Phellodendri chinensis by multivariate analysis approach. *Pharmacognosy Magazine* 12 (45): 41–49.
38 Wang, H., Sun, H., Zhang, A. et al. (2013). Rapid identification and comparative analysis of the chemical constituents and metabolites of Phellodendri amurensis cortex and Zhibai dihuang pill by ultra-performance liquid chromatography with quadrupole TOF-MS. *Journal of Separation Science* 36 (24): 3874–3882.
39 Sun, H., Wu, F., Zhang, A. et al. (2013). Profiling and identification of the absorbed constituents and metabolites of schisandra lignans by ultra-performance liquid chromatography coupled to mass spectrometry. *Biomedical Chromatography* 27 (11): 1511–1519.
40 Yin, Q., Wang, P., Zhang, A. et al. (2013). Ultra-performance LC-ESI/quadrupole-TOF MS for rapid analysis of chemical constituents of Shaoyao-Gancao decoction. *Journal of Separation Science* 36 (7): 1238–1246.
41 Wang, X., Zhang, A., Wang, P. et al. (2013). Metabolomics coupled with proteomics advancing drug discovery toward more agile development of targeted combination therapies. *Molecular & Cellular Proteomics* 12 (5): 1226–1238.
42 Zhou, X.H., Zhang, A.H., Wang, L. et al. (2016). Novel chinmedomics strategy for discovering effective constituents from ShenQiWan acting on ShenYangXu syndrome. *Chinese Journal of Natural Medicines* 14 (8): 561–581.

43 Liu, Q., Zhang, A., Wang, L. et al. (2016). High-throughput chinmedomics-based prediction of effective components and targets from herbal medicine AS1350. *Scientific Reports* 6: 38437.

44 Zhang, K., Yan, G., Zhang, A. et al. (2017). Recent advances in pharmacokinetics approach for herbal medicine. *RSC Advances* 7 (46): 28876–28888.

45 Sun, H., Wu, F., Zhang, A. et al. (2013). Pharmacokinetic study of schisandrin, schisandrol B, schisantherin A, deoxyschisandrin, and schisandrin B in rat plasma after oral administration of Shengmaisan formula by UPLC-MS. *Journal of Separation Science* 36 (3): 485–491.

46 Sun, H., Dong, T., Zhang, A. et al. (2013). Pharmacokinetics of hesperetin and naringenin in the Zhi Zhu wan, a traditional Chinese medicinal formulae, and its pharmacodynamics study. *Phytotherapy Research* 27 (9): 1345–1351.

47 Wang, P., Lv, H.t., Zhang, A.h. et al. (2013). Improved ultra-performance liquid chromatography with electrospray ionization quadrupole-time-of-flight high-definition mass spectrometry method for the rapid analysis of the chemical constituents of a typical medical formula: Liuwei Dihuang Wan. *Journal of Separation Science* 36 (21–22): 3511–3516.

48 Li, X., Sun, H., Zhang, A. et al. (2017). High-throughput LC-MS method for the rapid characterization of multiple chemical constituents and metabolites of Da-Bu-Yin-wan. *Journal of Separation Science* 40 (21): 4102–4112.

49 Han, Y., Zhang, A., Sun, H. et al. (2017). High-throughput ultra high performance liquid chromatography combined with mass spectrometry approach for the rapid analysis and characterization of multiple constituents of the fruit of Acanthopanax senticosus (Rupr. et Maxim.) Harms. *Journal of Separation Science* 40 (10): 2178–2187.

50 Liu, C., Zhang, A., Yan, G.L. et al. (2017). High-throughput ultra high performance liquid chromatography coupled to quadrupole time-of-flight mass spectrometry method for the rapid analysis and characterization of multiple constituents of Radix Polygalae. *Journal of Separation Science* 40 (3): 663–670.

51 Liu, X.Y., Zhang, A.H., Fang, H. et al. (2018). Serum metabolomics strategy for understanding the therapeutic effects of Yin-Chen-Hao-Tang against Yanghuang syndrome. *RSC Advances* 8 (14): 7403–7413.

52 Li, X.N., Zhang, A., Sun, H. et al. (2016). Rapid discovery of absorbed constituents and metabolites in rat plasma after the oral administration of Zi Shen Wan using high-throughput UHPLC-MS with a multivariate analysis approach. *Journal of Separation Science* 39 (24): 4700–4711.

53 Ren, J.L., Dong, H., Han, Y. et al. (2020). Network pharmacology combined with metabolomics approach to investigate the protective role and detoxification mechanism of Yunnan Baiyao formulation. *Phytomedicine: International Journal of Phytotherapy and Phytopharmacology* 77: 153266.

54 Zhang, A.H., Ma, Z.M., Kong, L. et al. (2020). High-throughput lipidomics analysis to discover lipid biomarkers and profiles as potential targets for evaluating efficacy of Kai-Xin-San against APP/PS1 transgenic mice based on UPLC-Q/TOF-MS. *Biomedical Chromatography* 34 (2): e4724.

55 Wang, X.J., Ren, J.L., Zhang, A.H. et al. (2019). Novel applications of mass spectrometry-based metabolomics in herbal medicines and its active ingredients: current evidence. *Mass Spectrometry Reviews* 38 (4–5): 380–402.

56 Gao, H.L., Zhang, A.H., Yu, J.B. et al. (2018). High-throughput lipidomics characterize key lipid molecules as potential therapeutic targets of Kaixinsan protects against Alzheimer's disease in APP/PS1 transgenic mice. *Journal of Chromatography. B, Analytical Technologies in the Biomedical and Life Sciences* 1092: 286–295.

57 Wang, X., Zhang, A., and Sun, H. (2012). Future perspectives of Chinese medical formulae: chinmedomics as an effector. *OMICS: A Journal of Integrative Biology* 16 (7–8): 414–421.

58 Wang, X., Zhang, A., Sun, H. et al. (2016). Discovery and development of innovative drug from traditional medicine by integrated chinmedomics strategies in the post-genomic era. *TrAC Trends in Analytical Chemistry* 76: 86–94.

59 Li, X.N., Zhang, A., Wang, M. et al. (2017). Screening the active compounds of Phellodendri Amurensis cortex for treating prostate cancer by high-throughput chinmedomics. *Scientific Reports* 7 (1): 1–15.

60 Wang, X., Zhang, A., Zhou, X. et al. (2016). An integrated chinmedomics strategy for discovery of effective constituents from traditional herbal medicine. *Scientific Reports* 6: 18997.

61 Zhang, A., Sun, H., Yan, G. et al. (2017). Recent developments and emerging trends of mass spectrometry for herbal ingredients analysis. *TrAC Trends in Analytical Chemistry* 94: 70–76.

62 Zhang, A.H., Sun, H., Yan, G.L. et al. (2019). Chinmedomics: a powerful approach integrating metabolomics with serum pharmacochemistry to evaluate the efficacy of traditional Chinese medicine. *Engineering* 5 (1): 60–68.

63 Sun, H., Zhang, A.H., Yang, L. et al. (2019). High-throughput chinmedomics strategy for discovering the quality-markers and potential targets for Yinchenhao decoction. *Phytomedicine* 54: 328–338.

64 Zhang, A.H., Yu, J.B., Sun, H. et al. (2018). Identifying quality-markers from Shengmai san protects against transgenic mouse model of Alzheimer's disease using chinmedomics approach. *Phytomedicine* 45: 84–92.

65 Wang, X.J., Zhang, A.H., Kong, L. et al. (2019). Rapid discovery of quality-markers from Kaixin San using chinmedomics analysis approach. *Phytomedicine* 54: 371–381.

66 Zhao, Q., Gao, X., Yan, G. et al. (2020). Chinmedomics facilitated quality-marker discovery of Sijunzi decoction to treat spleen qi deficiency syndrome. *Frontiers of Medicine* 14 (3): 335–356.

67 Lu, M., Li, K., He, H. et al. (2020). Systematic characterization of alkaloids in Eomecon chionantha Hance using ultrahigh-performance liquid chromatography-tandem quadrupole Exactive Orbitrap mass spectrometry with a four-step screening strategy. *Rapid Communications in Mass Spectrometry* 34 (19): e8880.

68 Wang, Y., Fan, Q., Xiang, J. et al. (2020). Structural characterization and discrimination of Paris polyphylla var. yunnanensis by a molecular networking strategy coupled with ultra-high-performance liquid chromatography with quadrupole time-of-flight mass spectrometry. *Rapid Communications in Mass Spectrometry* 34 (11): e8760.

69 Lyu, H., Chen, B., Xu, X. et al. (2020). Rapid simultaneous determination of fourteen antidepressants and thirteen antipsychotics in human plasma by using high-performance liquid chromatography tandem mass spectrometry with dynamic multiple reaction monitoring and its application to therapeutic drug monitoring. *Therapeutic Drug Monitoring*, https://doi.org/10.1097/FTD.0000000000000839.

70 Luo, Y., Wen, Q., Lai, C.J. et al. (2019). Characterization of polymeric phenolic acids and flavonoids in Clerodendranthi Spicati Herba using ultrahigh-performance liquid chromatography coupled to quadrupole time-of-flight tandem mass spectrometry with target and nontarget data mining strategy. *Rapid Communications in Mass Spectrometry* 33 (24): 1884–1893.

71 Xiong, H., Zhang, A.H., Zhao, Q.Q. et al. (2020). Discovery of quality-marker ingredients of Panax quinquefolius driven by high-throughput chinmedomics approach. *Phytomedicine: International Journal of Phytotherapy and Phytopharmacology* 74: 152928.

72 Xia, J.X., Zhao, B.B., Zan, J.F. et al. (2019). Simultaneous determination of phenolic acids and flavonoids in Artemisiae Argyi Folium by HPLC-MS/MS and discovery of antioxidant ingredients based on relevance analysis. *Journal of Pharmaceutical and Biomedical Analysis* 175: 112734.

14

Metabolomics Application in Herbal Medicine

Wanying Wang[1], *Aihua Zhang*[2], *Hui Sun*[2], *and Xijun Wang*[2]

[1] *Macau University of Science and Technology, State Key Laboratory of Quality Research in Chinese Medicine, Avenida Wai Long, Taipa, Macau, China*
[2] *Heilongjiang University of Chinese Medicine, National Chinmedomics Research Center, Functional Metabolomics Laboratory, Heping Road 24, Harbin, 150040, China*

14.1 Introduction

Herbal medicine (HM) has been applied in the clinic for thousands of years and treats diseases with multi-components acting on multi-target simultaneously through multiple pathways, generating a comprehensive overall effect [1–3]. The sources of the active constituents in HM are complicated, ranging from original compounds of the prescription to their metabolites of the drugs *in vivo* after administration. The identification of metabolites and absorbed constituents plays a key role in elucidating therapeutic effect and mechanism [4–7]. HM is usually prescribed in combination of several medicinal herbs at a certain mass ratio, playing a key role in the treatment of disease and the clinical prevention [8–13]. Elucidation of the efficacy of HM is important to understand the scientific value of HM and promotes the innovation and development of HM [14, 15]. However, due to complexity of formulation and vagueness of syndrome or disease, its efficacy is difficult to understand. Hence, it is necessary to establish an efficacy evaluation and quality control system of HM in a systematic approach.

During the last years, many technologies have been successfully applied in HM research. They have also become important tools in the field of HM and are widely applied to discover innovative drug, providing valuable methods to explore the function and essence of formulae in HM research. Metabolomics represents a global understanding of dynamic responses to the changes of both endogenous and exogenous factors and has many potential advantages for the investigation of complex systems [16–21]. It adopts a "top-down" strategy to reflect the function changes from the metabolic networks of a complete system caused by interventions in a holistic context [22, 23]. Metabolic profile by principal component analysis (Figure 14.1), tree cluster analysis (Figure 14.2), and horizontal cluster

Mass Spectrometry-Based Metabolomics in Clinical and Herbal Medicines: Strategies, Technologies, and Applications, First Edition. Edited by Aihua Zhang and Wanying Wang.

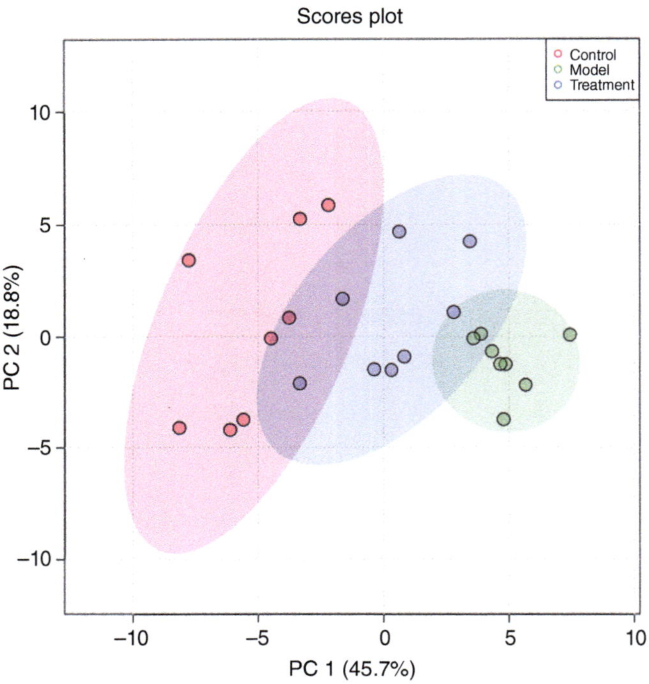

Figure 14.1 The principal component analysis for metabolic profile research that characterize the overall effect of HM.

analysis (Figure 14.3) can characterize the overall effect of HM, and metabolite fingerprint (Figure 14.4) and biomarkers (Figure 14.5) could finely express detailed changes in effects. Metabolite enrichment analysis (Figure 14.6), metabolic pathway analysis (Figure 14.7), and metabolic network analysis (Figure 14.8) can reveal the in-depth exploration of effective mechanisms. Consequently, its property agrees with the holistic thinking of HM and indicating that metabolomics has the huge potential to impact the understanding of the action mechanisms of HM. This chapter summarizes the applications of metabolomics in special HM issues.

14.2 Metabolomic Dissection of Herbal Medicine

HM pays attention to the whole parts of the body, and usually the satisfactory results can be achieved. Nowadays, there is some evidence that HM has increased popularity in the world [24, 25]. At present, using biomarkers to monitor the HM treatment is increasingly important to improve public health care. The key problem is how to establish new methods suitable for investigating HM. Fortunately, the rapid development of metabolomics appears to be similar to the system view of HM and provides a method for deep understanding of the action essence of HM

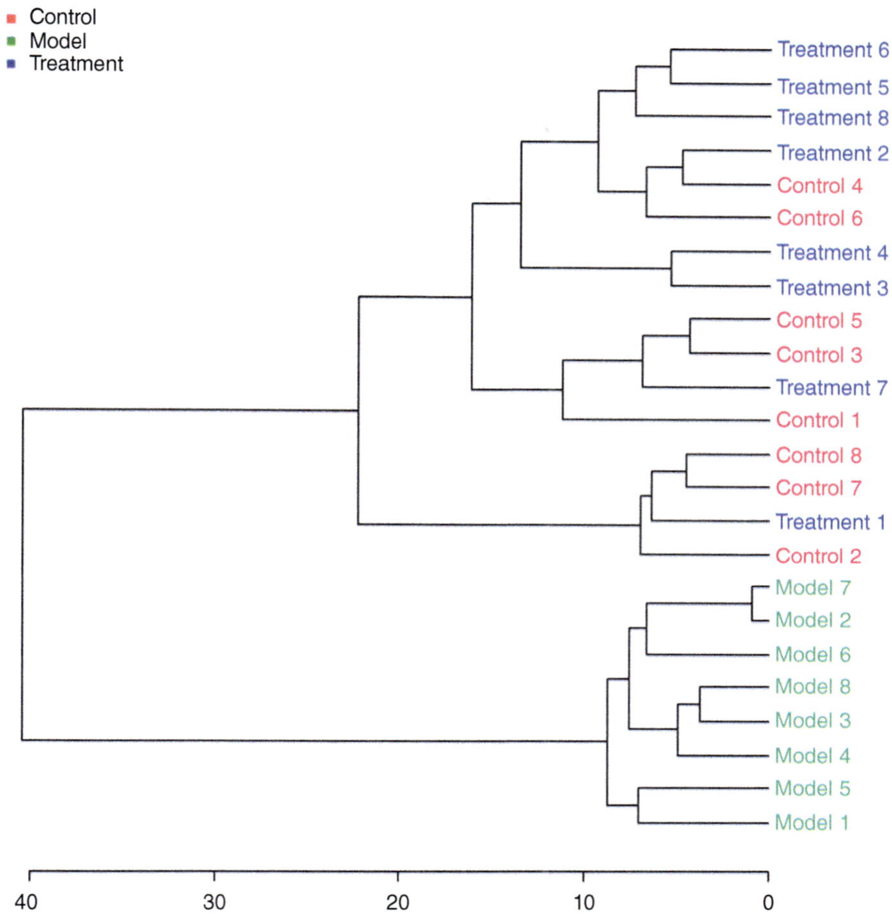

Figure 14.2 The tree cluster analysis for metabolic profile research.

[26–29]. Metabolomics has significantly increased in recent years and been used to explore the candidate metabolites, potential biomarkers, and effect pathways [30–34]. Using top-down systems, metabolomics could achieve the understanding of the interventions of HM at a global level. Its method and design resemble those of HM [35, 36]. The technology platforms of metabolomics could provide a more effective, direct, and rapid method for exploring the essence of HM efficacy [37]. Moreover, it has the potential to enable a dynamic picture of early-stage changes in diseases and thus provides an opportunity to develop biomarkers for early stage [38]. The metabolites could yield important information from HM treatment. Overall, the global metabolomics has great potential to explore the therapeutic effect of HM, and metabolomics integrated into HM would make it possible to clarify its essence.

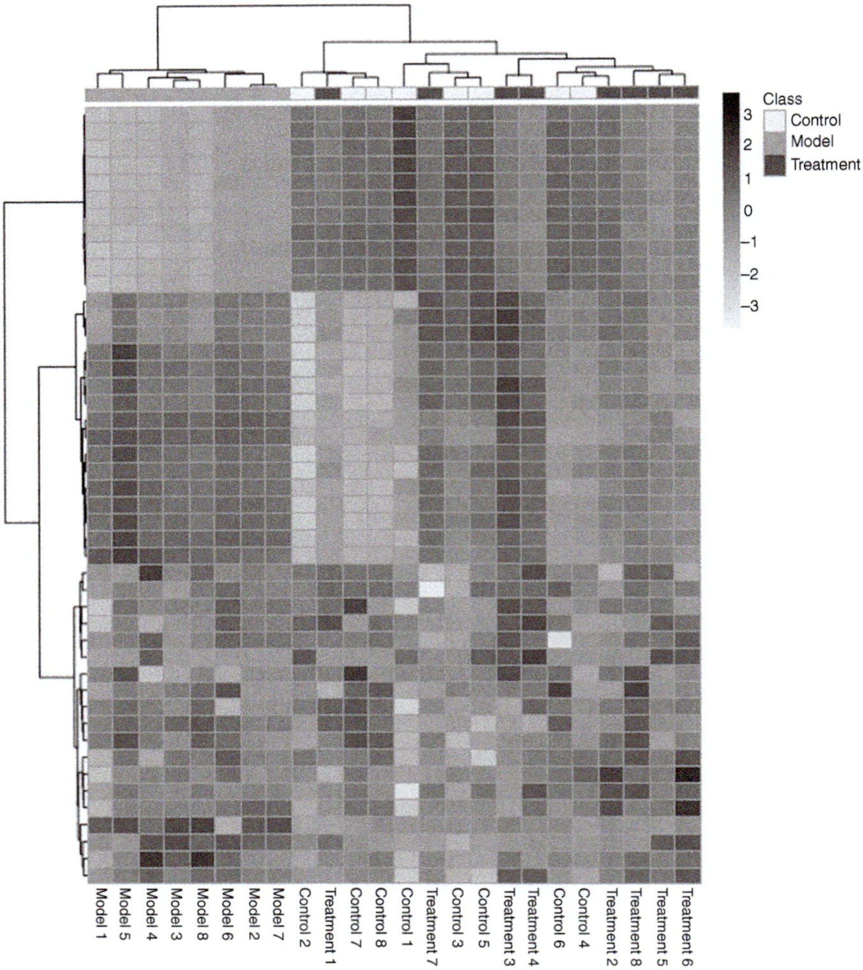

Figure 14.3 The horizontal cluster analysis for metabolic profile research.

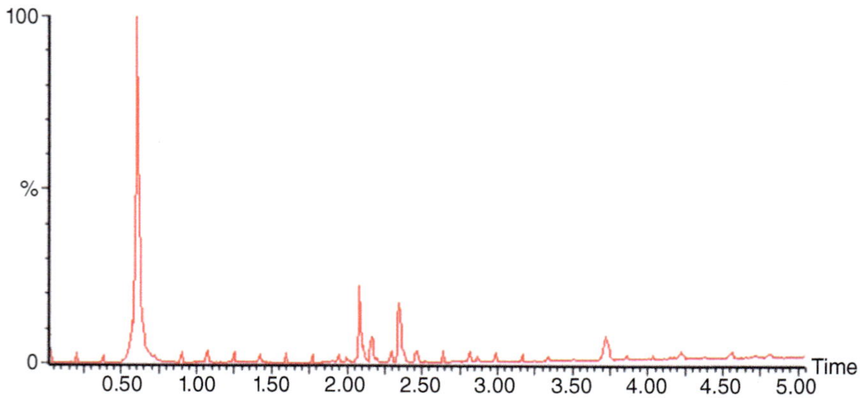

Figure 14.4 Metabolite fingerprint by high-performance liquid chromatography.

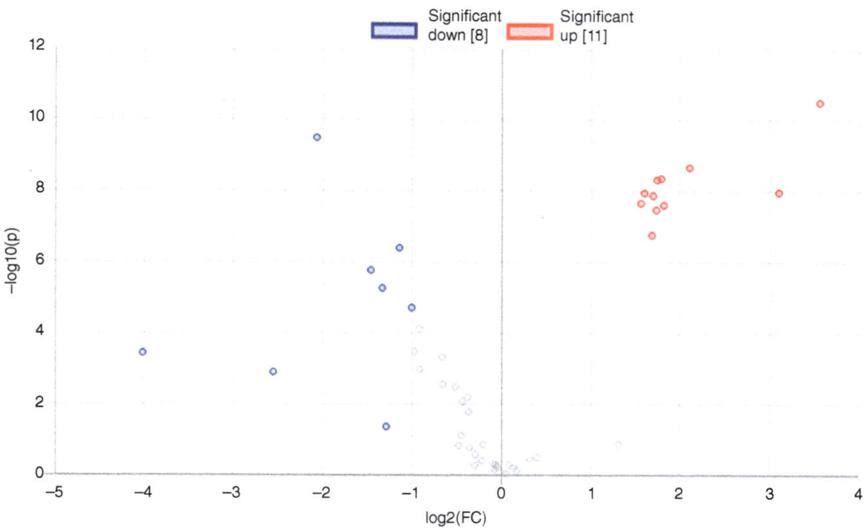

Figure 14.5 Volcano plot of metabolite biomarkers.

14.3 Application in Herbal Medicine

14.3.1 Effect Evaluations

In light of the advantages of metabolomics, many studies have applied this approach to explore the effect of HM. *Schisandra chinensis* (SC) is a well-known important HM that has been used to treat liver disease. The overall effects and mechanism of SC against alcoholic liver injury (ALI) were explored by metabolomics [39]. Principal component analysis was used to evaluate global metabolic alterations and therapeutic effects of SC. It showed that SC has therapeutic efficacy for ALI, as suggested by the recovery results of the inflammation and dysfunctions of amino acids and energy metabolism regulated by SC on ALI. *SangQiQingXuan* decoction is an herbal preparation exerting the therapeutic efficacy on high blood pressure. Metabolic profiling using ultra-performance liquid chromatography coupled with tandem mass spectrometry (MS) was analyzed for therapeutic efficacy and to reveal the optimal dosage [40]. It remarkably decreased systolic blood pressure, and the high dosage was with the best therapeutic effect where a total of 11 metabolites were markedly changed. Glycochenodeoxycholic acid, tryptophan betaine, and nicotinamide-*N*-oxide might be biomarkers that may be partially correlated with the protective effects of *SangQiQingXuan* and mainly involving in primary bile acid biosynthesis, cholesterol metabolism, and bile secretion.

A dynamic urinary metabolomics analysis based on MS was used to explore the distinctive metabolic patterns of blood stasis syndrome (BSS) development and to reveal the overall efficacy of *Danggui Sini* decoction [41]. A total of 21 metabolites were identified during the development of BSS, which are mainly involved in the riboflavin metabolism, tryptophan metabolism, and phenylalanine metabolism pathways. A receiver operating characteristic curve analysis showed

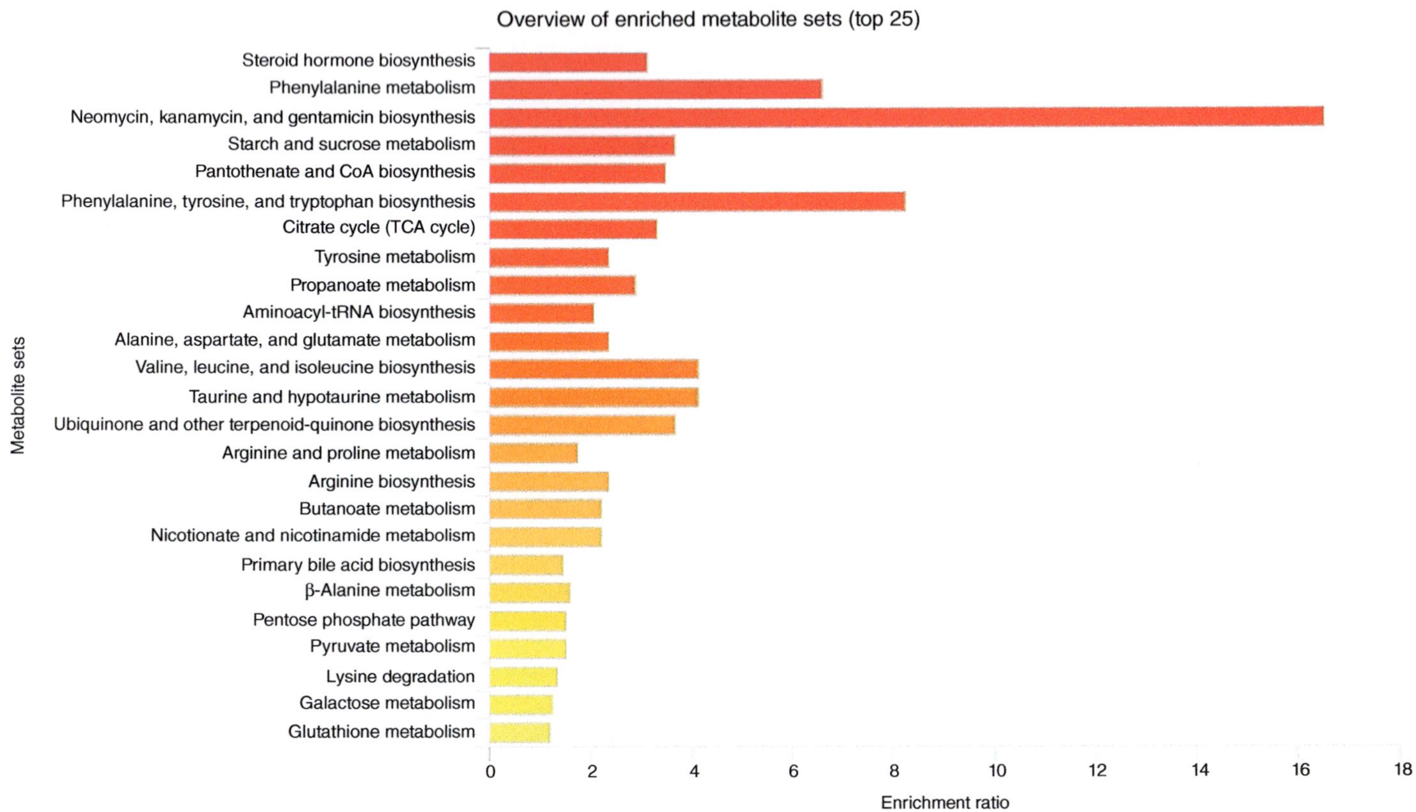

Figure 14.6 Metabolite enrichment analysis for metabolomics research.

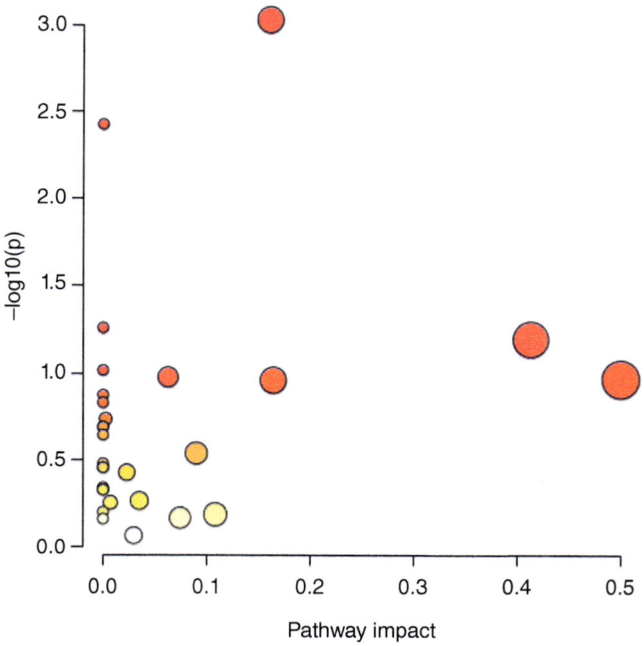

Figure 14.7 Metabolic pathway analysis of the biologically meaningful metabolites.

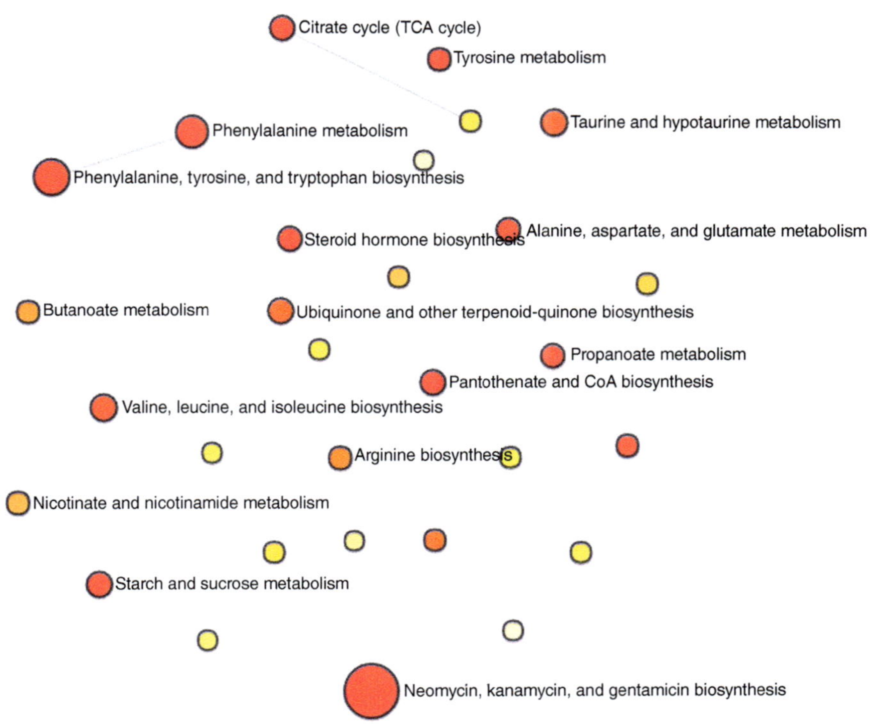

Figure 14.8 Metabolic network analysis of the biologically meaningful metabolites.

10 metabolites as potential biomarkers with an area under the curve (AUC) > 0.9. Metabolomics analysis was explored to explore the beneficial effect of an herbal decoction, *Jian-Pi-Yi-Shen* formula (JPYSF), in an acute kidney injury mouse model [42]. MS metabolomics was applied to analyze metabolic profiling variations in the kidney. The differentially expressed metabolites were identified in the kidney of acute kidney injury (AKI) mice responding to JPYSF treatment. The key pathways affected by JPYSF regulating renal metabolic disorders included alanine, aspartate, and glutamate metabolism, lysine biosynthesis, vitamin B6 metabolism, and butanoate metabolism.

14.3.2 Mechanism Explorations

The advantages of HM are difficult to understand due to the lack of the modern technologic approaches and the necessary molecular mechanism. Metabolomics is a promising new technique for the discovery of novel biomarkers for exploring the potential mechanisms of diverse diseases and assessing the therapeutic effects of HM. A traditional herb Epimedium has a therapeutic effect on osteoporosis (OP). The pharmacological effect of ethanol extract of epimedium (Ext-epi) on OP rat model was investigated to reveal the underlying mechanism by ultra-performance liquid chromatography (UPLC)/Q-time of flight (TOF)/MS metabolomics [43]. A total of 46 potential biomarkers were discovered and 16 potential biomarkers were significantly regulated after intervention with Ext-epi extracts solution. Metabolic pathway included glycerophospholipid and arachidonic acid metabolism as potential targets that are related to the efficacy. A comprehensive metabolomics was used to explore the underlying mechanisms of Suxiao Jiuxin pill (SJP) in an acute myocardial infarction rat model [44]. A total of 112 metabolites were identified from the metabolic profiles in different biological samples. It has comprehensive metabolic changes in galactose metabolism, fatty acid metabolism, and amino acid metabolism, while SJP regulated most of the differential metabolites, mainly affecting the fatty acid metabolism and amino acid metabolism. SJP could alter the metabolic activity of 12 potential biomarkers. It showed that the protective effect mechanism of SJP on cardiovascular disease was associated with the particular regulation of fatty acid and amino acid metabolism. Huangqi Jianzhong Tang (HQJZ) has a good efficacy in treating chronic atrophic gastritis (CAG). A metabolomic approach was applied to reveal the efficacy of HQJZ on the constructed CAG rats [45]. Thirty-five potential biomarkers related to CAG were detected and 32 metabolites were significantly regulated by HQJZ treatment. The metabolic pathways of valine, leucine and isoleucine, and tricarboxylic acid (TCA) cycle were the relevant pathways for HQJZ treatment. The main mechanism of HQJZ might be due to the balance of improvement of the immune system and oxidative stress, energy consumption, and inflammatory inhibition.

Metabolomics was used to investigate the pathological mechanism of post-menopausal osteoporosis and to evaluate the efficacy of total lignans in the stem of *Acanthophanax senticosus* (ASSL) on the ovariectomized osteoporosis model rats [46]. A total of 16 potential biomarkers were finally identified, and 10 biomarkers were found to be significantly regulated by ASSL, mainly including linoleic acid metabolism, unsaturated fatty acid biosynthesis, arachidonic acid metabolism, etc. It showed that the ASSL could affect the endogenous metabolite and provides a pharmacological mechanism basis of the ASSL. UPLC–MS combined with multivariate analysis was used for identification of lipid metabolites and metabolic pathway of diabetic retinopathy and discover the effect of *Keluoxin* capsule [47]. A total of 30 metabolite biomarkers were identified, and the related metabolic pathways included steroid hormone biosynthesis and arachidonic acid metabolism. *Keluoxin* capsule could regulate 22 biomarkers such as prostaglandin D_2, cortisol, γ-linolenic acid, etc., linked with PI3K/MAPK pathway. Serum metabolomics was carried out to reveal the effects of taxifolin on renal fibrosis [48]. It showed that the protective effect of taxifolin on renal fibrosis and a total of 32 metabolic biomarkers were identified. Of note, 27 biomarkers were regulated by taxifolin and participated in eight perturbed metabolic pathways.

14.3.3 Quality Control Analyses

Quality control of HM has always been a hot issue, and due to the ingredient complexity of HM, the current quality standards have problems that are difficult to guarantee the clinical efficacy. It is meaningful to choose the bioactive compounds as quality markers (Q-markers) for ensuring the efficacy to avoid clinical problems of HM. It needs to discover an efficient markers that can assess both quality and pharmacological mechanism of action. Metabolomics has provided a new method to discover quality markers related to the efficacy. MS metabolomics-based discovery method of bioactive compounds was a useful tool for screening the quality markers for HM. The targeted metabolomic profiles of 74 batches of *Sophora flavescens* were determined by UPLC/MS [49]. Multivariate statistical analysis screened six potential markers and were quantified as well as reverse-verified as the effective compounds. Antitumor activity of six flavonoids (kushenol N, kuraridin, trifolirhizin, kurarinone, norkurarinone, and genistein) were determined as the quality markers for *S. flavescens*. Metabolomics coupled with chinmedomics strategy were used to evaluate the effects of an herbal formula *Shengmai San* (SMS) in the treatment of APP/PS1 transgenic model and to discover the efficacy-related Q-markers [50]. SMS had a therapeutic effect on APP/PS1 mice, and 34 potential urine biomarkers were reversed by SMS treatment. A total of 17 *in vivo* constituents were detected, and 8 constituents were extremely correlated with protective effects of SMS and considered as potential Q-markers of SMS.

UPLC/MS was performed to identify the chemical profiling of *Guan-Xin-Jing* capsule (GXJC) and 148 compounds were tentatively identified [51]. Subsequently, the untargeted metabolomics analysis was performed, and a total of 11 compounds in GXJC were screened out as quality markers and lay the foundation for its quality control. Chinmedomics, an upgraded form of metabolomics, has been designed to discover and verify bioactive compounds of HM. In a study, chinmedomics were used to evaluate the efficacy of *Sijunzi* decoction (SJZD) against spleen qi deficiency syndrome (SQDS) to discover the potential Q-markers [52]. A total of 56 and 23 compounds were characterized *in vitro* and *in vivo*, respectively, and 10 constituents of them were associated with overall efficacy absorbed as effective ingredients. Chinmedomics has also discovered the Q-markers of *Panax quinquefolius* (American ginseng) such as pseudoginsenoside F11 and ginsenoside Rd, providing an effective and comprehensive method to clarify the significant ingredients from HM [53].

14.4 Conclusions

HM plays a great role in the treatment of diseases. Many HM and their related traditional treatments for diseases are used throughout the world that represent promising alternatives for the management of diseases. It needs to discover the bioactive compounds as Q-markers that can assess the quality control and effective mechanism. Metabolomics researches on the disease have contributed to explore biomarkers and to understand the progression of the disease at metabolic levels. In addition, in the last decade, a number of metabolomics studies have focused on screening the active ingredients and investigating the action mechanism of various HMs. Thus metabolomics is an effective approach to discover quality markers related to the efficacy.

Acknowledgments

This work was supported by grants from the National Key Research and Development Program of China (2018YFC1706103), Key Program of Natural Science Foundation of State (Grant No. 81830110, 81861168037, 81973745, 81302905), Natural Science Foundation of Heilongjiang Province (YQ2019H030), and Heilongjiang Touyan Innovation Team Program.

References

1 Wang, K.X., Gao, Y., Gong, W.X. et al. (2020). A novel strategy for decoding and validating the combination principles of Huanglian Jiedu decoction from multi-scale perspective. *Frontiers in Pharmacology* 11: 567088.

2 Ren, J.L., Zhang, A.H., and Wang, X.J. (2020). Traditional Chinese medicine for COVID-19 treatment. *Pharmacological Research* 155: 104743.

3 Wang, X., Zhang, A., and Sun, H. (2012). Future perspectives of Chinese medical formulae: chinmedomics as an effector. *OMICS: A Journal of Integrative Biology* 16 (7–8): 414–421.

4 Zhang, A., Sun, H., and Wang, X. (2018). Mass spectrometry-driven drug discovery for development of herbal medicine. *Mass Spectrometry Reviews* 37 (3): 307–320.

5 Xie, J., Zhang, A.H., Qiu, S. et al. (2019). Identification of the perturbed metabolic pathways associating with prostate cancer cells and anticancer affects of obacunone. *Journal of Proteomics* 206: 103447.

6 Zhang, A., Zou, D., Yan, G. et al. (2014). Identification and characterization of the chemical constituents of Simiao Wan by ultra high performance liquid chromatography with mass spectrometry coupled to an automated multiple data processing method. *Journal of Separation Science* 37 (14): 1742–1747.

7 Zhang, A.H., Wang, P., Sun, H. et al. (2013). High-throughput ultra-performance liquid chromatography-mass spectrometry characterization of metabolites guided by a bioinformatics program. *Molecular BioSystems* 9 (9): 2259–2265.

8 Zhang, A., Sun, H., Yuan, Y. et al. (2011). An in vivo analysis of the therapeutic and synergistic properties of Chinese medicinal formula Yin-Chen-Hao-Tang based on its active constituents. *Fitoterapia* 82 (8): 1160–1168.

9 Yan, G.L., Zhang, A.H., Sun, H. et al. (2013). An effective method for determining the ingredients of Shuanghuanglian formula in blood samples using high-resolution LC-MS coupled with background subtraction and a multiple data processing approach. *Journal of Separation Science* 36 (19): 3191–3199.

10 Zhang, A.H., Sun, H., Han, Y. et al. (2013). Ultraperformance liquid chromatography-mass spectrometry based comprehensive metabolomics combined with pattern recognition and network analysis methods for characterization of metabolites and metabolic pathways from biological data sets. *Analytical Chemistry* 85 (15): 7606–7612.

11 Wang, X., Zhang, A., Yan, G. et al. (2014). UHPLC-MS for the analytical characterization of traditional Chinese medicines. *TrAC Trends in Analytical Chemistry* 63: 180–187.

12 Sun, H., Liu, C., Zhang, A.H. et al. (2015). Rapid discovery and global characterization of multiple constituents from Kai-Xin-San using an integrated MS E data acquisition mode strategy based on ultra-performance liquid chromatography coupled to electrospray ionization/quadrupole-time-of-flight mass spectrometry. *Analytical Methods* 7 (1): 279–286.

13 Xue, C., Zhang, A., Sun, H. et al. (2014). An improved ultra-performance liquid chromatography-electrospray ionization/quadrupole-time-of-flight high-definition mass spectrometry method for determining ingredients of herbal Fructus corni in blood samples. *Pharmacognosy Magazine* 10 (40): 422–429.

14 Zhang, A.H., Sun, H., Yan, G.L. et al. (2019). Chinmedomics: a powerful approach integrating metabolomics with serum pharmacochemistry to evaluate the efficacy of traditional Chinese medicine. *Engineering* 5 (1): 60–68.

15 Zhang, A.H., Sun, H., Yan, G.L. et al. (2015). Chinmedomics: a new strategy for research of traditional Chinese medicine. *Zhongguo Zhong Yao Za Zhi* 40 (4): 569–576.

16 Zhang, A., Sun, H., Yan, G. et al. (2014). Metabolomics in diagnosis and biomarker discovery of colorectal cancer. *Cancer Letters* 345 (1): 17–20.

17 Zhang, A., Yan, G., Han, Y. et al. (2014). Metabolomics approaches and applications in prostate cancer research. *Applied Biochemistry and Biotechnology* 174 (1): 6–12.

18 Yang, Q., Zhang, A.H., Miao, J.H. et al. (2019). Metabolomics biotechnology, applications, and future trends: a systematic review. *RSC Advances* 9 (64): 37245–37257.

19 Wang, X., Zhang, A., Han, Y. et al. (2012). Urine metabolomics analysis for biomarker discovery and detection of jaundice syndrome in patients with liver disease. *Molecular & Cellular Proteomics* 11 (8): 370–380.

20 Zhang, A., Fang, H., Wang, Y. et al. (2017). Discovery and verification of the potential targets from bioactive molecules by network pharmacology-based target prediction combined with high-throughput metabolomics. *RSC Advances* 7 (81): 51069–51078.

21 Liang, Q., Liu, H., Xing, H. et al. (2016). High-resolution mass spectrometry for exploring metabolic signatures of sepsis-induced acute kidney injury. *RSC Advances* 6 (36): 29863–29868.

22 Wang, X., Wang, Q., Zhang, A. et al. (2013). Metabolomics study of intervention effects of Wen-Xin-Formula using ultra high-performance liquid chromatography/mass spectrometry coupled with pattern recognition approach. *Journal of Pharmaceutical and Biomedical Analysis* 74: 22–30.

23 Zhang, A.H., Sun, H., Yan, G.L. et al. (2014). Metabolomics study of type 2 diabetes using ultra-performance LC-ESI/quadrupole-TOF high-definition MS coupled with pattern recognition methods. *Journal of Physiology and Biochemistry* 70 (1): 117–128.

24 Tran, N., Pham, B., and Le, L. (2020). Bioactive compounds in anti-diabetic plants: from herbal medicine to modern drug discovery. *Biology* 9 (9): 252.

25 Sõukand, R., Mattalia, G., Kolosova, V. et al. (2020). Inventing a herbal tradition: the complex roots of the current popularity of Epilobium angustifolium in Eastern Europe. *Journal of Ethnopharmacology* 247: 112254.

26 Wang, X.J., Ren, J.L., Zhang, A.H. et al. (2019). Novel applications of mass spectrometry-based metabolomics in herbal medicines and its active ingredients: current evidence. *Mass Spectrometry Reviews* 38 (4–5): 380–402.

27 Zhang, A., Sun, H., Yan, G. et al. (2017). Recent developments and emerging trends of mass spectrometry for herbal ingredients analysis. *TrAC Trends in Analytical Chemistry* 94: 70–76.

28 Sun, H., Zhang, A., and Wang, X. (2012). Potential role of metabolomic approaches for Chinese medicine syndromes and herbal medicine. *Phytotherapy Research* 26 (10): 1466–1471.

29 Chu, H., Zhang, A., Han, Y. et al. (2016). Metabolomics approach to explore the effects of Kai-Xin-San on Alzheimer's disease using UPLC/ESI-Q-TOF mass spectrometry. *Journal of Chromatography B* 1015: 50–61.

30 Wang, X., Zhang, A., Wang, P. et al. (2013). Metabolomics coupled with proteomics advancing drug discovery toward more agile development of targeted combination therapies. *Molecular & Cellular Proteomics* 12 (5): 1226–1238.

31 Zhang, A., Zhou, X., Zhao, H. et al. (2017). Metabolomics and proteomics technologies to explore the herbal preparation affecting metabolic disorders using high resolution mass spectrometry. *Molecular BioSystems* 13 (2): 320–329.

32 Qiu, S., Zhang, A.H., Guan, Y. et al. (2020). Functional metabolomics using UPLC-Q/TOF-MS combined with ingenuity pathway analysis as a promising strategy for evaluating the efficacy and discovering amino acid metabolism as a potential therapeutic mechanism-related target for geniposide against alcoholic liver disease. *RSC Advances* 10 (5): 2677–2690.

33 Zhang, A.H., Ma, Z.M., Kong, L. et al. (2020). High-throughput lipidomics analysis to discover lipid biomarkers and profiles as potential targets for evaluating efficacy of Kai-Xin-San against APP/PS1 transgenic mice based on UPLC-Q/TOF-MS. *Biomedical Chromatography* 34 (2): e4724.
34 Gao, H.L., Zhang, A.H., Yu, J.B. et al. (2018). High-throughput lipidomics characterize key lipid molecules as potential therapeutic targets of Kaixinsan protects against Alzheimer's disease in APP/PS1 transgenic mice. *Journal of Chromatography B: Analytical Technologies in the Biomedical and Life Sciences* 1092: 286–295.
35 Wang, X., Sun, H., Zhang, A. et al. (2011). Potential role of metabolomics approaches in the area of traditional Chinese medicine: as pillars of the bridge between Chinese and Western medicine. *Journal of Pharmaceutical and Biomedical Analysis* 55 (5): 859–868.
36 Zhang, A., Sun, H., Wang, Z. et al. (2010). Metabolomics: towards understanding traditional Chinese medicine. *Planta Medica* 76 (17): 2026–2035.
37 Sun, H., Yang, L., Li, M.X. et al. (2018). UPLC-G2Si-HDMS untargeted metabolomics for identification of metabolic targets of Yin-Chen-Hao-Tang used as a therapeutic agent of dampness-heat jaundice syndrome. *Journal of Chromatography B: Analytical Technologies in the Biomedical and Life Sciences* 1081-1082: 41–50.
38 Yu, J., Kong, L., Zhang, A. et al. (2017). High-throughput metabolomics for discovering potential metabolite biomarkers and metabolic mechanism from the APPswe/PS1dE9 transgenic model of Alzheimer's disease. *Journal of Proteome Research* 16 (9): 3219–3228.
39 Su, L., Mao, J., Hao, M. et al. (2020). Integrated plasma and bile metabolomics based on an UHPLC-Q/TOF-MS and network pharmacology approach to explore the potential mechanism of *Schisandra chinensis*-protection from acute alcoholic liver injury. *Frontiers in Pharmacology* 10: 1543.
40 Wu, J., Li, L., Li, L. et al. (2020). Mechanism of SQQX Decoction's protective effect on SHR: a serum metabolomics-based analysis. *Evidence-based Complementary and Alternative Medicine* 2020: 8856943.
41 Huang, H., Wu, J., Lu, R. et al. (2020). Dynamic urinary metabolomics analysis based on UHPLC-Q-TOF/MS to investigate the potential biomarkers of blood stasis syndrome and the effects of Danggui Sini decoction. *Journal of Pharmaceutical and Biomedical Analysis* 179: 112986.
42 Li, Y., Liu, X., Liu, S. et al. (2020). Untargeted metabolomics reveals the protective effect of a traditional Chinese herbal decoction on cisplatin-induced acute kidney injury. *Evidence-based Complementary and Alternative Medicine* 2020: 8524132.
43 Zhao, J.F., Xu, J.Y., Xu, Y.E. et al. (2020). High-throughput metabolomics method for discovering metabolic biomarkers and pathways to reveal effects and molecular mechanism of ethanol extract from *Epimedium* against osteoporosis. *Frontiers in Pharmacology* 11: 1318.
44 Wu, G., Zhong, J., Chen, L. et al. (2020). Effects of the Suxiao Jiuxin pill on acute myocardial infarction assessed by comprehensive metabolomics. *Phytomedicine* 77: 153291.

45 Liu, Y., Jin, Z., Qin, X. et al. (2020). Urinary metabolomics research for Huangqi Jianzhong Tang against chronic atrophic gastritis rats based on ^{1}H NMR and UPLC-Q/TOF MS. *Journal of Pharmacy and Pharmacology* 72 (5): 748–760.

46 Zhang, A.H., Ma, Z.M., Sun, H. et al. High-throughput metabolomics evaluate the efficacy of total lignans from *Acanthophanax senticosus* stem against ovariectomized osteoporosis rat. *Frontiers in Pharmacology* 10: 553.

47 Ge, N., Kong, L., Zhang, A.H. et al. (2021). Identification of key lipid metabolites during metabolic dysregulation in the diabetic retinopathy disease mouse model and efficacy of Keluoxin capsule using an UHPLC-MS-based non-targeted lipidomics approach. *RSC Advances* 11 (10): 5491–5505.

48 Ren, L., Guo, H.N., Yang, J. et al. (2021). Dissecting efficacy and metabolic characteristic mechanism of *Taxifolin* on renal fibrosis by multivariate approach and ultra-performance liquid chromatography coupled with mass spectrometry-based metabolomics strategy. *Frontiers in Pharmacology* 11: 608511.

49 Chen, L., Huang, X., Wang, H. et al. (2020). Integrated metabolomics and network pharmacology strategy for ascertaining the quality marker of flavonoids for Sophora flavescens. *Journal of Pharmaceutical and Biomedical Analysis* 186: 113297.

50 Zhang, A.H., Yu, J.B., Sun, H. et al. (2018). Identifying quality-markers from Shengmai San protects against transgenic mouse model of Alzheimer's disease using chinmedomics approach. *Phytomedicine* 45: 84–92.

51 Zhang, G., Wang, Q., Liu, X. et al. (2020). An integrated approach to uncover quality markers of traditional Chinese medicine underlying chemical profiling, network target selection and metabolomics approach: Guan-Xin-Jing capsule as a model. *Journal of Pharmaceutical and Biomedical Analysis* 190: 113413.

52 Zhao, Q., Gao, X., Yan, G. et al. (2020). Chinmedomics facilitated quality-marker discovery of Sijunzi decoction to treat spleen qi deficiency syndrome. *Frontiers of Medicine* 14 (3): 335–356.

53 Xiong, H., Zhang, A.H., Zhao, Q.Q. et al. (2020). Discovery of quality-marker ingredients of Panax quinquefolius driven by high-throughput chinmedomics approach. *Phytomedicine* 74: 152928.

15

Mass Spectrometry-Based Metabolomics Insights into the Mode of Action of Natural Products

Junling Ren, Aihua Zhang, Hui Sun, and Xijun Wang

Heilongjiang University of Chinese Medicine, National Chinmedomics Research Center, Functional Metabolomics Laboratory, Department of Pharmaceutical Analysis, Heping Road 24, Harbin 150040, China

15.1 Introduction

Compounds extracted from natural sources such as plants, animals, and microorganisms are defined as natural products (NPs). Many NPs have pharmacological or physiological activities, which are important sources for drug development [1], for example, artemisinin (*Artemisia annua* L.) for malaria; reserpine (*Rauwolfia serpentina* L.), an antihypertensive drug; digoxin (*Digitalis purpurea* L.) for heart failure; Sennae Folium (the dried lobule of *Cassia angustifolia* Vahl or *Cassia acutifolia* Delile) as laxatives; etc. (Figure 15.1). Among all the new molecular entities approved by the US Food and Drug Administration, 25% were derived from plant NPs [2]. Herbal medicine (HM), as a treasure house for NPs, is a gift of nature to humans, up to 80% of the population in many Asian and African countries relies on HMs for primary health care [3]. It is a vital source of inspiration to discover potential new drugs based on their ability to create unique and diverse chemical structures [4, 5]. HMs have therapeutical effects on a variety of diseases, but most mechanisms are unknown primarily because active ingredients are ambiguous. The chemical composition of HMs are very diverse; although the plant kingdom served an important source of chemical entities, valuable information in HM is still waiting to be discovered by modern medical methods.

Recently, metabolomics benefits from the rapid development of mass spectrometry (MS) and has carried out extensive research on various aspects of NPs. Like other booming technologies, it still experiences a series of optimization and methodology development that are essential for improvement [6, 7]. Thus, MS-based metabolomics is a valuable tool and a new opportunity for HMs in the drug discovery field. Here, we highlighted the advantages of MS-based metabolomics, current research status and challenges of NPs, and the application of MS in the analysis of herbal components, pharmacodynamic evaluation, quality markers (Q-markers) discovery, toxicity explores, and potential impact on drug discovery are also elaborated.

Mass Spectrometry-Based Metabolomics in Clinical and Herbal Medicines: Strategies, Technologies, and Applications, First Edition. Edited by Aihua Zhang and Wanying Wang.

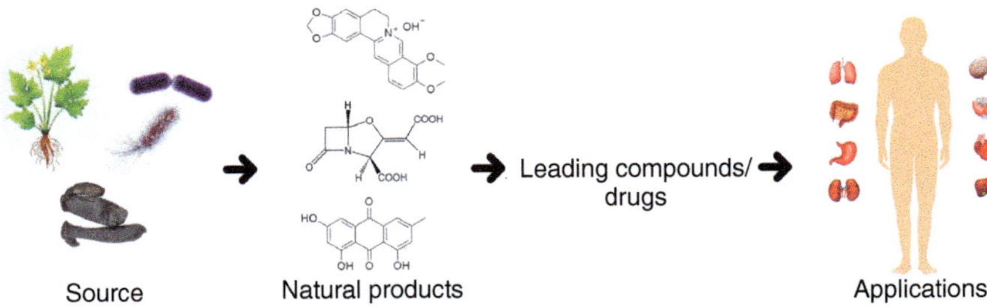

Figure 15.1 Natural products as the main leading compounds or drug sources for the treatment of a variety of diseases.

15.2 Metabolomics: An Overview

Metabolomics, defined as the study of small molecules (<1000 Da) produced or consumed within biological systems, is a booming omics field of post-genomic research that provides a holistic approach to explore the molecular changes caused by a variety of factors in the entire organism [6, 8, 9]. Metabolites are regarded as more representative of phenotypes because they exhibit faster changes than that of genetic materials [10, 11]. Metabolomics has been widely used and considered to be a useful technique for biomarker identification [12, 13], disease diagnosis [14, 15], drug discovery [16], drug mechanism study [17, 18], and drug safety evaluation [19, 20].

15.2.1 Comprehensive Metabolomics for Metabolites Detection

Untargeted metabolomics is characterized from a large quantity of metabolites across a sample set, without prior knowledge of any components. Scanning mode was usually employed to identify all detected peaks and to obtain the statistical analysis with the annotated and non-annotated peak information. Comprehensive analysis of global metabolites supplies a better approach for the detection of novel metabolites. The detection of these metabolites was usually carried out with MS and nuclear magnetic resonance (NMR) [21, 22]. Thus, numerous studies have become popular to analyze the same sample with MS and NMR [23, 24].

15.2.2 Understanding Pharmacological Effects

For thousands of years, people realized that some plants have medicinal properties. Although the chemical constituents of NPs have often been identified, their pharmacological effects are not well understood. Metabolomics, preferred as an integral, dynamic, and comprehensive, coincides with the systematic perspective of HM, thus widely applied to the research of HMs [25]. It is also an ideal tool to bridge the gap between HMs and molecular pharmacology [26]. In recent years, metabolomics techniques have been employed to study pharmacological effects. Obacunone, a highly oxygenated triterpenoids NP, has been commonly found in Citrus. Xie et al. adopted a functional metabolomics strategy to understand the pharmacological effects of obacunone on prostate cancer and showed that obacunone effectually controls proliferation and promotes apoptosis in 22RV1 prostate cancer cells, which were related with nicotinate and tryptophan metabolism, nicotinamide metabolism, phenylalanine metabolism, and ascorbate metabolism [27]. A serum metabolomics strategy was also used to explore the hepatoprotective activity of geniposide, a kind of iridoid glycoside NP compound, and it showed that the MS-based metabolomics was a potentially powerful tool to discover amino acid metabolism as mechanism-related target against alcoholic liver disease [28].

15.2.3 Discovery of Biomarkers and Targets

Metabolomics as a fast, sensitive, and non-invasive tool investigates metabolic changes in biological systems and provides metabolic fingerprints of small molecules related to disease [29]. The characteristic of biomarkers and pathway analysis has a great potential to classify diseases. The number of such metabolomics research is increasing exponentially in virtue of the urgent need to discover sensitive and specific biomarkers that can be used for early diagnosis of diseases to improve the survival rate. Metabolomics combined with multivariate statistical analysis allowed us to study the perturbations of metabolic pathways under different physiological conditions [30]. In addition, metabolomics can indicate potential targets for therapeutic intervention, especially promising enzymatic targets. Understanding the molecular targets is important for toxicology studies and the optimization of lead compounds [31], all of which are important sections of drug discovery and development.

15.2.4 Screen Bioactive Natural Products

The discovery and development model of bioactive NPs has changed from the phytochemistry determination to the use of advanced analytical techniques, high-throughput genomics, and metabolomics as well as high-content screening [32]. Metabolomics provides an efficacious method to discover active ingredients from HMs. An untargeted metabolomics analysis was capable to predict the action modes of NP constituents from the extract library [33]. Fourteen compounds were screened out based on the MS information from the NP extract library, and quinocinnolinomycins, a new family of NP, has a unique carbon skeleton that can cause endoplasmic reticulum stress.

15.3 Superiority of MS-Based Metabolomics

Electrospray ionization (ESI) accelerated the commercialization of MS, has been accepted in the analytical field rapidly, and has been gradually applied into metabolomics and NPs. In view of the power of MS-based metabolomics, metabolomics is becoming increasingly popular in molecular medicines, drug development, HMs, and other biotechnology fields. It can directly indicate phenotype and variation compared with other omics techniques.

15.3.1 Enhanced MS Coverage for Small-Molecule Metabolites

Continuous improvement in analytical technology has also contributed significantly to expand the coverage of polar and nonpolar metabolites. For instance, gas chromatography (GC) × GC–MS, which is derived from the basis of GC and increased peak sensitivity and capacity, can detect approximately sevenfold more metabolites than GC–MS in the same serum sample [34]. Moreover, the emergence of ultra performance liquid chromatography (UPLC)–MS can produce about 20% more metabolite detection than that of high performance liquid chromatography (HPLC)–MS [35]. In ion mobility–mass spectrometry (IM–MS) platform, the peak capacity can be raised by threefold compared with individual MS analysis [36]. The updated MS technology leads to a broader domain of metabolites.

15.3.2 MS-Based Metabolomics Guided Biochemical Analysis

As the incidence of various diseases increases, the burden of disease in society is aggravated. At present, the diagnosis of most diseases still depends on traditional technologies, such as biopsy and computed tomography, which are usually invasive and frequent checks cause body damage. The maximum potential of metabolomics for medical care lies in the detection and identification of biomarkers [37]. MS-based metabolomics provides a great potential for noninvasive diagnosis and supplies unique insights into the disturbance of metabolic pathways [38]. The continuous development of MS-based metabolomics has promoted the discovery of disease biomarkers and the study of metabolic pathways.

15.3.3 Identify Molecular Mechanisms by Integrating MS with Databases

Metabolite identification is still time-consuming and challenging. For the identification of "known" metabolites, which often search in accessible databases, it is accurate for "known" metabolites identification [39]. By contrast, the identification of the "unknown" metabolites is one of the major challenges in metabolomics. In view of this problem, the database has undergone tremendous evolution. Kyoto encyclopedia of Genes and Genomes (HMDB) is the most complete collection of human metabolites and metabolism [40]. The most recent version (HMDB 4.0) contains about 114 265 metabolites to greatly expand the recognition of the metabolites. Other widely used database, such as METLIN [41], NIST [42], and MassBank [43], also aided the metabolite identification. There were also other databases, such as Serum Metabolome databases [44] and ReSpect database, which can be conducive to identification of phytochemicals [45].

15.4 Mass Spectrometry Techniques Employed in Metabolomics

Despite the commonly employed analytical techniques in metabolomics, MS and NMR, in most cases, MS is preferred due to its high resolution, incomparable sensitivity, and structural specificity. In practical applications, MS is combined with different separation techniques such as GC, liquid chromatography (LC), capillary electrophores (CE), or ion mobility spectrometry (IMS) for a pre-separation [46]. The vigorous development of mass spectrometry imaging (MSI) technology has widened the cognition of the spatial distribution of not only metabolites but also NPs, to add one more dimension to expand cognition.

15.4.1 Gas Chromatography–Mass Spectrometry

The earliest research of metabolite profiling study was published in the early 1970s by Devaux. GC–MS was used to analyze multi-component of steroids and urinary drug metabolites [47]. GC–MS, as a repeatable, inexpensive, and high peak-capacity method, is especially suitable for the detection of thermo-stable volatile compounds, which requires complex and time-consuming sample preparation, and in this process, the structure of analytes might change. Wei et al. developed a GC–MS method to investigate the urine metabolites characteristics of systemic lupus erythematosus and therapeutic effects of Jiedu Quyu Ziyin prescription [48]. A total of 11 metabolites were identified and related to amino acid metabolism and energy metabolism. After treatment with Jiedu Quyu Ziyin prescription, the levels of metabolites were close to the control group to achieve the therapeutic effect.

15.4.2 Liquid Chromatography–Mass Spectrometry

LC–MS has a higher sensitivity than GC–MS, allowing analysis of thermally unstable nonvolatile compounds. Polar sugars, various nonpolar lipids, and non-aromatic organic acids can be detected with LC–MS. Another reason that LC–MS has such an overwhelming popularity is largely due to a series of interfaces that operate at ESI and atmospheric pressure chemical ionization (APCI). LC–MS supplies a robust analytical platform for the analysis of small molecules, thus being extensively used to study metabolomics.

Yin Chen Hao Tang (YCHT) is a classic HM used for the treatment of dampness–heat jaundice syndrome. In order to explore the therapeutic effect of YCHT, Sun et al. used UPLC–MS-based technology to reveal metabolite alteration during the treatment [49]. As a result, 22 biomarkers and 8 metabolic pathways were identified. The biomarker–protein interaction network was also set up to predict potential targets. Zhu et al. adopted UPLC–MS to explore the effect mechanism and targets of *Huanglian Jiedu* decoction [50]. Nineteen biomarkers were characterized and related with 10 metabolic pathways. The glutamate/γ-aminobutyric acid (GABA)–glutamine cycle regulation and metabolic stress modulation were recognized as the main pathways to relate with the pharmacological action of *Huanglian Jiedu* decoction. These studies showed that UPLC–MS-based metabolomics can be used to comprehensively elucidate the mechanisms and targets of HMs.

15.4.3 Capillary Electrophoresis–Mass Spectrometry

CE is usually 1 order of magnitude higher than that in conventional HPLC [51]. It is especially suitable to separate charged and polar compounds [52]. CE–MS could get similar or even better separation efficiency than that of GC–MS or LC–MS.

Kimura et al. performed a CE–MS metabolic profiling of end-stage kidney disease with plasma from a prospective cohort with 112 patients [53]. A total of 218 metabolites were detected, and 16 of which have predictive values that relate to the high risk of end-stage kidney disease. The measurement of these metabolites might facilitate chronic kidney disease management for predicting the risk of progression to end-stage kidney disease.

15.4.4 Ion Mobility–Mass Spectrometry

IMS is a gas-phase electrophoretic technique that can rapidly separate ions according to size, shape, and charge state [54]. Collisional cross section (CCS) values directly reflect ionic size and configuration and can facilitate metabolite identification [55]. The combination of IM with MS provides not only mass information but also multidimensional and structural resolution within milliseconds to improve identification and coverage depth of metabolites [55]. Lipids are the main components of cell membranes and play important biological roles in life. Bowman et al. broadly explored IM–MS to separate lipid isomers, with a 75% success rate among four main types of phospho- and glycerolipids [56]. Chen et al. developed a combinational strategy for accelerating lipid identification and quantification by IM–MS [57].

15.4.5 Mass Spectrometry Imaging

For the analysis of biological tissues within LC–MS-based metabolomics, the first step is sample homogenization to extract metabolites [58]. MSI is emerging in the field of "spatial metabolomics." It can visualize the location, distribution, and production of metabolites in an intact biological samples [59]. Matrix-assisted laser desorption/ionization (MALDI) MSI is currently the most mature imaging technique. Tissues are sectioned into specified thickness, transferred to the MALDI plate, covered with matrix, and laser raster the tissue, and MS spectra is then obtained. After processing, peak intensities can be reconstructed into two-dimensional images; thus, metabolite changes can be displayed intuitively. MSI techniques have opened up new avenues in the life sciences and have become powerful supplements to classical metabolomics [60]. Sarabia et al. established a high-resolution MALDI MSI platform to reveal the detailed spatial distribution of metabolites in the root of barley seedlings in response to salinity stress [61]. As a response to salt stress, phosphatidylcholine showed an overall reduction in salt-treated roots. It allows the researchers to study plants and their metabolites during growth from a spatial perspective.

15.4.6 New Challenges in MS-Based Metabolomics

Although metabolomics is a mature technology, it still lacks standardized processes, including collection of samples to metabolite identification, and data obtained from different laboratories cannot be shared and verified. Standardization process should be well established to enhance the repeatability of experimental data. Due to the multitudinous type and number of metabolites detected, coupled with the lack of standards, the concentration of metabolites is quite different, and all of these lead to the quantitative difficulties. In addition, the identification of metabolites is largely dependent on database, and when the identification of the metabolites is not loaded in the database, it is often difficult to determine, which is not conducive to the discovery of new metabolites. Because each analytical method has its own advantages, more information about metabolites can be obtained through the combined use of various analytical techniques.

15.5 Challenges of Natural Products

Although some NPs have been successfully transformed into drugs, in recent years, NPs have also been controversial and challenged. The lack of modern scientific and clinical evidence of safety, efficacy, and mechanism of action also made some NPs not widely accepted [62]. The identified bioactive ingredients faced with mass production issues, limited supply of raw materials, and other issues. HM as the main source of NP has been used to prevent and treat diseases for thousands of years. It has accumulated the pharmacological information and medication experience to form an integrated and unique theoretical system. The multiple active components of herbs or formulae could provide better therapeutic effects. Such complexity in HMs brings great challenges to modernization of HMs. These challenges have seriously hampered the integration of NPs into modern health-care system, which need to be solved in the future studies.

15.6 Herbal Ingredients: Current Research and Applications

Artemisinin, a famous NP found in Artemisiae Annuae Herba, brought a breakthrough in the treatment of tropical parasitic diseases; its known bioactive ingredients from HMs played a vital role in drug discovery [63]. Therefore, Tu Youyou won the 2015 Nobel Prize in medicine [64].

15.6.1 Various Analytical Strategies for Herbal Ingredients

HMs are usually comprised of multi-components to exert their efficacy, and systematic identification of herbal ingredients is the basis for the research of pharmacological effect [65]. The diversity of the chemical structure of HMs determines that the analysis of herbal ingredients is a time-consuming and laborious process [66]. Thus, it is imperative to develop sensitive methods for HM analysis. MS offers the mass of compounds, and structural information can be gleaned from the MS/MS fragments. NMR provides information about the chemical connectivity through NMR-active isotopes [67]. If sufficient sample is available, MS has a much higher level of sensitivity than that of NMR [24, 68]. The coverage of metabolites was remarkably increased by combining these two methods. Hybrid MS and NMR methods are expected as an additional or even an alternative to the existing protocols in MS-based analysis of herbal ingredients.

15.6.2 Effective Constituents *In Vitro* and *In Vivo*

The complexity of chemical components of HMs increases the difficulty of screen bioactive constituents. The bioactive components can be absorbed into the blood first and then undergo complex metabolic and distribution processes before the therapeutic effect is achieved [69]. Therefore, *in vivo* and *in vitro* chemical component analysis is the basis to discover the active NPs. *Acanthopanacis senticosi* Radix et Rhizoma seu Caulis (dried root and rhizome, or stem of *Acanthopanax senticosus* [Rupr. Et Maxim.] Harms) has progressively declined to cause an ecological crisis. An optimized UPLC–MS method was developed to analyze effective constituents in leaf extracts [70]. In a study, 131 chemical compounds were identified *in vitro* and 21 compounds were absorbed into blood; among them, 12 prototypes and 9 metabolites were characterized. It showed that UPLC–MS could rapidly and comprehensively analyze the chemical components of HMs *in vivo*. *Kai-Xin-San* is mainly used to treat neurasthenia and Alzheimer's disease [71, 72]. Wang et al. had applied UPLC–MS to screen the effective constituents *in vitro* and *in vivo* [73]. A total of 108 compounds were identified *in vitro* and 69 components were characterized *in vivo*, of which were 47 prototype and 22 metabolites.

15.7 MS in Natural Product

Modern analytical technologies allow qualitative and quantitative measurement of a large number of NPs in complex systems. Among these techniques, MS is preferred because of its unparalleled sensitivity, ultra-high resolution, and structural specificity.

15.7.1 Qualitative Analysis

Because an HM might contain hundreds of NPs, it is necessary to determine as many compounds as possible to discover bioactive ingredients, quality control of medicinal materials, and clinical applications. Qualitative analysis of herbal ingredients includes confirmation of known compounds and clarification of unknown structures. HM prescriptions gained international popularity in health promotion and adjuvant therapy [74, 75]. The identification of chemical constituents of prescription helps to clarify the basis of the therapeutic effect and to ensure quality. UPLC–MS was established for ingredient analysis of *A. Senticosi* Radix et Rhizoma Seu Caulis fruit [76]. A total of 104 compounds were identified, including flavones, lignans, phenolic acids, etc., and 7 of them were the first time reported. Polygalae Radix has the effect to improve intelligence, eliminate phlegm, and disperse swelling of the carbuncle [77]. Liu et al. employed UPLC–Q-TOF–MS to characterize the constituents of *Polygalae* Radix; 35 compounds were affirmed, and 10 of them were the first discovery [78]. This qualitative analysis study will be conducive to the clinical application and will lead to compound discovery of *Polygalae* Radix. Li et al. investigated the multiple constituents in the 80% methanol extract of processed products of Ligustri Lucidi Fructus, and luteolin-7-*O*-rutinoside, secologanoside, and oleoside were firstly detected in Ligustri Lucidi Fructus [79].

15.7.2 Quantitative Analysis

Quantitative determination of the bioactive ingredients is also an important aspect of the quality evaluation of HMs. Because of the alteration of HMs contents in different species, origins, and harvest seasons, it is essential to quantitatively analyze the main components to guarantee the safety and effectiveness of each HM [80]. The continuous development of MS makes the quantitative analysis of HMs simple and feasible. An HPLC–MS method was successfully established for quantitative analysis of 16 bioactive components varied among 18 samples from different regions and revealed the quality difference of *Cyclocarya paliurus* [81]. *Bupi Yishen Formula* is clinically applied for the recovery and delay of chronic kidney disease. After the chemical component analysis, seven main ingredients were simultaneously analyzed, which provided the chemical basis for the evaluation of active components in *Bupi Yishen Formula* [82]. *Xuebijing Injection* is an HM formula widely used in the treatment of sepsis. Zuo et al. employed ultra-high performance liquid chromatography (UHPLC)–MS to rapidly confirm and quantify 30 bioactive components [83]. Moreover, senkyunolide I and hydroxysafflor yellow A were selected as quality markers.

15.7.3 Mechanism Analysis of Action

With MS advantages, researchers can bring insight into the therapeutic effect from single NP or multi-component and multi-target of HMs, as well as to elucidate the mechanism of drug action. Metabolomic analysis suggested berberine could regulate the disordered metabolism [84]. Li et al. investigated the dynamic changes of metabolite profiles in cell and mouse models of prostate cancer to reveal a therapeutic mechanism of berberine [85]. Seven metabolites were identified with MS, which are involved in linoleic acid metabolism, arginine and proline metabolism, purine metabolism, and retinol metabolism. An MS-based metabolomics platform was established by Zhang et al. to elucidate the potential mechanism of geniposide toward alcohol-induced liver injury [86]. Geniposide reduced alcohol-induced liver injury mainly by multiple amino acid metabolism, pyrimidine, and purine metabolism. With thorough analysis of biomarkers and clinical chemistry, geniposide was considered to reduce oxidative stress and correct neurotransmitter abnormality. Moreover, by down-regulation of the expression of *N*-acylethanolamine-hydrolyzing acid amidase in synovial tissue, geniposide could further decrease the content of palmitoylethanolamines [87], which plays important roles in inflammatory stimulation [88]. Scoparone is a representative ingredient of *Artemisia capillaris* Thunb, which has an effect on various liver diseases [89]. Fang et al. studied the mechanism of scoparone and its efficacy on metabolic levels with MS-based metabolomics [90]. Differential metabolites were characterized for the therapeutic effects of scoparone on *Yang Huang* syndrome. Thirty-three biomarkers, accurately identified with MS, played vital roles in the regulation of key metabolic pathways related to disease.

15.8 MS-Based Metabolomics in Bioactive Ingredients

In response to the growing interest in NPs, various strategies have improved the discovery of bioactive ingredients. Metabolomics combined with MS can nicely be applied to such cases to speed up the identification of bioactive ingredients and can capture differences in the content of active ingredients (Figure 15.2).

15.8.1 Reveal Efficacy of HMs

Metabolomics can not only detect the metabolites but can also determine the metabolic pathways and molecular targets to clarify the efficacy of HMs from a deeper perspective. Zhang et al. explored the phenotypic characteristics of the metabolic signatures in the "Kidney yang deficiency syndrome" (KYDS) with *Nanshi Oral Liquid* (NOL), for interpreting the therapeutic effect of NOL on KYDS [91]. The 3-methyldioxyindole, 4,6-dihydroxyquinoline, 3-metanephrine, taurine, kynurenic acid, tyramine, etc. were significantly changed and related to disturbances of steroid–hormone biosynthesis, lysine degradation, hypotaurine metabolism, phenylalanine metabolism, and tyrosine metabolism. It indicated that NOL can

Figure 15.2 Application of MS-based metabolomics analysis in bioactive ingredients.

alleviate the kidney impairment, and that those biomarkers were also regulated after NOL treatment.

15.8.2 High-Throughput Analysis of HMs

The constant development of MS-metabolomics has invigorated a new interest in high-throughput analysis of HMs because it breaks through the previous restrictions of separation and extraction of complex mixtures in one analysis. The AS1350 has a protective effect against KYDS, whereas the therapeutic mechanism, bioactive ingredients, and potential targets of AS1350 are not identified. Liu et al. employed MS-based metabolomics to elucidate the previously mentioned issues [92]. Forty-eight biomarkers were identified, which were associated with amino acid metabolism, alpha-linolenic acid metabolism, phospholipid metabolism, fatty acid metabolism, and sphingolipid metabolism. The correlation coefficient between herbal components *in vivo* and biomarkers was both analyzed, and nine active components were identified. MS-based metabolomics was established by Liu et al. to explore the key biomarkers, pathways, mechanism, bioactive compounds, and potential targets of *Nanshi Capsule* in the treatment of KDYS [93]. Fifty-five urine and serum biomarkers were identified, which relate to 15 metabolic pathways, and 8 active components were characterized. Potential target-prediction showed that 11 potential targets were correlated with these bioactive compounds.

15.8.3 Discovery of Quality Markers

HMs have wide sources and multiple producing areas that result in significant differences in quality, especially the number of active ingredients [94]. It is noteworthy that the quality of HMs determines its effectiveness and safety. The establishment of qualitative and quantitative determination methods is the basis of quality standard research. In order to improve the consistency and controllability of HMs, the concept "Q-marker" was proposed to help establish quality standards for herbs [95]. Since the concept of Q-marker has been put forward, the strategy of MS-based metabolomics was especially widely used to discover Q-markers. Ginseng has the effect to strengthen energy, enhance intelligence and body function, and promote longevity [96]. Huang et al. employed UHPLC–Q-TOF MS and multivariate statistical analysis to obtain the metabolite data and discover a Q-marker [97]. It indicated that 2- to 6-year-old ginsengs can be well separated in the OPLS-DA model. Thirty-six Q-markers were discovered as Q-markers. *Sijunzi* decoction (SJZD) is used for the treatment of spleen *qi* deficiency syndrome (SQDS). Chinmedomics strategy has been used for the discovery of bioactive ingredients from HM (Figure 15.3). Zhao et al. applied chinmedomics strategy to evaluate the efficacy and discover the potential Q-marker for its quality control [98]. Finally, SJZD can regulated a large proportion of biomarkers of SQDS, and 23 compounds of SJZD were discovered *in vivo*, of these bioactive components, ginsenoside Ro, malonyl-ginsenoside Rb2, dehydrotumulosic acid, glycyrrhizic acid, dihydroxy lanostene-triene-21-acid, glycyrrhetinic acid, isoglabrolide, and 2-atractylenolide were identified as Q-markers for the quality control of SJZD.

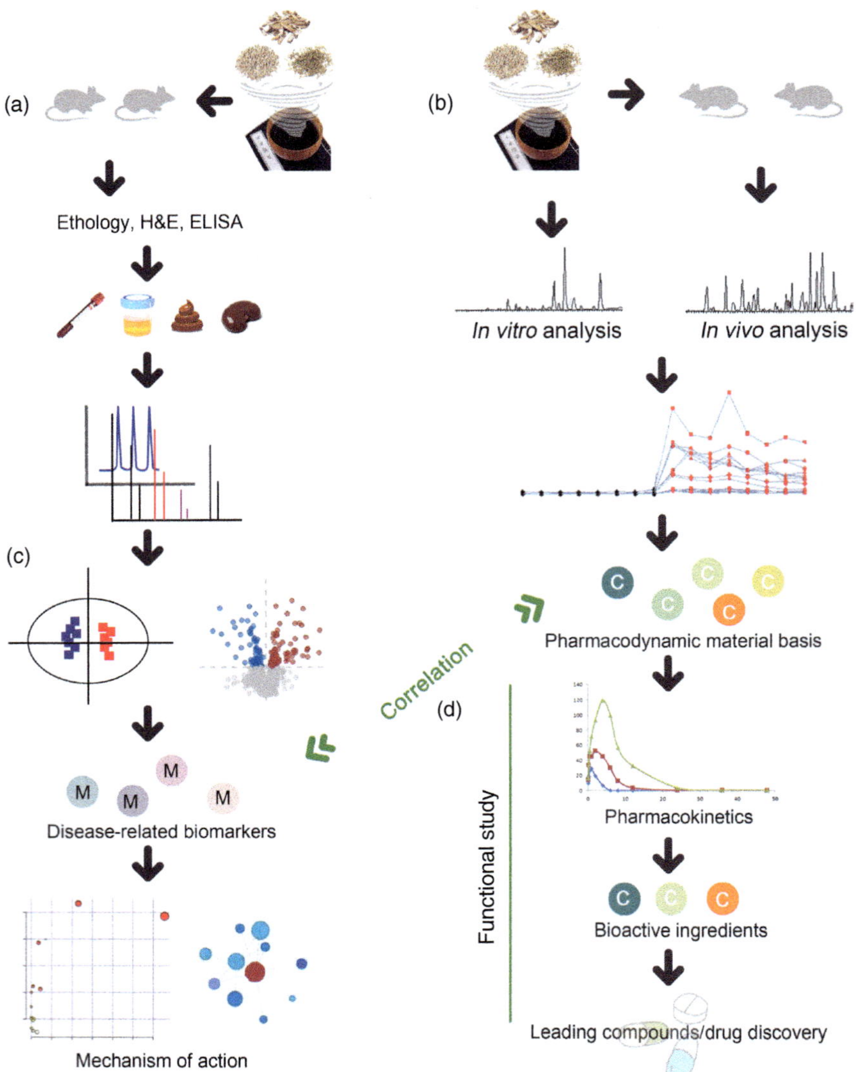

Figure 15.3 Schematic diagram of chinmedomics strategy for the discovery of bioactive ingredients from herbal medicine. (a) Classical clinical methods (e.g. Enzyme-linked immuno sorbent assay (ELISA), ethology, H&E) were used to evaluate the efficacy of herbal medicine; in addition, MS-based metabolomics strategy was employed to depict metabolic profiling and identify biomarkers, disturbed metabolic pathways, and potential targets, thus exploring the action mechanism. (b) UPLC–MS combined with the corresponding databases were used to identify the chemical constituents of herbal medicines *in vitro*. After animal models were administrated with herbal medicines, compounds *in vivo* were studied by UPLC–MS combined with the corresponding databases to identify effective components. (c) The correlation between chemical components *in vivo* and diseases-related biomarkers was analyzed to discover pharmacodynamic material basis. (d) Through pharmacokinetic and functional studies, candidate leading compounds were finally screened out.

15.8.4 Toxicity Explorations

The safety issues of NPs have been raised, including the preparation methods, administration schemes, the possibility of interaction with other herbs and conventional drugs, and assurances that herbal products are authentic and do not contain toxins or pollutants [99]. If the toxic ingredients, concentration, and mechanism of action are uncertain, the use of NPs might be unhealthy, or even fatal. Therefore, the effective methods for assessing NP toxicity were required, and metabolomics seems to be ideal for this purpose. Great efforts have been made to investigate the toxicity of aristolochic acid by employing the strategy of metabolomics. It was suggested that energy metabolism, gut microbiota, and purine metabolism were associated with aristolochic acid-induced nephrotoxicity [100]. Besides, by inhibiting amino acids metabolism, β-oxidation of fatty acids, glucose metabolism, and aristolochic acid can also induce testicular toxicity [101]. Aconiti Kusnezoffii Radix, the dried tuberous root of *Aconitum kusnezoffii* Reichb., possessed highly bioactive as well as highly toxic character for medicinal use. Yunnan Baiyao is being extensively used around the world, which contains Aconiti Kusnezoffii Radix, and suffered to the controversy. From the perspective of metabolomics, five core toxicity biomarkers were regulated by *Yunnan Baiyao*, and lysine degradation was considered to be the core metabolic pathway of detoxification of *Yunnan Baiyao* [19].

15.8.5 Potential Impacts on Drug Discovery

Metabolomics has certified its value in the discovery process of NPs and provided molecular data for drug discovery. YCHT has been used to treat liver disorders [102]. First, MS-based metabolomics was employed to clarify treatment efficacy. After treatment with YCHT, the contents of the differential metabolites were closed to control group and indicated the therapeutic effect of YCHT, and pathway analysis showed that pyrimidine metabolism and primary bile acid biosynthesis might relate with YCHT effect [103]. It also showed remarkable effects to prevent *liver* injury progression in dampness–heat jaundice syndrome [104]. Second, by establishing an UPLC–MS platform, 45 chemical compounds in YCHT and 21 compounds in serum after oral administration of YCHT were identified [105], and 8 ingredients have hepatoprotective activities. Third, a sensitive UPLC–MS method was used to monitor the pharmacokinetic time-course of 21 compounds absorbed *in vivo* [106]. 6,7-Dimethoxycoumarin, geniposide, and rhein were delineated as bioactive ingredients. Last, Qiu et al. dissected geniposide efficacy with metabolomics [107]. Zhang et al. also presented a lipidomics strategy to determine the changes of lipids after scoparone exposure in primary hepatocytes, and key lipids were identified that elucidated scoparone effects [108]. These findings demonstrated that scoparone can regulate the lipid alteration to achieve protective effects on liver cells. Zhang et al. also explored the mechanism of scoparone protective function against alcohol-induced hepatotoxicity [65]. Based on the previously mentioned studies, the active ingredients in YCHT were discovered, and the pharmacokinetics and pharmacological effects of these active ingredients were explored. It not only preliminarily confirmed the pharmaceutical action of these bioactive components but also affirmed the role of MS-based metabolomics in NP discovery.

15.9 Conclusions and Future Perspective

Current evidence shows that NPs could be beneficial for the treatment of various diseases. However, the complexities present in chemical compositions of plant source are currently an obstacle for the progression of NPs, which involve unclear bioactive compounds, mechanisms of action, undetermined targets for therapy, non-specific features for drug metabolism, etc. To overcome these issues, metabolomics could solve these scientific difficulties with NPs from a metabolic perspective.

In recent years, MS-based metabolomics was widely applied for the analysis of NPs *in vivo* and *in vitro*. The progress of analytical techniques along with the development of data analysis methodologies has led to the extensive application of MS-based metabolomics in HM fields. Although metabolomics is a relatively new field of science, it has exhibited considerable potential in NPs research or even revolutionized the study of NPs. MS-based metabolomics has certified its value to enhance discovery of bioactive ingredients, as well as to provide molecular data for the development of NP discovery in a high-throughput screening processes. In addition to characterizing the bioactive ingredients and understanding the mechanism of action, metabolomics also confers a powerful tool to assess quality of HMs to improve the safety.

Despite the rapid development of MS coupled with GC, LC, or other separation technologies, there are still some problems that need to be solved in future NPs research. The main challenges that impact MS-based metabolomics on NP ingredients are the technical and biological challenges to improve coverage and our understanding of active ingredient functions, respectively. Isotope labeling is a powerful way to improve annotation of NPs and to provide useful information about pathway elucidation [109]. Although MS-metabolomics holds great promise in the research of NPs, expanded analytical abilities are urgently needed to obtain a fully integrated analysis of metabolic networks with a true biological significance. MS-based metabolomics is still in a continuously growing period and has the ability to become a valuable tool for NP research.

Acknowledgment

This study was supported by the National Natural Science Foundation of China (Grant No. 81903818, 81973745, 81302905), the University Nursing Program for Young Scholars with Creative Talents in Heilongjiang Province (UNPYSCT-2020224, UNPYSCT-2015118), and Foundation of Heilongjiang University of Chinese Medicine (2018bs02, 2018jc01).

References

1 Baker, D.D., Chu, M., Oza, U., and Rajgarhia, V. (2007). The value of natural products to future pharmaceutical discovery. *Natural Product Reports* 24 (6): 1225–1244.

2 Patridge, E., Gareiss, P., Kinch, M.S., and Hoyer, D. (2016). An analysis of FDA-approved drugs: natural products and their derivatives. *Drug Discovery Today* 21 (2): 204–207.

3 Commisso, M., Strazzer, P., Toffali, K. et al. (2013). Untargeted metabolomics: an emerging approach to determine the composition of herbal products. *Computational and Structural Biotechnology Journal* 4: e201301007.

4 Xie, T., Song, S., Li, S. et al. (2015). Review of natural product databases. *Cell Proliferation* 48 (4): 398–404.

5 Yuliana, N.D., Khatib, A., Choi, Y.H., and Verpoorte, R. (2011). Metabolomics for bioactivity assessment of natural products. *Phytotherapy Research* 25 (2): 157–169.

6 Jang, C., Chen, L., and Rabinowitz, J.D. (2018). Metabolomics and isotope tracing. *Cell* 173 (4): 822–837.

7 Rinschen, M., M., Ivanisevic, J., Giera, M., and Siuzdak, G. (2019). Identification of bioactive metabolites using activity metabolomics. *Nature Reviews Molecular Cell Biology* 20 (6): 353–367.

8 Newgard, C.B. (2017). Metabolomics and metabolic diseases: where do we stand? *Cell Metabolism* 25 (1): 43–56.

9 Wishart, D.S. (2019). Metabolomics for investigating physiological and pathophysiological processes. *Physiological Reviews* 99 (4): 1819–1875.

10 Cui, L., Lu, H., and Lee, Y.H. (2018). Challenges and emergent solutions for LC-MS/MS based untargeted metabolomics in diseases. *Mass Spectrometry Reviews* 37 (6): 772–792.

11 Hu, L., Liu, J., Zhang, W. et al. (2020). Functional metabolomics decipher biochemical functions and associated mechanisms underlie small-molecule metabolism. *Mass Spectrometry Reviews* 39 (5, 6): 417–433.

12 Guijas, C., Montenegro-Burke, J.R., Warth, B. et al. (2018). Metabolomics activity screening for identifying metabolites that modulate phenotype. *Nature Biotechnology* 36 (4): 316–320.

13 Montaner, J., Ramiro, L., Simats, A. et al. (2020). Multilevel omics for the discovery of biomarkers and therapeutic targets for stroke. *Nature Reviews Neurology* 16 (5): 247–264.

14 Costello, E. (2018). A metabolomics-based biomarker signature discriminates pancreatic cancer from chronic pancreatitis. *Gut* 67 (1): 2–3.

15 Badhwar, A., McFall, G.P., Sapkota, S. et al. (2020). A multiomics approach to heterogeneity in Alzheimer's disease: focused review and roadmap. *Brain: A Journal of Neurology* 143 (5): 1315–1331.

16 Wishart, D.S. (2016). Emerging applications of metabolomics in drug discovery and precision medicine. *Nature Reviews Drug Discovery* 15 (7): 473–484.

17 Beyer, B.A., Fang, M., Sadrian, B. et al. (2018). Metabolomics-based discovery of a metabolite that enhances oligodendrocyte maturation. *Nature Chemical Biology* 14 (1): 22–28.

18 Matsui, T., Omuro, H., Liu, Y.F. et al. (2017). Astrocytic glycogen-derived lactate fuels the brain during exhaustive exercise to maintain endurance capacity. *Proceedings of the National Academy of Sciences of the United States of America* 114 (24): 6358–6363.

19 Ren, J.L., Dong, H., Han, Y. et al. (2020). Network pharmacology combined with metabolomics approach to investigate the protective role and detoxification mechanism of Yunnan Baiyao formulation. *Phytomedicine: International Journal of Phytotherapy and Phytopharmacology* 77: 153266.

20 Ren, J.L., Sun, H., Dong, H. et al. (2019). A UPLC-MS-based metabolomics approach to reveal the attenuation mechanism of Caowu compatibility with Yunnan Baiyao. *RSC Advances* 9: 8926–8933.

21 Vavricka, C.J., Hasunuma, T., and Kondo, A. (2020). Dynamic metabolomics for engineering biology: accelerating learning cycles for bioproduction. *Trends in Biotechnology* 38 (1): 68–82.

22 Xu, W.J., Wen, H., Kim, H.S. et al. (2018). Observation of acetyl phosphate formation in mammalian mitochondria using real-time in-organelle NMR metabolomics. *Proceedings of the National Academy of Sciences of the United States of America* 115 (16): 4152–4157.

23 Gathungu, R.M., Kautz, R., Kristal, B.S. et al. (2020). The integration of LC-MS and NMR for the analysis of low molecular weight trace analytes in complex matrices. *Mass Spectrometry Reviews* 39 (1, 2): 35–54.

24 Bingol, K. and Bruschweiler, R. (2015). NMR/MS translator for the enhanced simultaneous analysis of metabolomics mixtures by NMR spectroscopy and mass spectrometry: application to human urine. *Journal of Proteome Research* 14 (6): 2642–2648.

25 Wang, M., Chen, L., Liu, D. et al. (2017). Metabolomics highlights pharmacological bioactivity and biochemical mechanism of traditional Chinese medicine. *Chemico-Biological Interactions* 273: 133–141.

26 Qiu, J. (2007). 'Back to the future' for Chinese herbal medicines. *Nature Reviews Drug Discovery* 6 (7): 506–507.

27 Xie, J., Zhang, A.H., Qiu, S. et al. (2019). Identification of the perturbed metabolic pathways associating with prostate cancer cells and anticancer affects of obacunone. *Journal of Proteomics* 206: 103447.

28 Qiu, S., Zhang, A.H., Guan, Y. et al. (2020). Functional metabolomics using UPLC-Q/TOF-MS combined with ingenuity pathway analysis as a promising strategy for evaluating the efficacy and discovering amino acid metabolism as a potential therapeutic mechanism-related target for geniposide against alcoholic liver disease. *RSC Advances* 10: 2677–2690.

29 Fessenden, M. (2016). Metabolomics: small molecules, single cells. *Nature* 540 (7631): 153–155.

30 Han, Y., Sun, H., Zhang, A. et al. (2020). Chinmedomics, a new strategy for evaluating the therapeutic efficacy of herbal medicines. *Pharmacology & Therapeutics* 216: 107680.

31 Isgut, M., Rao, M., Yang, C. et al. (2018). Application of combination high-throughput phenotypic screening and target identification methods for the discovery of natural product-based combination drugs. *Medicinal Research Reviews* 38 (2): 504–524.

32 Newman, D. (2017). Screening and identification of novel biologically active natural compounds. *F1000Research* 6: 783.

33 Kurita, K.L., Glassey, E., and Linington, R.G. (2015). Integration of high-content screening and untargeted metabolomics for comprehensive functional annotation of natural product libraries. *Proceedings of the National Academy of Sciences of the United States of America* 112 (39): 11999–12004.

34 Miyazaki, T., Okada, K., Yamashita, T., and Miyazaki, M. (2017). Two-dimensional gas chromatography time-of-flight mass spectrometry-based serum metabolic fingerprints of neonatal calves before and after first colostrum ingestion. *Journal of Dairy Science* 100 (6): 4354–4364.

35 Nordstrom, A., O'Maille, G., Qin, C., and Siuzdak, G. (2006). Nonlinear data alignment for UPLC-MS and HPLC-MS based metabolomics: quantitative analysis of endogenous and exogenous metabolites in human serum. *Analytical Chemistry* 78 (10): 3289–3295.

36 Arthur, K.L., Turner, M.A., Reynolds, J.C., and Creaser, C.S. (2017). Increasing peak capacity in nontargeted omics applications by combining full scan field asymmetric waveform ion mobility spectrometry with liquid chromatography-mass spectrometry. *Analytical Chemistry* 89 (6): 3452–3459.

37 Nicholson, J.K., Holmes, E., Kinross, J.M. et al. (2012). Metabolic phenotyping in clinical and surgical environments. *Nature* 491 (7424): 384–392.

38 Khamis, M.M., Adamko, D.J., and El-Aneed, A. (2017). Mass spectrometric based approaches in urine metabolomics and biomarker discovery. *Mass Spectrometry Reviews* 36 (2): 115–134.

39 Bingol, K., Bruschweiler-Li, L., Li, D. et al. (2016). Emerging new strategies for successful metabolite identification in metabolomics. *Bioanalysis* 8 (6): 557–573.

40 Wishart, D.S., Feunang, Y.D., Marcu, A. et al. (2018). HMDB 4.0: the human metabolome database for 2018. *Nucleic Acids Research* 46 (D1): D608–D617.

41 Guijas, C., Montenegro-Burke, J.R., Domingo-Almenara, X. et al. (2018). METLIN: a technology platform for identifying knowns and unknowns. *Analytical Chemistry* 90 (5): 3156–3164.

42 Simon-Manso, Y., Lowenthal, M.S., Kilpatrick, L.E. et al. (2013). Metabolite profiling of a NIST Standard Reference Material for human plasma (SRM 1950): GC-MS, LC-MS, NMR, and clinical laboratory analyses, libraries, and web-based resources. *Analytical Chemistry* 85 (24): 11725–11731.

43 Horai, H., Arita, M., Kanaya, S. et al. (2010). MassBank: a public repository for sharing mass spectral data for life sciences. *Journal of Mass Spectrometry* 45 (7): 703–714.

44 Psychogios, N., Hau, D.D., Peng, J. et al. (2011). The human serum metabolome. *PLoS One* 6 (2): e16957.

45 Sawada, Y., Nakabayashi, R., Yamada, Y. et al. (2012). RIKEN tandem mass spectral database (ReSpect) for phytochemicals: a plant-specific MS/MS-based data resource and database. *Phytochemistry* 82: 38–45.

46 Ren, J.L., Zhang, A.H., Kong, L., and Wang, X.J. (2018). Advances in mass spectrometry-based metabolomics for investigation of metabolites. *RSC Advances* 8: 22335.

47 Devaux, P., Horning, M., and Horning, E. (1971). Benzyloxime derivatives of steroids. A new metabolic profile procedure for human urinary steroids human urinary steroids. *Analytical Letters* 4: 151–160.

48 Wei, H., Miao, H., Yun, Y. et al. (2014). Validation of an LC-MS/MS method for quantitative analysis of the 5 bioactive components of Wuzhi capsule in human plasma samples. *Therapeutic Drug Monitoring* 36 (6): 781–788.

49 Sun, H., Yang, L., Li, M.X. et al. (2018). UPLC-G2Si-HDMS untargeted metabolomics for identification of metabolic targets of Yin-Chen-Hao-Tang used as a therapeutic agent of dampness-heat jaundice syndrome. *Journal of Chromatography B: Analytical Technologies in the Biomedical and Life Sciences* 1081, 1082: 41–50.

50 Zhu, B., Cao, H., Sun, L. et al. (2018). Metabolomics-based mechanisms exploration of Huang-Lian Jie-Du decoction on cerebral ischemia via UPLC-Q-TOF/MS analysis on rat serum. *Journal of Ethnopharmacology* 216: 147–156.

51 Kuehnbaum, N.L. and Britz-McKibbin, P. (2013). New advances in separation science for metabolomics: resolving chemical diversity in a post-genomic era. *Chemical Reviews* 113 (4): 2437–2468.

52 Wolfender, J.L., Marti, G., Thomas, A., and Bertrand, S. (2015). Current approaches and challenges for the metabolite profiling of complex natural extracts. *Journal of Chromatography A* 1382: 136–164.

53 Kimura, T., Yasuda, K., Yamamoto, R. et al. (2016). Identification of biomarkers for development of end-stage kidney disease in chronic kidney disease by metabolomic profiling. *Scientific Reports* 6: 26138.

54 Hinz, C., Liggi, S., and Griffin, J.L. (2018). The potential of ion mobility mass spectrometry for high-throughput and high-resolution lipidomics. *Current Opinion in Chemical Biology* 42: 42–50.

55 Zhang, X., Quinn, K., Cruickshank-Quinn, C. et al. (2018). The application of ion mobility mass spectrometry to metabolomics. *Current Opinion in Chemical Biology* 42: 60–66.

56 Bowman, A.P., Abzalimov, R.R., and Shvartsburg, A.A. (2017). Broad separation of isomeric lipids by high-resolution differential ion mobility spectrometry with tandem mass spectrometry. *Journal of the American Society for Mass Spectrometry* 28 (8): 1552–1561.

57 Chen, X., Yin, Y., Zhou, Z. et al. (2020). Development of a combined strategy for accurate lipid structural identification and quantification in ion-mobility mass spectrometry based untargeted lipidomics. *Analytica Chimica Acta* 1136: 115–124.

58 Sheikh, K.D., Khanna, S., Byers, S.W. et al. (2011). Small molecule metabolite extraction strategy for improving LC/MS detection of cancer cell metabolome. *Journal of Biomolecular Techniques* 22 (1): 1–4.

59 Parrot, D., Papazian, S., Foil, D., and Tasdemir, D. (2018). Imaging the unimaginable: desorption electrospray ionization–imaging mass spectrometry (DESI–IMS) in natural product research. *Planta Medica* 84 (9, 10): 584–593.

60 Pareek, V., Tian, H., Winograd, N., and Benkovic, S.J. (2020). Metabolomics and mass spectrometry imaging reveal channeled de novo purine synthesis in cells. *Science* 368 (6488): 283–290.

61 Sarabia, L.D., Boughton, B.A., Rupasinghe, T. et al. (2018). High-mass-resolution MALDI mass spectrometry imaging reveals detailed spatial distribution of metabolites and lipids in roots of barley seedlings in response to salinity stress. *Metabolomics: Official Journal of the Metabolomic Society* 14 (5): 63.

62 Izzo, A.A., Hoon-Kim, S., Radhakrishnan, R., and Williamson, E.M. (2016). A critical approach to evaluating clinical efficacy, adverse events and drug interactions of herbal remedies. *Phytotherapy Research* 30 (5): 691–700.

63 Zhang, A.H., Sun, H., Yan, G.L., and Wang, X.J. (2017). Recent developments and emerging trends of mass spectrometry for herbal ingredients analysis. *Trends in Analytical Chemistry* 94: 70–76.

64 Liu, W. and Liu, Y. (2016). Youyou Tu: significance of winning the 2015 Nobel Prize in Physiology or Medicine. *Cardiovascular Diagnosis and Therapy* 6 (1): 1–2.

65 Zhang, A., Sun, H., and Wang, X. (2014). Urinary metabolic profiling of rat models revealed protective function of scoparone against alcohol induced hepatotoxicity. *Scientific Reports* 4: 6768.

66 Zhang, A., Sun, H., and Wang, X. (2018). Mass spectrometry-driven drug discovery for development of herbal medicine. *Mass Spectrometry Reviews* 37 (3): 307–320.

67 Bradshaw, J., Butina, D., Dunn, A.J. et al. (2001). A rapid and facile method for the dereplication of purified natural products. *Journal of Natural Products* 64 (12): 1541–1544.

68 Bingol, K., Bruschweiler-Li, L., Yu, C. et al. (2015). Metabolomics beyond spectroscopic databases: a combined MS/NMR strategy for the rapid identification of new metabolites in complex mixtures. *Analytical Chemistry* 87 (7): 3864–3870.

69 Wang, X., Zhang, A., Zhou, X. et al. (2016). An integrated chinmedomics strategy for discovery of effective constituents from traditional herbal medicine. *Scientific Reports* 6: 18997.

70 Zhang, Y., Zhang, A., Zhang, Y. et al. (2016). Application of ultra-performance liquid chromatography with time-of-flight mass spectrometry for the rapid analysis of constituents and metabolites from the extracts of *Acanthopanax senticosus* Harms leaf. *Pharmacognosy Magazine* 12 (46): 145–152.

71 Chu, H., Zhang, A., Han, Y. et al. (2016). Metabolomics approach to explore the effects of Kai-Xin-San on Alzheimer's disease using UPLC/ESI-Q-TOF mass spectrometry. *Journal of Chromatography B: Analytical Technologies in the Biomedical and Life Sciences* 1015, 1016: 50–61.

72 Wang, X.J., Zhang, A.H., Kong, L. et al. (2019). Rapid discovery of quality-markers from Kaixin San using chinmedomics analysis approach. *Phytomedicine: International Journal of Phytotherapy and Phytopharmacology* 54: 371–381.

73 Wang, X., Liu, J., Yang, X. et al. (2018). Development of a systematic strategy for the global identification and classification of the chemical constituents and metabolites of Kai-Xin-San based on liquid chromatography with quadrupole time-of-flight mass spectrometry combined with multiple data-processing approaches. *Journal of Separation Science* 41 (12): 2672–2680.

74 Wu, H., Guo, J., Chen, S. et al. (2013). Recent developments in qualitative and quantitative analysis of phytochemical constituents and their metabolites using liquid chromatography-mass spectrometry. *Journal of Pharmaceutical and Biomedical Analysis* 72: 267–291.

75 Yang, B., Dong, W., Zhang, A. et al. (2011). Ultra-performance liquid chromatography coupled with electrospray ionization/quadrupole-time-of-flight mass spectrometry for rapid analysis of constituents of Suanzaoren decoction. *Journal of Separation Science* 34 (22): 3208–3215.

76 Han, Y., Zhang, A., Sun, H. et al. (2017). High-throughput ultra high performance liquid chromatography combined with mass spectrometry approach for the rapid analysis and characterization of multiple constituents of the fruit of *Acanthopanax senticosus* (Rupr. et Maxim.) Harms. *Journal of Separation Science* 40 (10): 2178–2187.

77 Feng, G.F., Liu, S., Pi, Z.F. et al. (2018). Studies on the chemical and intestinal metabolic profiles of Polygalae Radix by using UHPLC-IT-MS(n) and UHPLC-Q-TOF-MS method coupled with intestinal bacteria incubation model in vitro. *Journal of Pharmaceutical and Biomedical Analysis* 148: 298–306.

78 Liu, C., Zhang, A., Yan, G.L. et al. (2017). High-throughput ultra high performance liquid chromatography coupled to quadrupole time-of-flight mass spectrometry method for the rapid analysis and characterization of multiple constituents of Radix Polygalae. *Journal of Separation Science* 40 (3): 663–670.

79 Li, H., Yao, W., Liu, Q. et al. (2017). Application of UHPLC-ESI-Q-TOF-MS to identify multiple constituents in processed products of the herbal medicine Ligustri Lucidi Fructus. *Molecules* 22 (5): 689–703.

80 Zhao, J., Ma, S.C., and Li, S.P. (2018). Advanced strategies for quality control of Chinese medicines. *Journal of Pharmaceutical and Biomedical Analysis* 147: 473–478.

81 Cao, Y., Fang, S., Yin, Z. et al. (2017). Multicomponent quantitative analysis for the quality evaluation of *Cyclocarya paliurus* leaves by HPLC-Q-TOF-MS. *Molecules* 22 (11): 1927–1943.

82 Zhang, J., Xu, W., Wang, P. et al. (2018). Chemical analysis and multi-component determination in Chinese medicine preparation Bupi Yishen formula using ultra-high performance liquid chromatography with linear ion trap-orbitrap mass spectrometry and triple-quadrupole tandem mass spectrometry. *Frontiers in Pharmacology* 9: 568.

83 Zuo, L., Sun, Z., Hu, Y. et al. (2017). Rapid determination of 30 bioactive constituents in XueBiJing injection using ultra high performance liquid chromatography-high resolution hybrid quadrupole-orbitrap mass spectrometry coupled with principal component analysis. *Journal of Pharmaceutical and Biomedical Analysis* 137: 220–228.

84 Qin, X., Jiang, M., Zhao, Y. et al. (2020). Berberine protects against diabetic kidney disease via promoting PGC-1alpha-regulated mitochondrial energy homeostasis. *British Journal of Pharmacology* 177 (16): 3646–3661.

85 Li, X., Zhang, A., Sun, H. et al. (2017). Metabolic characterization and pathway analysis of berberine protects against prostate cancer. *Oncotarget* 8 (39): 65022–65041.

86 Zhang, T., Zhang, A., Qiu, S. et al. (2016). High-throughput metabolomics approach reveals new mechanistic insights for drug response of phenotypes of geniposide towards alcohol-induced liver injury by using liquid chromatography coupled to high resolution mass spectrometry. *Molecular BioSystems* 13 (1): 73–82.

87 Zhan, X., Wu, H., and Wu, H. (2020). Joint synovial fluid metabolomics method to decipher the metabolic mechanisms of adjuvant arthritis and Geniposide intervention. *Journal of Proteome Research* 19 (9): 3769–3778.

88 Impellizzeri, D., Esposito, E., Di Paola, R. et al. (2013). Palmitoylethanolamide and luteolin ameliorate development of arthritis caused by injection of collagen type II in mice. *Arthritis Research & Therapy* 15 (6): R192.

89 Liu, X. and Zhao, X. (2017). Scoparone attenuates hepatic stellate cell activation through inhibiting TGF-beta/Smad signaling pathway. *Biomedicine & Pharmacotherapy* 93: 57–61.

90 Fang, H., Zhang, A., Yu, J. et al. (2016). Insight into the metabolic mechanism of scoparone on biomarkers for inhibiting Yanghuang syndrome. *Scientific Reports* 6: 37519.

91 Zhang, A., Liu, Q., Zhao, H. et al. (2016). Phenotypic characterization of nanshi oral liquid alters metabolic signatures during disease prevention. *Scientific Reports* 6: 19333.

92 Liu, Q., Zhang, A., Wang, L. et al. (2016). High-throughput chinmedomics-based prediction of effective components and targets from herbal medicine AS1350. *Scientific Reports* 6: 38437.

93 Liu, Q., Zhao, H.W., Zhang, A.H. et al. (2016). Chinmedomics strategy to discover effective constituents and elucidate action mechanism of Nanshi capsule against kidney-yang deficiency syndrome. *Zhongguo Zhong yao za zhi = Zhongguo zhongyao zazhi = China Journal of Chinese Materia Medica* 41 (15): 2901–2914.

94 Wong, M. (2010). Herbs and spices: unexpected sources of antiepileptogenic drug treatments? *Epilepsy Currents* 10 (1): 21–23.

95 Liu, C., Chen, S., Zhang, T. et al. (2016). A new concept on quality marker of Chinese materia medica: quality control for Chinese medicinal products. *Chinese Traditional and Herbal Drugs* 47: 1443–1457.

96 He, Y., Yang, J., Lv, Y. et al. (2018). A review of ginseng clinical trials registered in the WHO international clinical trials registry platform. *BioMed Research International* 2018: 1843142.

97 Huang, B.M., Zha, Q.L., Chen, T.B. et al. (2018). Discovery of markers for discriminating the age of cultivated ginseng by using UHPLC-QTOF/MS coupled with OPLS-DA. *Phytomedicine: International Journal of Phytotherapy and Phytopharmacology* 45: 8–17.

98 Zhao, Q., Gao, X., Yan, G. et al. (2020). Chinmedomics facilitated quality-marker discovery of Sijunzi decoction to treat spleen qi deficiency syndrome. *Frontiers of Medicine* 14 (3): 335–356.

99 Chen, D.Q., Chen, H., Chen, L. et al. (2016). Metabolomic application in toxicity evaluation and toxicological biomarker identification of natural product. *Chemico-Biological Interactions* 252: 114–130.

100 Hu, X., Shen, J., Pu, X. et al. (2017). Urinary time- or dose-dependent metabolic biomarkers of aristolochic acid-induced nephrotoxicity in rats. *Toxicological Sciences: An Official Journal of the Society of Toxicology* 156 (1): 123–132.

101 Cui, Y., Han, J., Ren, J. et al. (2019). Untargeted LC-MS-based metabonomics revealed that aristolochic acid I induces testicular toxicity by inhibiting amino acids metabolism, glucose metabolism, beta-oxidation of fatty acids and the TCA cycle in male mice. *Toxicology and Applied Pharmacology* 373: 26–38.

102 Wang, X., Zhang, A., Wang, P. et al. (2013). Metabolomics coupled with proteomics advancing drug discovery toward more agile development of targeted combination therapies. *Molecular & Cellular Proteomics* 12 (5): 1226–1238.

103 Sun, H., Zhang, A.H., Zou, D.X. et al. (2014). Metabolomics coupled with pattern recognition and pathway analysis on potential biomarkers in liver injury and hepatoprotective effects of yinchenhao. *Applied Biochemistry and Biotechnology* 173 (4): 857–869.

104 Cao, H.X., Sun, H., Jiang, X.G. et al. (2009). Comparative study on the protective effects of Yinchenhao Decoction against liver injury induced by alpha-naphthylisothiocyanate and carbon tetrachloride. *Chinese Journal of Integrative Medicine* 15 (3): 204–209.

105 Wang, X., Sun, W., Sun, H. et al. (2008). Analysis of the constituents in the rat plasma after oral administration of Yin Chen Hao Tang by UPLC/Q-TOF-MS/MS. *Journal of Pharmaceutical and Biomedical Analysis* 46 (3): 477–490.

106 Wang, X., Sun, H., Zhang, A. et al. (2011). Pharmacokinetics screening for multi-components absorbed in the rat plasma after oral administration traditional Chinese medicine formula Yin-Chen-Hao-Tang by ultra performance liquid chromatography-electrospray ionization/quadrupole-time-of-flight mass spectrometry combined with pattern recognition methods. *Analyst* 136 (23): 5068–5076.

107 Qiu, S., Zhang, A., Zhang, T. et al. (2017). Dissect new mechanistic insights for geniposide efficacy on the hepatoprotection using multiomics approach. *Oncotarget* 8 (65): 108760–108770.

108 Zhang, A., Qiu, S., Sun, H. et al. (2016). Scoparone affects lipid metabolism in primary hepatocytes using lipidomics. *Scientific Reports* 6: 28031.

109 Wang, Z. and Jones, A.D. (2014). Profiling of stable isotope enrichment in specialized metabolites using liquid chromatography and multiplexed nonselective collision-induced dissociation. *Analytical Chemistry* 86 (21): 10600–10607.

16

Mass Spectrometry-Driven Lipidomics for Biomarker, Molecular Mechanism, and Therapy

Qiang Yang, Shi Qiu, Aihua Zhang, and Xijun Wang

Heilongjiang University of Chinese Medicine, National Chinmedomics Research Center, Functional Metabolomics Laboratory, Heping Road 24, Harbin 150040, China

16.1 Introduction

Lipids are a basic component of biofilms and a metabolite of the body, playing a crucial role in cellular energy storage, composition, and signaling [1–3]. Lipids mainly consist of fats and oils. Fat is an essential part of an animal's body, which is mainly concerned with storing and supplying energy, maintaining body temperature, and protecting internal organs. Lipids include phospholipids, sugar lipids, cholesterol, and its esters. Phospholipids, as the main component of biofilm, mainly participate in maintaining the stability of biofilm. Glycolipids are complex lipids based on carbohydrate residues, whose role is mainly involved in signal recognition and immunity [4–6]. Cholesterol and cholesterol ester are precursors of hormones, vitamins, and pigments and are mainly engaged in signal transmission and enzyme activators [7–9]. Lipidomics is a new discipline for systematic analysis of intact lipids. By comparing changes in lipid metabolism networks under different physiological conditions, it can identify key lipid biomarkers in metabolic regulation and ultimately reveal lipids in various life activities' mechanism of action. In the last decade, with the advancement of mass spectrometry (MS) technology, it has been widely used in the identification of disease lipid biomarkers, disease diagnosis, drug targets, and the discovery of lead compounds [10–12]. This chapter summarizes biotechnologies of lipidomics based on mass spectrometry and highlights the latest applications of lipidomics as a powerful tool for the discovery of lipid biomarkers and disease diagnosis (Figure 16.1).

16.2 Lipidomics Method

With the deepening of research in biology and medicine, the rapid development of omics technology from the very beginning to the present shows the importance of omics technology [13–15]. At the same time, the existing science and technology

Mass Spectrometry-Based Metabolomics in Clinical and Herbal Medicines: Strategies, Technologies, and Applications, First Edition. Edited by Aihua Zhang and Wanying Wang.

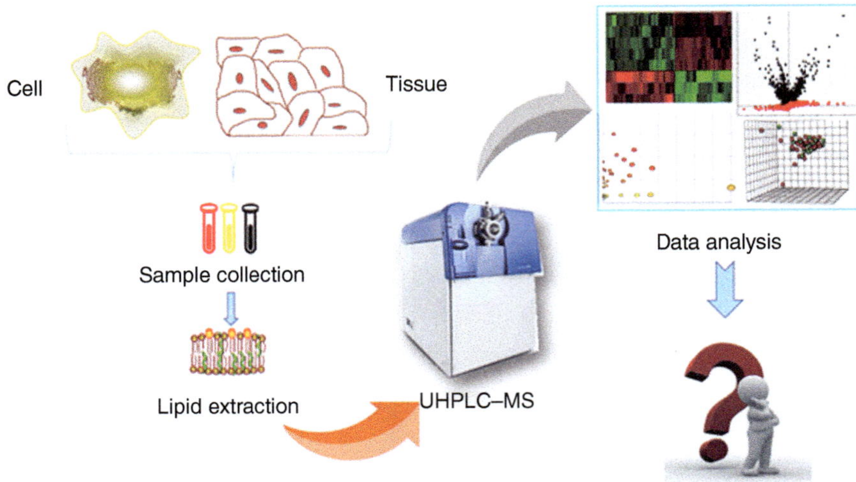

Figure 16.1 A flowchart of lipidomics analysis technology.

cannot provide a clear, accurate, and comprehensive explanation of the mechanism of disease, but can only provide some references for disease prevention and treatment. Fortunately, omics technology has been developed in recent years, playing an important role in elucidating the pathogenesis and mechanism of some diseases, the study of disease biomarkers, and the role of drug components in the metabolic processes in the body [16, 17]. Metabolomics technology is closely related to the fields studied by other group schools, often intermingling in the process of research, and more is the exchange of multiple disciplines. As one of the important branches of metabolomics technology, the lipidomics technology gradually shows its unique vitality in the research of disease lipid biomarkers and disease diagnosis [18–20]. Lipidomics technology has been gradually applied to various fields of medical research, such as the research of metabolic syndrome, vascular diseases, and neuropsychiatric diseases, and has achieved very significant effects.

In the process of cell–molecule research, scientists often divide the direction of analysis and elaboration into targeted and non-targeted. Non-target therapy has gradually become a more common method in daily medical care, and often drug therapy belongs to non-targeted therapy. Targeted therapy, as a research hotspot in recent years, often focuses on one point to achieve the goals of remission and treatment of diseases [21, 22]. According to the research direction, metabolomics can be divided into targeted metabolomics and non-targeted metabolomics. The lipidomics technology is also divided into targeted lipidomics and non-targeted lipidomics in the research process. It will become more important to clearly and accurately clarify changes in lipids in the body.

16.2.1 Targeted Lipidomics

Targeted lipidomics technology is mainly used for accurate qualitative and quantitative analysis of one or more types of target lipids. It has high sensitivity and high specificity for low abundance lipids and specific pathways in metabolic pathways [23]. Lipid testing has significant advantages and is often used for the analysis or validation of key metabolic pathways or targets. Based on the targeted lipidomics method of liquid chromatography–mass spectrometry (LC–MS)/MS and multivariate statistical analysis, quantitative research on lysophosphatidylcholine, including the evaluation of multiple lysophosphatidylcholine indexes, can be used to distinguish different cancer types [24]. The targeted lipidomics plays an important role in the early prediction of type 2 diabetes and depression [25, 26].

16.2.2 Non-targeted Lipidomics

Non-targeted lipidomics is the unbiased analysis of changes in overall lipids in biological samples, which reflect the changes and trends of lipids in samples to the greatest extent, so as to discover the biomarkers for further diagnosis of diseases. Psoriasis is a joint chronic inflammatory skin disease that is complicated by the interaction of genes with the environment. Although a series of omics studies have been performed, the role of lipid metabolites is unclear. Non-targeted lipidomics techniques were used to explore the lipid profile of psoriasis and to characterize its lipid metabolites. The differential expression of some phospholipids in blood lipids was revealed, which provided new ideas for further exploring the role of lipids in psoriasis [27]. Untargeted lipidomics was also used to assess the effect of vitamin D3 on postprandial lipid status in obese and vitamin D-deficient women. After taking vitamin D3, plasma sphingomyelin levels decrease. It showed that sphingomyelin conversion may be due to an increase in neutral sphingomyelinase activity, which affects a series of physiological changes, potentially increasing the likelihood of cardiovascular disease [28]. Non-targeted lipidomics has a wide range of applications in the diagnosis of diseases. The main purpose is to explain the changes in disease lipids and explore the biomarkers and provide directions for in-depth research on the mechanism of action.

16.3 Sample Preparation

The process of lipidomics sample preparation is an important step in lipidomics research. Preparation of lipid samples includes extraction, pretreatment, and separation processes. Sample extraction is the first stage in sample preparation. In lipidomics research, liquid-phase extraction and solid-phase extraction are mostly used for preliminary separation and extraction of samples. Different extraction methods have different advantages and disadvantages. To choose the correct extraction method according to the nature of the sample is a prerequisite for lipid research [29–31].

Samples usually require a large amount of pretreatment before they can be analyzed. Due to the complexity and variety of lipid molecules and the large differences in chemical properties, this results in significant differences in ionization efficiency during the detection process and ion suppression effects. Lipid internal standard

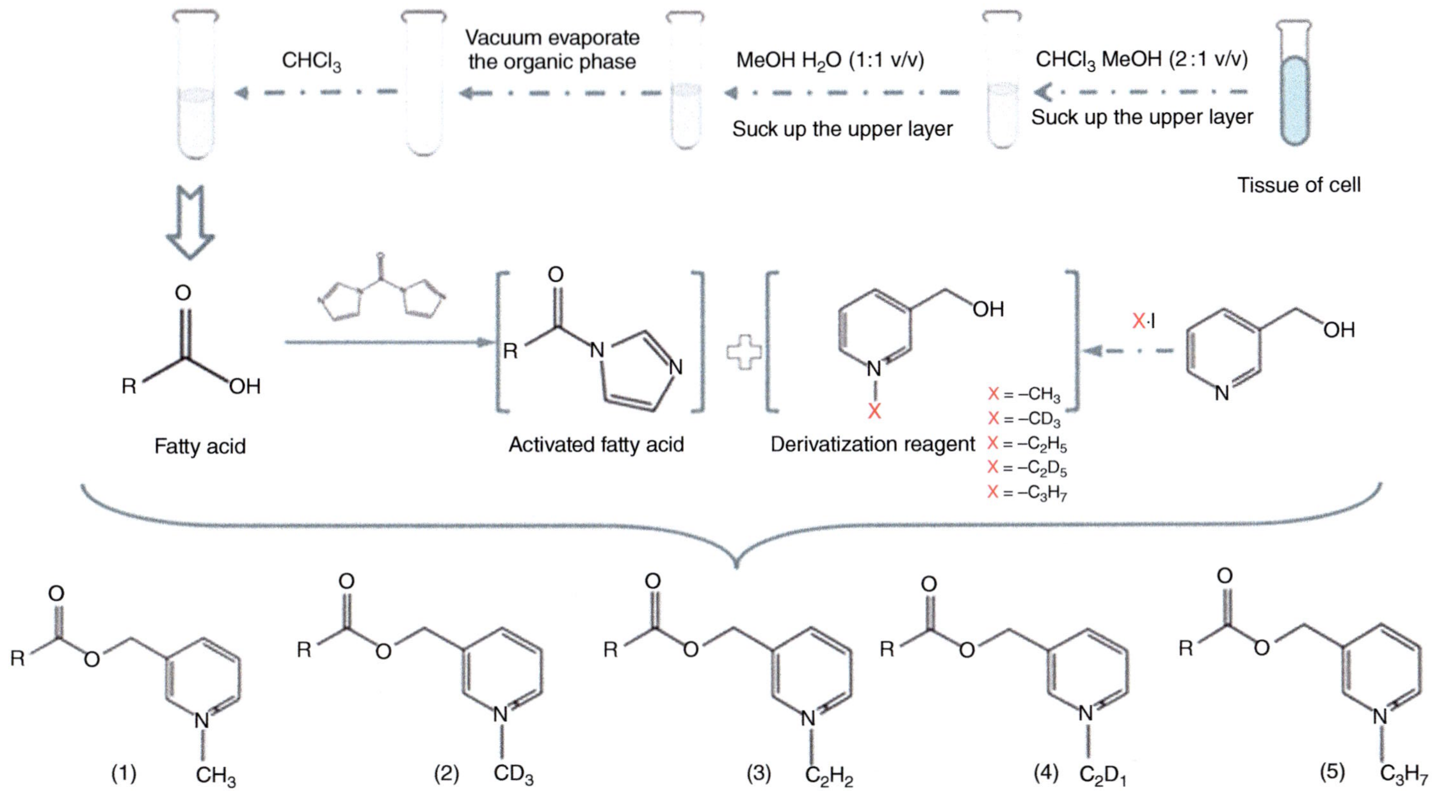

Figure 16.2 Fatty acid extraction and derivatization process.

molecules for quantitative analysis are not easy to obtain, and there are a large number of isomers, which lead to difficulties in qualitative and quantitative analysis of lipids. Choosing the right derivatization method is one of the important ways to solve the previously mentioned problems. Lipid derivatization is mainly targeted at double bonds and lipid polar head groups in lipids. Different derivatization methods can be applied to different positions of lipids. However, the ultimate purpose of derivatization is to improve the ionization efficiency, detection sensitivity, and stability of the compound structure and even to distinguish isomers of lipids (Figure 16.2) [32–34].

The chemical derivatization method can significantly improve the lipidomics analysis performance of the "shotgun lipidomics" or "LC–MS," improve the chromatographic resolution of lipids in the LC–MS analysis, and improve the MS in the shotgun method. The detected mass-to-charge ratio distribution improves MS ionization efficiency and MS/MS structure analysis capabilities [35, 36]. However, while chemical derivatization methods are introduced, it is easy to introduce new interference. How to choose the best derivatizing agent is very important in the pretreatment process of lipid analysis.

16.4 Analytical Techniques in Lipidomics

Mass spectrometry is gradually applied to the research of biological samples because of its high sensitivity and high throughput. In practical applications, chromatographic technology is often combined with mass spectrometry to facilitate the separation of lipid compounds and reduce matrix interference. It has become a common method in the process of lipidomics research.

16.4.1 Liquid Chromatography–Mass Spectrometry

LC–MS is widely used in lipidomics research. It is a more comprehensive analytical method in the study of lipidomics. Among them, many liquid chromatography separation mechanisms such as normal-phase, reversed-phase, or hydrophilic chromatography can be applied to the separation of different types of lipids [37, 38]. For example, normal-phase chromatography, which separates lipids with molecular polarities, can not only separate phospholipids with different polar head groups but can also separate phospholipids with the same polar head groups. However, due to its long equilibration time, the reproducibility is poor, and the mobile phase used is volatile, which easily causes changes in polarity and the sample retention time to drift. Reversed-phase chromatography is based on the hydrophobicity of the fatty acid chains in the lipid structure. The elution sequence is related to the length and number of double bonds in the fatty acid chains. It is often used for targeted lipidomics analysis of fatty acids, phospholipids, etc. Hydrophilic interaction chromatography has the characteristics of low mobile phase volatility, high separation efficiency, and good compatibility with mass spectrometry. It has been successfully used in the analysis of glycerophospholipids such as cyclic phosphatidic acid, lysophosphatidic acid, and lysophosphatidylcholine in serum [16, 39].

At present, the most advanced lipomics technology is the research of lipomics based on MS technology. Because it is often used to determine lipid isomers, chromatography–mass spectrometry is the preferred method for lipid analysis. UHPLC–MS is one representative of LC–MS method.

16.4.2 Gas Chromatography–Mass Spectrometry

In lipidomics research, gas chromatography (GC) is usually used in combination with chemical ionization mass spectrometry or electron ionization mass spectrometry. Due to its high separation efficiency, high peak capacity, and good reproducibility, it is mainly used for the separation of volatile and thermally stable compounds [40, 41]. However, analysis of some nonvolatile lipids requires derivatization. This step is time consuming and easily causes sample changes or introduces new interference, which limits the detection of complex lipids by GC–MS and high-throughput non-targeted lipidomics analysis [42]. At present, GC–MS is mainly applied to the targeted analysis of fatty acids and relatively polar lipids.

16.4.3 Supercritical Fluid Chromatography–Mass Spectrometry

In recent years, supercritical fluid chromatography–mass spectrometry has shown great potential for rapid and comprehensive separation and analysis of lipid metabolites. Supercritical carbon dioxide is the most commonly used mobile phase in supercritical fluid chromatography due to its chemical inertness, non-flammability, low toxicity, low price, low viscosity, and high diffusion coefficient [43, 44]. In supercritical carbon dioxide, mobile phase modifiers are usually used to adjust the solubility of the analytes and improve the elution capacity of the mobile phase. The supercritical carbon dioxide can be miscible with most polar and nonpolar organic solvents, it is especially suitable for the separation and analysis of lipid components. It found that the MS method can quickly separate 30 nonpolar and polar fats within six minutes and has good sensitivity to polar lipids and the highest sample throughput in a short time. The method is suitable for fat group analysis of large sample groups in clinical research [45–47].

16.4.4 Ion Mobility–Mass Spectrometry

Ion mobility–mass spectrometry (IMS) is a new two-dimensional mass spectrometry (2D-MS) technology that combines ion mobility separation and mass spectrometry. The principle of ion mobility separation is based on the fact that ions collide with buffer gas in a drift tube, and ions can be separated according to their size and shape [48, 49]. IMS combines the characteristics of ion mobility technology, which is sensitive, fast, and can provide ion structure information, and mass spectrometry, which can provide accurate quality information [50, 51]. It shows more and more advantages in the analysis of compound isomers and biological macromolecular interaction [52]. This method can realize the separation of lipid isomers and has a wide application prospect in the study of lipidomics. The study of lipid metabolism provides new ideas for the diagnosis and pathophysiology of diseases. The combination of high-resolution mass spectrometry technology and high-performance chromatography opens up new methods for lipid analysis. The addition of orthogonal IMS to the LC–MS analysis process improves the resolution of complex lipids and makes it easier to distinguish isomers, and lipid identification and characterization are more accurate [53–55].

16.4.5 Mass Spectrometry Imaging

Mass spectrometry imaging technology is a method of imaging using a mass spectrometer that analyzes the standard molecular weight of biomolecules by measuring the mass-to-charge ratio under the control of special mass spectrometry imaging software. It can simultaneously analyze the spatial distribution characteristics of hundreds of molecules on the same tissue section. Mass spectrometry imaging technique is widely used in lipid analysis. It can provide the spatial distribution information of lipid molecules in biological samples for visual analysis and dynamic data acquisition. It does not require labeling and complex sample pretreatment because its detection sensitivity limits its application [56–58].

16.4.6 Shotgun Lipidomics

Shotgun lipidomics has become one of the most powerful and useful methods for direct and comprehensive analysis of lipid components in various biological samples due to its high accuracy and high precision. Now, electrospray ionization–mass spectrometry (ESI–MS) has become the most widely used lipid analysis method in direct injection mass spectrometry. According to ESI–MS, Han and coworkers used shotgun lipidomics based on ESI–MS and carried out product ion scanning, precursor ion scanning, neutral drop scanning, and selective ionization [59, 60]. Monitoring lipidomics combined with multidimensional mass spectrometry shotgun method is widely used in current studies of lipidomics. However, without chromatographic separation, ion suppression still exists, and some low-concentration lipid compounds are difficult to detect. At the same time, the difference between the isomers is also a major challenge for the shotgun method.

Some scholars identify and quantify the blood lipid levels of patients with systemic lupus erythematosus based on the shotgun lipomics platform of multidimensional mass spectrometry. It showed that the patient's lipid composition, including diacylphosphatidylethanolamine, lysozyme, ceramide, etc., had significant changes. The changes in the components of diacylphosphatidylethanolamine and ceramide can be used as potential biomarkers for the early diagnosis and/or prognosis of systemic lupus erythematosus [61].

16.5 Lipidomics for Biomarker Discovery

Lipidomics technology is often used in drug efficacy research and drug development. The action process and curative effect of some drugs *in vivo* are still worth studying [62–64]. The appearance of lipomics can systematically explain the changes of lipid *in vivo* after administration and can screen significant biomarkers, which provides a theoretical basis for its clinical application and development. Some scholars have studied the therapeutic effects of bortezomib and the degree of side effects through lipidomics methods, and then a total of 359 lipid metabolites were found in the serum, of which several low levels of glycerophospholipids, sphingolipids, and cholesterol esters are closely related to side effects. Among them, metabolites related to platelet activating factor biosynthesis and cholesterol metabolism play an important role [65].

Lipidomics is widely used in the process of discovering biomarkers of diseases, mainly involving cancer, pulmonary fibrosis, Alzheimer's disease (AD), and other diseases [11, 14, 66–68]. The discovery of disease biomarkers has certain predictive effects on disease monitoring and diagnosis. Yan et al. used ultra-performance liquid chromatography–mass spectrometry (UPLC–MS) to detect and identify lipid changes in patients with idiopathic pulmonary fibrosis. Through statistical analysis, a total of 62 special lipids were identified in patients, of which 6 were considered potential biomarkers of idiopathic pulmonary fibrosis [69]. Lipidomics technology has also been used to discover the potential biomarkers for the early diagnosis of non-small cell lung cancer. The preliminary identification and verification of phosphatidylcholine and phosphatidylethanolamine are potential biomarkers for early diagnosis of non-small cell lung cancer. Patients and healthy patients have a better distinction, and its content level can be regulated in the early glyceride metabolism of non-small cell lung cancer [70]. Recent studies have shown that the pathogenesis of AD is closely related to lipid metabolism disorders. Some of the adjustable lipid sets are related to potential biomarkers and clinical symptoms of diseases. The pathogenesis of AD is mainly amyloid-β plaques and tau tangles and mitochondrial dysfunction [71–75]. The study of AD is still a problem worthy of attention, in which lipid metabolism disorder is a key factor affecting amyloid-β plaques. Studies have found that four lipids such as polyunsaturated fatty acids are related to cognitive decline, and monounsaturated lipid metabolism plays an important role in early AD, while polyunsaturated lipid metabolism is related to late AD [76]. Lipids are closely related to many diseases. With the deepening of lipidomics research, we have a deeper understanding of the body's life mechanism and disease treatment. Analysis of patient plasma will become an effective means of clinical discovery and disease monitoring. It can help in the early detection and diagnosis of diseases by identifying biomarkers and contribute to the development of personalized treatments.

16.6 Application of MS-Based Lipidomics

16.6.1 Efficacy Analysis

The efficacy evaluation is an essential requirement for discovering the effective constituents and therapeutic targets. Therefore, based on the lipidomics techniques, the innovative methods such as functional lipidomics (funclipidomics) could be used for evaluating the efficacy and revealing the effective mechanism of drug, herbal medicine, traditional medicine, etc. and seek to elucidate the therapeutic properties with modern techniques. After the drug enters into the body, under the influence of various internal environments, it will undergo biotransformation in a variety of ways and then exert its efficacy. In pharmacological research, lipidomics technology focuses on monitoring the characteristics of lipid changes after the drug enters the body and elucidating the lipid action mechanism. Chang et al. used a liposome method to describe the lipid metabolism of patients with non-alcoholic fatty liver before and after berberine treatment. It was found that berberine can change the content of circulating ceramide, thereby improving the symptoms of fatty liver [77]. Studies have also found that *Shenling Baizhu San* has a protective effect on non-alcoholic fatty liver, because it changes certain liver lipids in high-fat diet rats. The mechanism of action is mainly related to the regulation of glycerophospholipid and glycerolipid metabolism [78].

16.6.2 Food Science

Lipids play a vital role in maintaining human health and internal environment stability. Lipidomics mainly studies molecular lipids in cells, tissues, and biological fluids and has made rapid progress in the past decade. In recent years, lipidomics technology has also been gradually applied to nutrition research. It can effectively explain the interaction between diet, nutrition, and human metabolism [79, 80]. It can also be used to optimize the impact of food processing on dietary value and for food related to the assessment of health impact that has broad research prospects. Lipids, as one of the main ingredients of food, are an important component of biomolecules [81]. Lipidomics is widely used in the field of food science, among which it is more prominent in the inspection of diet, agricultural products, and meat [82–84]. Dairy products are an important part of food, in which lipid and protein components in dairy products are the main sources of nutrition, and the fat content of dairy products is the key to the quality control of dairy products. Using lipidomics technology as the core to explore the types and content of lipids in dairy products, it provides a theoretical basis for identifying the quality of dairy products and coordinating the nutritional ratio [85, 86]. Coffee is widely used in food and drink because it contains a large number of active substances. Kuang et al. had analyzed the effect of coffee intake on lipid by using ion mobility spectrometry. It was found that coffee intake could lead to the decrease of specific lysophosphatidylcholine species and even had certain effects on the metabolism of glycerophosphatidylcholine [87]. Lipids in cereals are important for human nutrition, but because of the complex lipid structure, exploring lipids in cereals is a major challenge. Hammann et al. used GC–MS to analyze the lipids in grains and found that there were significant differences in morphology of alkylresorcinol and tocopherol among different grains, showing lipidomics and food science close relationships [88].

16.6.3 Environment

Lipids play a key role in the process of cellular tolerance to environmental stress and adaptation to environmental changes. Changes in lipid composition and changes in some lipid synthesis pathways reflect the response mechanism and adaptability of organisms to changes in the external environment. Lipidomics technology has broad application prospects in environmental stress tolerance and environmental pollutant toxicity evaluation [89–91]. Some scholars have studied the lipid changes of pankeratitis, an inflammatory disease of the aquatic organism *Oreochromis mossambicus* under environmental stress. It showed that the content of 13 lipids related to inflammation and apoptosis has increased, and ceramides that are sensitive to the degree of disease can be a potential biomarker of pan-inflammation [92].

16.6.4 Clinical Biomarker Discovery

Processing clinical samples, analyzing the changes in lipid metabolism, and determining their clinical biomarkers are helpful to explain the pathogenesis and principles of the disease. In clinical disease research, lipidomics technology is often used in polycystic ovary syndrome, stroke, cancer, malnutrition, etc. [7, 93, 94]. Lipidomics analysis technology reveals the correlation between abnormal lipid metabolism and diseases, and searching for potential biomarkers has been a research hotspot in recent years. Alcoholic liver disease is an important health problem, and its pathogenesis is still unclear. The changes of the lipid metabolites may provide new ideas for the study of the mechanism of alcoholic liver disease [95]. Wu et al. used ultra-high performance LC–MS to analyze the lipidomics of plasma samples, and identified 35 differentially expressed lipids as potential biomarkers related to antenatal depression in cesarean section women Metabolites, including cholesterol sulfate and PC(18 : 2(2E,4E/0 : 0)) may be their potential biomarkers [96]. In the future, the analysis of patients' plasma will become an effective means of clinical discovery and monitoring of diseases. It can assist in the early detection and diagnosis of diseases by identifying biomarkers, which will help the development of personalized treatment.

16.6.5 Pharmacological Research

Hyperlipidemia refers to high levels of blood lipids in the human body, which can cause a series of diseases that endanger human health such as coronary atherosclerosis and coronary heart disease. UPLC–HDMS (high resolution mass spectrometry) technology was used to explore the effect and mechanism of *Poria* on the treatment of hyperlipidemia induced by high-fat diet. It found that *Poria* treatment can significantly improve hyperlipemia and related abnormalities of lipid metabolites [97]. The characteristics of lipid medium changes in acute diisopropyl phosphorofluoridate (DFP) poisoning rat model study. Lipidomics studies have found that inhibiting cyclooxygenase (COX) to block the formation of prostaglandins and inhibiting enzyme-soluble cyclohydrolase to stabilize anti-inflammatory lipid mediators can be used as potential therapeutic targets [98].

16.6.6 Toxicology Analysis

Lipidomics is mainly used in the study of toxicology to explain the changes in lipids when substances produce toxic and side effects on the body and play a certain role in elaborating the toxicological mechanism and exploring treatment methods in depth. The lipidomics technology in toxicological research can deeply explore the role of lipid molecules in disease occurrence. Studies have found that troxerutin can effectively reduce the levels of serum triglycerides, phosphatidylcholine, and phosphatidylethanolamine in patients with nephropathy, thereby reducing nephrotoxicity in mice with nephropathy caused by ochratoxin A (OTA) [99]. Using lipomics to investigate the changes of bisphenol on lipid metabolism and spleen toxicity in mice, it found that bisphenol exposure caused morphological changes such as splenomegaly and led to the disorder of immune cell liposomes, which laid the foundation for further research in toxicology [100].

16.7 Future Perspectives

With its advanced analysis platform, scientific research methods and data processing technology, lipidomics plays an important role in clinical diagnosis, especially in the early diagnosis of chronic and occult diseases. Lipidomics explores the biological functions of the body from the perspective of overall lipid metabolism, which is consistent with the overall action characteristics of multiple components and multiple pathways in traditional medicine [101]. Therefore, research based on lipomics can actively reveal the complex effects of herbal medicine on organisms. With the deepening of research, lipidomics has gradually shown its unique advantages in the research field of traditional Chinese medicine (TCM) syndrome. Through the study of TCM syndrome, the common changes of different lipid metabolites were discovered in human physiological changes (Figure 16.3). Abnormal biomarkers explain

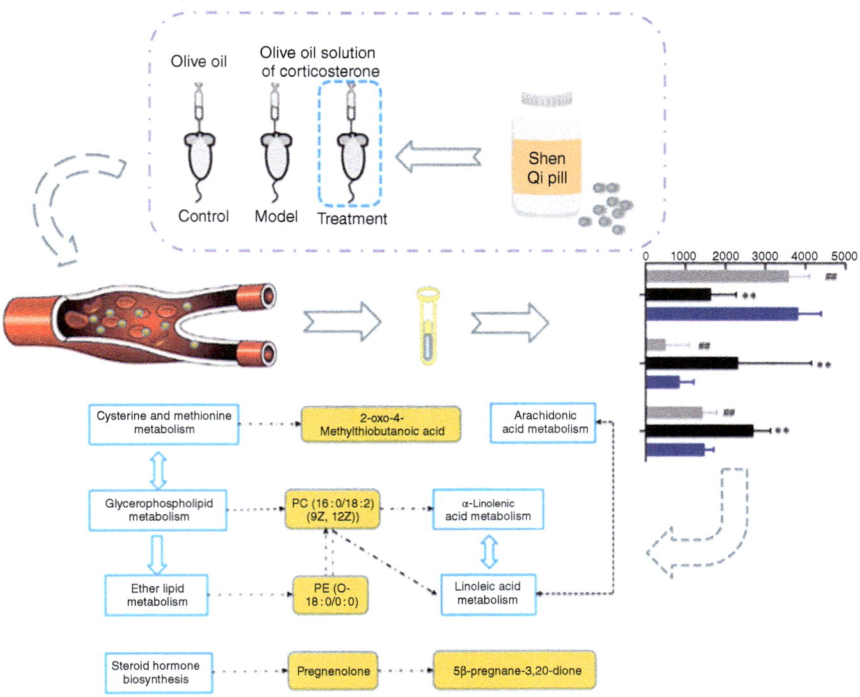

Figure 16.3 A lipidomics study for TCM syndromes.

the theory of TCM syndrome differentiation [2, 102–104]. Lipidomics will play an irreplaceable role in the research of TCM [105–109].

The current lipidomics research is still at the screening level of biomarkers, and further exploration is needed for the detailed analysis of the relationship between lipid metabolism pathway abnormalities and disease. There are still challenges in the process of lipidomics analysis. Firstly, current lipidomics analysis technology is limited, the existing lipid compound database is not perfect, and the commercial lipid reference substance is scarce. Comprehensive qualitative and quantitative analysis has caused certain difficulties. Secondly, as an important part of systems biology, lipidomics needs to integrate data mining methods in the field of statistics with lipidomics data analysis, which is important for analysts. Lipidomics research based on mass spectrometry technology still has majestic vitality, providing strong and powerful theoretical support for the development of the medical industry. Therefore, it is still necessary to develop more complete lipidomics analysis technologies, enrich lipidomics databases, strengthen the commercialization of lipid reference substances, and improve the ability of analysts to analyze and mine lipidomics data.

16.8 Conclusions

More and more studies have shown that lipids are not only the skeletal components and energy storage of biofilms, but they also participate in many important functions of cells. Lipidomics is a system capable of comprehensive analysis, identification, and understanding of lipid structure and function and reveals the relationship between lipid metabolism and cells, organs, and the human body. With its accuracy and precision, lipidomics technology has been gradually applied in clinical, pharmacology, food, and other fields. Lipidomics enables a large-scale understanding of the role of lipids in life as a whole and may lead to the discovery of new biomarkers that could provide new insights into future research of disease or drug efficacy.

Acknowledgments

This work was supported by grants from the Key Program of Natural Science Foundation of State (Grant No. 81973745, 81302905), Natural Science Foundation of Heilongjiang Province (YQ2019H030), and Heilongjiang Touyan Innovation Team Program.

References

1 Welte, M.A. and Gould, A.P. (2017). Lipid droplet functions beyond energy storage. *Biochimica et Biophysica Acta – Molecular and Cell Biology of Lipids* 1862: 1260–1272, https://doi.org/10.1016/j.bbalip.2017.07.006.

2 Zhang, H., Zhang, A., Zhou, X. et al. (2018). High-throughput lipidomics reveal mirabilite regulating lipid metabolism as anticancer therapeutics. *RSC Advances* 8 (62): 35600–35610.

3 Nielson, J.R. and Rutter, J.P. (2018). Lipid-mediated signals that regulate mitochondrial biology. *Journal of Biological Chemistry* 293: 7517–7521, https://doi.org/10.1074/jbc.R117.001655.

4 Iwabuchi, K. (2018). Gangliosides in the immune system: role of glycosphingolipids and glycosphingolipid-enriched lipid rafts in immunological functions. *Methods in Molecular Biology* 1804: 83–95, https://doi.org/10.1007/978-1-4939-8552-4_4.

5 Ge, N., Kong, L., Zhang, A.H. et al. (2021). Identification of key lipid metabolites during metabolic dysregulation in the diabetic retinopathy disease mouse model and efficacy of Keluoxin capsule using an UHPLC–MS-based non-targeted lipidomics approach. *RSC Advances* 11: 5491.

6 Kentala, H., Weber-Boyvat, M., and Olkkonen, V.M. (2016). OSBP-related protein family: mediators of lipid transport and signaling at membrane contact sites. *International Review of Cell and Molecular Biology* 321: 299–340, https://doi.org/10.1016/bs.ircmb.2015.09.006.

7 Hyotylainen, T., Ahonen, L., Poho, P., and Oresic, M. (2017). Lipidomics in biomedical research-practical considerations. *Biochimica et Biophysica Acta – Molecular and Cell Biology of Lipids* 1862: 800–803, https://doi.org/10.1016/j.bbalip.2017.04.002.

8 Yang, L., Li, M., Shan, Y. et al. (2016). Recent advances in lipidomics for disease research. *Journal of Separation Science* 39: 38–50, https://doi.org/10.1002/jssc.201500899.

9 Shahine, A., Van Rhijn, I., Cheng, T.Y. et al. (2017). A molecular basis of human T cell receptor autoreactivity toward self-phospholipids. *Science Immunology* 2, https://doi.org/10.1126/sciimmunol.aao1384: 1384–1406.

10 Rogers, M.A., Liu, J., Song, B.L. et al. (2015). Acyl-CoA: cholesterol acyltransferases (ACATs/SOATs): enzymes with multiple sterols as substrates and as activators. *Journal of Steroid Biochemistry and Molecular Biology* 151: 102–107, https://doi.org/10.1016/j.jsbmb.2014.09.008.

11 Zhang, A.H., Ma, Z.M., Kong, L. et al. (2020). High-throughput lipidomics analysis to discover lipid biomarkers and profiles as potential targets for evaluating efficacy of Kai–Xin–San against APP/PS1 transgenic mice based on UPLC–Q/TOF–MS. *Biomedical Chromatography* 34 (2): e4724.

12 Koeberle, A. (2016). Target identification and lead discovery by functional lipidomics. *Future Medicinal Chemistry* 8: 2169–2171, https://doi.org/10.4155/fmc-2016-0182.

13 Hasin, Y., Seldin, M., and Lusis, A. (2017). Multi-omics approaches to disease. *Genome Biology* 18: 83, https://doi.org/10.1186/s13059-017-1215-1.

14 Gao, H.L., Zhang, A.H., Yu, J.B. et al. (2018). High-throughput lipidomics characterize key lipid molecules as potential therapeutic targets of Kaixinsan protects against Alzheimer's disease in APP/PS1 transgenic mice. *Journal of Chromatography B: Analytical Technologies in the Biomedical and Life Sciences* 1092: 286–295.

15 Sun, Y.V. and Hu, Y.J. (2016). Integrative analysis of multi-omics data for discovery and functional studies of complex human diseases. *Advances in Genetics* 93: 147–190, https://doi.org/10.1016/bs.adgen.2015.11.004.

16 Liang, Q., Zhu, Y., Liu, H. et al. (2017). High-throughput lipidomics enables discovery of the mode of action of huaxian capsule impacting the metabolism of sepsis. *RSC Advances* 7 (71): 44990–44996.

17 Zhang, A., Sun, H., Yan, G. et al. (2016). Mass spectrometry-based metabolomics: applications to biomarker and metabolic pathway research. *Biomedical Chromatography* 30: 7–12, https://doi.org/10.1002/bmc.3453.

18 Liang, Q., Liu, H., Jiang, Y. et al. (2016). Discovering lipid phenotypic changes of sepsis-induced lung injury using high-throughput lipidomic analysis. *RSC Advances* 6: 38233–38237.

19 Liang, Q., Zhang, T., Jiang, Y. et al. (2016). Untargeted lipidomics study of coronary artery disease by FUPLC–Q–TOF–MS. *Analytical Methods* 8: 1229–1334.

20 Zhang, A., Qiu, S., Sun, H. et al. (2016). Scoparone affects lipid metabolism in primary hepatocytes using lipidomics. *Scientific Reports* 6: 28031.

21 Gong, L.L. et al. (2020). Discovery of metabolite profiles of metabolic syndrome using untargeted and targeted LC–MS based lipidomics approach. *Journal of Pharmaceutical and Biomedical Analysis* 177: 112848, https://doi.org/10.1016/j.jpba.2019.112848.

22 Lee, H.C. and Yokomizo, T. (2018). Applications of mass spectrometry-based targeted and non-targeted lipidomics. *Biochemical and Biophysical Research Communications* 504: 576–581, https://doi.org/10.1016/j.bbrc.2018.03.081.

23 Hyötyläinen, T. and Orešič, M. (2016). Bioanalytical techniques in nontargeted clinical lipidomics. *Bioanalysis* 8: 351–364, https://doi.org/10.4155/bio.15.244.

24 Zhang, Q., Xu, H., Liu, R. et al. (2019). A novel strategy for targeted lipidomics based on LC-tandem-MS parameters prediction, quantification, and multiple statistical data mining: evaluation of lysophosphatidylcholines as potential cancer biomarkers. *Analytical Chemistry* 91: 3389–3396, https://doi.org/10.1021/acs.analchem.8b04715.

25 Brunkhorst-Kanaan, N.s, Klatt-Schreiner, K., Hackel, J. et al. (2019). Targeted lipidomics reveal derangement of ceramides in major depression and bipolar disorder. *Metabolism, Clinical and Experimental* 95: 65–76, https://doi.org/10.1016/j.metabol.2019.04.002.

26 Lu, J., Lam, S.M., Wan, Q. et al. (2019). High-coverage targeted lipidomics reveals novel serum lipid predictors and lipid pathway dysregulation antecedent to type 2 diabetes onset in normoglycemic Chinese adults. *Diabetes Care* 42: 2117–2126, https://doi.org/10.2337/dc19-0100.

27 Zeng, C., Wen, B., Hou, G. et al. (2017). Lipidomics profiling reveals the role of glycerophospholipid metabolism in psoriasis. *GigaScience* 6: 1–11, https://doi.org/10.1093/gigascience/gix087.

28 Fernández-Arroyo, S., Hernández-Aguilera, A., de Vries, M.A. et al. (2019). Effect of vitamin D3 on the postprandial lipid profile in obese patients: a non-targeted lipidomics study. *Nutrients* 11, https://doi.org/10.3390/nu11051194: 1194–1205.

29 Jurowski, K., Kochan, K., Walczak, J. et al. (2017). Comprehensive review of trends and analytical strategies applied for biological samples preparation and storage in modern medical lipidomics: state of the art. *TrAC Trends in Analytical Chemistry* 86: 276–289.

30 Gil, A., Zhang, W., Wolters, J.C. et al. (2018). One- vs two-phase extraction: re-evaluation of sample preparation procedures for untargeted lipidomics in plasma samples. *Analytical and Bioanalytical Chemistry* 410: 5859–5870, https://doi.org/10.1007/s00216-018-1200-x.

31 López-Bascón, M.A., Calderón-Santiago, M., Sánchez-Ceinos, J. et al. (2018). Influence of sample preparation on lipidomics analysis of polar lipids in adipose tissue. *Talanta* 177: 86–93, https://doi.org/10.1016/j.talanta.2017.09.017.

32 Ryan, E. and Reid, G.E. (2016). Chemical derivatization and ultrahigh resolution and accurate mass spectrometry strategies for "shotgun" lipidome analysis. *Accounts of Chemical Research* 49: 1596–1604, https://doi.org/10.1021/acs.accounts.6b00030.

33 Meckelmann, S.W., Hellhake, S., Steuck, M. et al. (2017). Comparison of derivatization/ionization techniques for liquid chromatography tandem mass spectrometry analysis of oxylipins. *Prostaglandins & Other Lipid Mediators* 130: 8–15, https://doi.org/10.1016/j.prostaglandins.2017.02.003.

34 Tie, C., Hu, T., Jia, Z.X., and Zhang, J.L. (2016). Derivatization strategy for the comprehensive characterization of endogenous fatty aldehydes using HPLC-multiple reaction monitoring. *Analytical Chemistry* 88: 7762–7768, https://doi.org/10.1021/acs.analchem.6b01756.

35 Yang, K., Dilthey, B.G., and Gross, R.W. (2016). Shotgun lipidomics approach to stabilize the regiospecificity of monoglycerides using a facile low-temperature derivatization enabling their definitive identification and quantitation. *Analytical Chemistry* 88: 9459–9468, https://doi.org/10.1021/acs.analchem.6b01862.

36 Zhao, X.E., Zhu, S., and Liu, H. (2020). Recent progresses of derivatization approaches in the targeted lipidomics analysis by mass spectrometry. *Journal of Separation Science* 43: 1838–1846, https://doi.org/10.1002/jssc.201901346.

37 Jurowski, K., Kochan, K., Walczak, J. et al. (2017). Analytical techniques in lipidomics: state of the art. *Critical Reviews in Analytical Chemistry* 47: 418–437, https://doi.org/10.1080/10408347.2017.1310613.

38 Liang, Q., Wang, C., Li, B. et al. (2015). Lipidomics analysis based on liquid chromatography mass spectrometry for hepatocellular carcinoma and intrahepatic cholangiocarcinoma. *RSC Advances* 5: 63711–63718.

39 Taraji, M., Haddad, P.R., Amos, R.I.J. et al. (2018). Chemometric-assisted method development in hydrophilic interaction liquid chromatography: a review. *Analytica Chimica Acta* 1000: 20–40, https://doi.org/10.1016/j.aca.2017.09.041.

40 Salivo, S., Beccaria, M., Sullini, G. et al. (2015). Analysis of human plasma lipids by using comprehensive two-dimensional gas chromatography with dual detection and with the support of high-resolution time-of-flight mass spectrometry for structural elucidation. *Journal of Separation Science* 38: 267–275, https://doi.org/10.1002/jssc.201400844.

41 Strączyński, G. and Ligor, T. (2018). Comprehensive gas chromatography: food and metabolomocs applications. *Critical Reviews in Analytical Chemistry* 48: 176–185, https://doi.org/10.1080/10408347.2017.1390426.

42 Chiu, H.H. and Kuo, C.H. (2020). Gas chromatography–mass spectrometry-based analytical strategies for fatty acid analysis in biological samples. *Journal of Food and Drug Analysis* 28: 60–73, https://doi.org/10.1016/j.jfda.2019.10.003.

43 Shulaev, V. and Isaac, G. (2018). Supercritical fluid chromatography coupled to mass spectrometry – a metabolomics perspective. *Journal of Chromatography B: Analytical Technologies in the Biomedical and Life Sciences* 1092: 499–505, https://doi.org/10.1016/j.jchromb.2018.06.021.

44 Pauk, V. and Lemr, K. (2018). Forensic applications of supercritical fluid chromatography – mass spectrometry. *Journal of Chromatography B: Analytical Technologies in the Biomedical and Life Sciences* 1086: 184–196, https://doi.org/10.1016/j.jchromb.2018.04.015.

45 Lísa, M. and Holčapek, M. (2015). High-throughput and comprehensive lipidomic analysis using ultrahigh-performance supercritical fluid chromatography–mass spectrometry. *Analytical Chemistry* 87: 7187–7195, https://doi.org/10.1021/acs.analchem.5b01054.

46 Lísa, M., Cífková, E., Khalikova, M. et al. (2017). Lipidomic analysis of biological samples: comparison of liquid chromatography, supercritical fluid chromatography and direct infusion mass spectrometry methods. *Journal of Chromatography A* 1525: 96–108, https://doi.org/10.1016/j.chroma.2017.10.022.

47 Wolrab, D., Chocholoušková, M., Jirásko, R. et al. (2020). Validation of lipidomic analysis of human plasma and serum by supercritical fluid chromatography-mass spectrometry and hydrophilic interaction liquid chromatography–mass spectrometry. *Analytical and Bioanalytical Chemistry* 412: 2375–2388, https://doi.org/10.1007/s00216-020-02473-3.

48 Gabelica, V. and Marklund, E. (2018). Fundamentals of ion mobility spectrometry. *Current Opinion in Chemical Biology* 42: 51–59, https://doi.org/10.1016/j.cbpa.2017.10.022.

49 Romero, K.I. and Fernandez-Maestre, R. (1992). Ion mobility spectrometry: the diagnostic tool of third millennium medicine. *Revista da Associação Médica Brasileira* 64: 861–868. https://doi.org/10.1590/1806-9282.64.09.861 (2018).

50 Hernández-Mesa, M., Ropartz, D., García-Campaña, A.M. et al. (2019). Ion mobility spectrometry in food analysis: principles, current applications and future trends. *Molecules* 24, https://doi.org/10.3390/molecules24152706.

51 Ewing, M.A., Glover, M.S., and Clemmer, D.E. (2016). Hybrid ion mobility and mass spectrometry as a separation tool. *Journal of Chromatography. A* 1439: 3–25, https://doi.org/10.1016/j.chroma.2015.10.080.

52 Mairinger, T., Causon, T.J., and Hann, S. (2018). The potential of ion mobility-mass spectrometry for non-targeted metabolomics. *Current Opinion in Chemical Biology* 42: 9–15, https://doi.org/10.1016/j.cbpa.2017.10.015.

53 Zandkarimi, F. and Brown, L.M. (2019). Application of ion mobility mass spectrometry in lipidomics. *Advances in Experimental Medicine and Biology* 1140: 317–326, https://doi.org/10.1007/978-3-030-15950-4_18.

54 Hinz, C., Liggi, S., and Griffin, J.L. (2018). The potential of ion mobility mass spectrometry for high-throughput and high-resolution lipidomics. *Current Opinion in Chemical Biology* 42: 42–50, https://doi.org/10.1016/j.cbpa.2017.10.018.

55 Paglia, G. and Astarita, G. (2017). Metabolomics and lipidomics using traveling-wave ion mobility mass spectrometry. *Nature Protocols* 12: 797–813, https://doi.org/10.1038/nprot.2017.013.

56 Vaysse, P.M., Heeren, R.M.A., Porta, T., and Balluff, B. (2017). Mass spectrometry imaging for clinical research – latest developments, applications, and current limitations. *Analyst* 142: 2690–2712, https://doi.org/10.1039/c7an00565b.

57 Ferguson, C.N., Fowler, J.W.M., Waxer, J.F. et al. (2019). Mass spectrometry-based tissue imaging of small molecules. *Advances in Experimental Medicine and Biology* 1140: 99–109, https://doi.org/10.1007/978-3-030-15950-4_5.

58 Porta Siegel, T., Hamm, G., Bunch, J. et al. (2018). Mass spectrometry imaging and integration with other imaging modalities for greater molecular understanding of biological tissues. *Molecular Imaging and Biology* 20: 888–901, https://doi.org/10.1007/s11307-018-1267-y.

59 Hu, C., Duan, Q., and Han, X. (2020). Strategies to improve/eliminate the limitations in shotgun lipidomics. *Proteomics* 20: e1900070, https://doi.org/10.1002/pmic.201900070.

60 Hu, C., Wang, C., He, L., and Han, X. (2019). Novel strategies for enhancing shotgun lipidomics for comprehensive analysis of cellular lipidomes. *Trends in Analytical Chemistry* 120, https://doi.org/10.1016/j.trac.2018.11.028.

61 Lu L, Hu C, Zhao Y, et al. (2018). Shotgun lipidomics revealed altered profiles of serum lipids in systemic lupus erythematosus closely associated with disease activity. *Biomolecules* 8, https://doi.org/10.3390/biom8040105.

62 Admasu, T.D., Chaithanya Batchu, K., Barardo, D. et al. (2018). Drug synergy slows aging and improves healthspan through IGF and SREBP lipid signaling. *Developmental Cell* 47: 67–79.e65, https://doi.org/10.1016/j.devcel.2018.09.001.

63 Pati, S., Angel, P., Drake, R.R. et al. (2019). Lipidomic changes in the rat hippocampus following cocaine conditioning, extinction, and reinstatement of drug-seeking. *Brain and Behavior: A Cognitive Neuroscience Perspective* 9: e01451, https://doi.org/10.1002/brb3.1451.

64 García-Cañaveras, J.C., Peris-Díaz, M.D., Alcoriza-Balaguer, M.I. et al. (2017). A lipidomic cell-based assay for studying drug-induced phospholipidosis and steatosis. *Electrophoresis* 38: 2331–2340, https://doi.org/10.1002/elps.201700079.

65 Maekawa, K., Ri, M., Nakajima, M., Sekine, A. et al. (2019). Serum lipidomics for exploring biomarkers of bortezomib therapy in patients with multiple myeloma. *Cancer Science* 110: 3267–3274, https://doi.org/10.1111/cas.14178.

66 Hu, C., Wang, Y., Fan, Y. et al. (2015). Lipidomics revealed idiopathic pulmonary fibrosis-induced hepatic lipid disorders corrected with treatment of baicalin in a murine model. *The AAPS Journal* 17: 711–722, https://doi.org/10.1208/s12248-014-9714-4.

67 Kulkarni, Y.M., Dutta, S., Iyer, A.K.V. et al. (2018). A Lipidomics approach to identifying key lipid species involved in VEGF-inhibitor mediated attenuation of bleomycin-induced pulmonary fibrosis. *Proteomics. Clinical Applications* 12: e1700086, https://doi.org/10.1002/prca.201700086.

68 Yan, F., Zhao, H., and Zeng, Y. (2018). Lipidomics: a promising cancer biomarker. *Clinical and Translational Medicine* 7: 21, https://doi.org/10.1186/s40169-018-0199-0.

69 Yan, F., Wen, Z., Wang, R. et al. (2017). Identification of the lipid biomarkers from plasma in idiopathic pulmonary fibrosis by lipidomics. *BMC Pulmonary Medicine* 17: 174, https://doi.org/10.1186/s12890-017-0513-4.

70 Chen Y, Ma Z, Shen X, et al. (2018). Serum lipidomics profiling to identify biomarkers for non-small cell lung cancer. *BioMed Research International* 2018: 5276240, https://doi.org/10.1155/2018/5276240.

71 Pinheiro, L. and Faustino, C. (2019). Therapeutic strategies targeting amyloid-β in Alzheimer's disease. *Current Alzheimer Research* 16: 418–452, https://doi.org/10.2174/1567205016666190321163438.

72 Gosztyla, M.L., Brothers, H.M., and Robinson, S.R. (2018). Alzheimer's amyloid-β is an antimicrobial peptide: a review of the evidence. *Journal of Alzheimer's Disease* 62: 1495–1506, https://doi.org/10.3233/jad-171133.

73 Naseri, N.N., Wang, H., Guo, J. et al. (2019). The complexity of tau in Alzheimer's disease. *Neuroscience Letters* 705: 183–194, https://doi.org/10.1016/j.neulet.2019.04.022.

74 Chong, F.P., Ng, K.Y., Koh, R.Y., and Chye, S.M. (2018). Tau proteins and tauopathies in Alzheimer's disease. *Cellular and Molecular Neurobiology* 38: 965–980, https://doi.org/10.1007/s10571-017-0574-1.

75 Shoshan-Barmatz, V., Nahon-Crystal, E., Shteinfer-Kuzmine, A., and Gupta, R. (2018). VDAC1, mitochondrial dysfunction, and Alzheimer's disease. *Pharmacological Research* 131: 87–101, https://doi.org/10.1016/j.phrs.2018.03.010.

76 Barupal, D.K., Baillie, R., Fan, S. et al. (2019). Sets of coregulated serum lipids are associated with Alzheimer's disease pathophysiology. *Alzheimers Dement (Amst)* 11: 619–627, https://doi.org/10.1016/j.dadm.2019.07.002.

77 Chang, X., Wang, Z., Zhang, J. et al. (2016). Lipid profiling of the therapeutic effects of berberine in patients with nonalcoholic fatty liver disease. *Journal of Translational Medicine* 14: 266, https://doi.org/10.1186/s12967-016-0982-x.

78 Deng, Y., Pan, M., Nie, H. et al. (2019). Lipidomic analysis of the protective effects of Shenling Baizhu San on non-alcoholic fatty liver disease in rats. *Molecules* 24, https://doi.org/10.3390/molecules24213943.

79 Luque de Castro, M.D. and Quiles-Zafra, R. (2020). Lipidomics: an omics discipline with a key role in nutrition. *Talanta* 219: 121197, https://doi.org/10.1016/j.talanta.2020.121197.

80 Newsholme, P. (2016). Overview: metabolomics and lipidomics in nutrition and metabolism research. *Essays in Biochemistry* 60: 407, https://doi.org/10.1042/ebc20160047.

81 Ibanez, C., Mouhid, L., Reglero, G., and Ramirez de Molina, A. (2017). Lipidomics insights in health and nutritional intervention studies. *Journal of Agricultural and Food Chemistry* 65: 7827–7842, https://doi.org/10.1021/acs.jafc.7b02643.

82 Liu, Z., Rochfort, S., and Cocks, B. (2018). Milk lipidomics: what we know and what we don't. *Progress in Lipid Research* 71: 70–85, https://doi.org/10.1016/j.plipres.2018.06.002.

83 Trivedi DK, Hollywood KA, Rattray NJ, et al. (2016). Meat, the metabolites: an integrated metabolite profiling and lipidomics approach for the detection of the adulteration of beef with pork. *Analyst* 141: 2155–2164, https://doi.org/10.1039/c6an00108d.
84 Mi S, Shang K, Jia W, et al. (2018). Characterization and discrimination of Taihe black-boned silky fowl (Gallus gallus domesticus Brisson) muscles using LC/MS-based lipidomics. *Food Research International* 109: 187–195, https://doi.org/10.1016/j.foodres.2018.04.038.
85 Wang, L., Li, X., Liu, L. et al. (2020). Comparative lipidomics analysis of human, bovine and caprine milk by UHPLC–Q–TOF–MS. *Food Chemistry* 310: 125865, https://doi.org/10.1016/j.foodchem.2019.125865.
86 Argov-Argaman, N., Glasser, T., Muklada, H. et al. (2020). Lipidome changes, with a focus on phospholipids, due to feeding systems and processing in goat milk. *Food Chemistry* 340: 127938, https://doi.org/10.1016/j.foodchem.2020.127938.
87 Kuang A, Erlund I, Herder C, et al. (2018). Lipidomic response to coffee consumption. *Nutrients* 10, https://doi.org/10.3390/nu10121851.
88 Hammann, S., Korf, A., Bull, I.D. et al. (2019). Lipid profiling and analytical discrimination of seven cereals using high temperature gas chromatography coupled to high resolution quadrupole time-of-flight mass spectrometry. *Food Chemistry* 282: 27–35, https://doi.org/10.1016/j.foodchem.2018.12.109.
89 Higashi, Y. and Saito, K. (2019). Lipidomic studies of membrane glycerolipids in plant leaves under heat stress. *Progress in Lipid Research* 75: 100990, https://doi.org/10.1016/j.plipres.2019.100990.
90 Hunter, J.E., Brandsma, J., Dymond, M.K. et al. (2018). Lipidomics of Thalassiosira pseudonana under phosphorus stress reveal underlying phospholipid substitution dynamics and novel Diglycosylceramide substitutes. *Applied and Environmental Microbiology* 84, https://doi.org/10.1128/aem.02034-17.
91 Lu S, Fadlalla T, Tang S, et al. (2019). Genome-wide analysis of phospholipase D gene family and profiling of phospholipids under abiotic stresses in Brassica napus. *Plant & Cell Physiology* 60: 1556–1566, https://doi.org/10.1093/pcp/pcz071.
92 Koelmel, J.P., Ulmer, C.Z., Fogelson, S. et al. (2019). Lipidomics for wildlife disease etiology and biomarker discovery: a case study of pansteatitis outbreak in South Africa. *Metabolomics* 15: 38, https://doi.org/10.1007/s11306-019-1490-9.
93 Sethi, S. and Brietzke, E. (2017). Recent advances in lipidomics: analytical and clinical perspectives. *Prostaglandins & Other Lipid Mediators* 128-129: 8–16, https://doi.org/10.1016/j.prostaglandins.2016.12.002.
94 Jové M, Pradas I, Naudí A, et al. (2018). Lipidomics reveals altered biosynthetic pathways of glycerophospholipids and cell signaling as biomarkers of the polycystic ovary syndrome. *Oncotarget* 9: 4522–4536, https://doi.org/10.18632/oncotarget.23393.
95 Warner, D.R., Liu, H., Ghosh Dastidar, S. et al. (2018). Ethanol and unsaturated dietary fat induce unique patterns of hepatic ω-6 and ω-3 PUFA oxylipins in a mouse model of alcoholic liver disease. *PLoS One* 13: e0204119, https://doi.org/10.1371/journal.pone.0204119.

96 Wu, Z., Zhao, P., Long, Z. et al. (2019). Biomarker screening for antenatal depression in women who underwent caesarean section: a matched observational study with plasma lipidomics. *BMC Psychiatry* 19: 259, https://doi.org/10.1186/s12888-019-2241-1.

97 Miao, H., Zhao, Y.H., Vaziri, N.D. et al. (2016). Lipidomics biomarkers of diet-induced hyperlipidemia and its treatment with Poria cocos. *Journal of Agricultural and Food Chemistry* 64: 969–979, https://doi.org/10.1021/acs.jafc.5b05350.

98 Yang, J., Bruun, D.A., Wang, C. et al. (2019). Lipidomes of brain from rats acutely intoxicated with diisopropylfluorophosphate identifies potential therapeutic targets. *Toxicology and Applied Pharmacology* 382: 114749, https://doi.org/10.1016/j.taap.2019.114749.

99 Yang, X., Xu, W., Huang, K. et al. (2019). Precision toxicology shows that troxerutin alleviates ochratoxin A-induced renal lipotoxicity. *The FASEB Journal* 33: 2212–2227, https://doi.org/10.1096/fj.201800742R.

100 Zhao C, Yong T, Zhang Y, et al. (2020). Evaluation of the splenic injury following exposure of mice to bisphenol S: a mass spectrometry-based lipidomics and imaging analysis. *Environment International* 135: 105378, https://doi.org/10.1016/j.envint.2019.105378.

101 Ren, J.L., Zhang, A.H., Kong, L. et al. (2020). Analytical strategies for the discovery and validation of quality-markers of traditional Chinese medicine. *Phytomedicine* 67: 153165, https://doi.org/10.1016/j.phymed.2019.153165.

102 Sun H, Zhang AH, Yang L, et al. (2019). High-throughput chinmedomics strategy for discovering the quality-markers and potential targets for Yinchenhao decoction. *Phytomedicine* 54: 328–338, https://doi.org/10.1016/j.phymed.2018.04.015.

103 LLiW., Zhang A., Zhou X., et al. High-throughput liquid chromatography mass-spectrometry-driven lipidomics discover metabolic biomarkers and pathways as promising targets to reveal the therapeutic effects of the Shenqi pill. *RSC Advances* 10: 2347–2358.

104 Liu, C., Zong, W., Zhang, A. et al. (2018). Lipidomic characterisation discovery for coronary heart disease diagnosis based on high-throughput ultra-performance liquid chromatography and mass spectrometry. *RSC Advances* 8: 647–654.

105 Wang, X.J., Ren, J.L., Zhang, A.H. et al. (2019). Novel applications of mass spectrometry-based metabolomics in herbal medicines and its active ingredients: current evidence. *Mass Spectrometry Reviews* 38: 380–402, https://doi.org/10.1002/mas.21589.

106 Zhang, A.H., Yu, J.B., Sun, H. et al. (2018). Identifying quality-markers from Shengmai San protects against transgenic mouse model of Alzheimer's disease using chinmedomics approach. *Phytomedicine* 45: 84–92, https://doi.org/10.1016/j.phymed.2018.04.004.

107 Zhang, A.H., Ma, Z.M., Sun, H. et al. (2019). High-throughput metabolomics evaluate the efficacy of total lignans from acanthophanax senticosus stem against ovariectomized osteoporosis rat. *Frontiers in Pharmacology* 10: 553, https://doi.org/10.3389/fphar.2019.00553.

108 Fang, H., Zhang, A.H., Sun, H. et al. (2019). High-throughput metabolomics screen coupled with multivariate statistical analysis identifies therapeutic targets in alcoholic liver disease rats using liquid chromatography-mass spectrometry. *Journal of Chromatography B: Analytical Technologies in the Biomedical and Life Sciences* 1109: 112–120, https://doi.org/10.1016/j.jchromb.2019.01.017.

109 Zhou, X.H., Zhang, A.H., Wang, L. et al. (2016). Novel chinmedomics strategy for discovering effective constituents from ShenQiWan acting on ShenYangXu syndrome. *Chinese Journal of Natural Medicines* 14: 561–581, https://doi.org/10.1016/s1875-5364(16)30067-x.

17

Potential Application of Mass Spectrometry-Based Lipidomics for Herbal Medicine

Ying Cai, Aihua Zhang, Hui Sun, and Xijun Wang

Heilongjiang University of Chinese Medicine, National Chinmedomics Research Center, Functional Metabolomics Laboratory, Department of Pharmaceutical Analysis, Heping Road 24, Harbin 150040, China

17.1 Introduction

Lipids have many other essential biological functions. Recent studies have shown that lipids play an irreplaceable role in regulating cell metabolism, maintaining homeostasis in the body, and regulating immunity [1–3]. More and more experiments have shown that the fluctuation of lipid level is closely related to the status of many diseases [4, 5]. The occurrence and the development of diabetes, neurodegenerative diseases, metabolic syndrome, and other diseases are highly correlated with abnormal lipid metabolism [6–9]. Therefore, a comprehensive and systematic characterization and analysis of lipids in biological systems can provide rich and valuable information for the pathological study of diseases and the explanation of the pathogenesis of diseases. The molecular structure of lipids is different, which makes lipids have rich biological functions. In the field of biology, the further study of lipids has a broad prospect, which also promotes the continuous development of lipidomics as an independent subject. The objective of lipidomics is to comprehensively analyze lipids in biological systems. With the development of mass spectrometry (MS), the changes of lipidome and the details of various biochemical pathways could be fully described [10, 11].

Emerging analytical strategies based on mass spectrometry, such as liquid chromatography (LC) coupled MS, MS imaging, and ion mobility MS, have promoted the development and application of lipidomics. Qualitative and quantitative identification and dynamic analysis of lipid metabolism network changes using lipidomics combined with mass spectrometry and bioinformatics could explain in detail the physiological significance of lipid changes at the molecular level. This chapter will show the current research status of lipidomics based on mass spectrometry technology according to different analysis strategies such as non-targeting, targeting, and pseudotargeting and will summarize the findings of new technologies and methods in disease discovery in recent years, as well as the potential application of mass spectrometry-driven lipidomics for herbal medicine.

Mass Spectrometry-Based Metabolomics in Clinical and Herbal Medicines: Strategies, Technologies, and Applications, First Edition. Edited by Aihua Zhang and Wanying Wang.

17.2 Lipidomics Method

MS-based lipidomics approaches are the most powerful and effective tools for analysis of lipidome with high sensitivity and specificity. Advances in mass spectroscopy-based lipidomics platforms have provided the opportunities for comprehensive analysis of lipids in biological samples.

17.2.1 Sample Preparation

17.2.1.1 Classical Method

The structure of lipids is diverse, the types are diverse, and the content of lipids varies greatly *in vivo*. Consequently, the extraction process is limited by numerous conditions. Instead of being free in the matrix, lipids are mostly embedded in the matrix. Therefore, it is necessary to minimize the interference of other small-molecular polar metabolites in the substrate to lipid analysis. Differences in extraction rates between lipid classes may lead to deviations from downstream analysis, thus affecting the reliability of the results [12]. As a consequence, different lipid extraction methods should be adopted for different lipid structures.

The single solvent extraction method, namely, protein precipitation method, adds excessive organic solvent to the matrix, vortexes the protein precipitation and centrifugation, and takes the supernatant for the purpose of separating and removing the protein in the sample. Some researchers used four different water-soluble organic solvents to prepare samples and conducted lipid mass spectrometry analysis. Then, according to the results of lipid analysis, the optimal lipid preparation method was optimized [13]. The results showed that acetonitrile (ACN) caused lipid decomposition due to insufficient protein denaturation and the efficiency of lipid extraction was not satisfactory. Alcohol is very adequate for LC/MS sample pretreatment, in which isopropyl alcohol (IPA) is suitable for the extraction of hydrophobic lipids such as triacylglycerol (TG), and methanol is suitable for the extraction of hydrophilic lipids. In addition, some studies have found that ethanol extraction of lipids is the simplest extraction method and has the advantage of high coverage rate. Compared with the liquid and liquid extraction method using chloroform, single-phase protein deposition method avoids the use of toxic solvents, and the operation is simpler [14]. Besides, there have been good results in the four key parts of sample preparation, such as simplicity, lipid coverage, reproducibility, and recovery. Single-phase extraction may be a suitable alternative to biphasic extraction for large-scale global lipid analysis [15]. Alshehry method is a new single-phase extraction method, using methanol/1-butanol as the solvent for extraction. Some scholars investigated four different lipid extraction methods and used LC–MS to analyze the extracted lipids. It showed that the Alshehry method has unique advantages in polar lipids extracted and is extremely reproducible, safer, and more environment-friendly since it doesn't need chloroform [16].

Liquid–liquid extraction is the most commonly used method for lipid extraction, which can be widely used for the extraction of lipid from various substrates. Folch extraction, Bligh–Dyer extraction, Matyash extraction, BUME extraction, and other methods are commonly used for sample pretreatment in lipidomics analysis. Some researchers used direct infusion mass spectrometry and LC–MS technology to prepare three different types of samples consisting of human serum and urine. Non-targeted methods investigated three different extraction methods including the improved Matyash method, original Matyash method, and stepwise Bligh and Dyer for the sake of optimizing the best sample preparation method. The modified Matyash method shows excellent performance when the analysis requires a higher yield and reproducibility [17]. The use of biphasic lipid extraction method can be used to investigate both the water layer and the organic layer, which has a unique advantage for non-targeted studies to increase lipid coverage [18].

17.2.1.2 Green Technologies

Although Folch and Bligh–Dyer methods have been widely used as the gold standard for lipid extraction, chloroform, as a toxic solvent used in the sample preparation operation of the liquid–liquid extraction method, may cause health damage. In the last decade, many scholars and researchers have been committed to developing green lipid extraction technology and have tried to find different solvents to replace the highly toxic chloroform because of the high safety risks in traditional extraction methods. In recent years, it has been found that ethyl acetate may be a good substitute for chloroform [19]. Some researchers have tried to use ethanol and ethyl acetate as alternative solvents for methanol and chloroform in Folch or Bligh–Dyer methods. The results showed that this method could be used to extract fatty acids and other lipids and had good extraction ability [20].

Microwave-assisted extraction (MAE) uses microwave energy to heat the solvent in contact with the sample to separate the targeted lipids from the sample. This method has short extraction time, high extraction efficiency, and a small amount of extraction reagent. Recently, NaCl solution was considered as a solvent for MAE and for the extraction of lipid [21, 22]. The results showed that the method has a higher yield and can extract high-quality lipids. This method is cheap, green, and environmentally friendly, can quickly extract high-quality lipids, and has a good application prospect. However, the extraction conditions should be optimized according to the targeted lipids extracted [23], and factors such as temperature rise may lead to lipid degradation.

Ultrasound-assisted extraction could improve the extraction efficiency by using ultrasonic energy to promote the transfer of lipid molecules between immiscible phases. It is suitable for various types of lipid extraction. Some scholars have studied and compared the extraction efficiency of three extraction methods. The experimental results showed that compared with Soxhlet and chloroform/methanol two-phase extraction, ultrasound-assisted extraction not only guaranteed the extraction efficiency but could also greatly reduce the extraction time and save solvent, making it a cheap and environmentally friendly extraction technology [24]. This method has the advantages of simple operation, short extraction time, high extraction efficiency, good repeatability, small amount of extraction reagent, easy regulation of temperature and pressure, and other extraction conditions and is often used to extract total lipids from human serum and plasma.

A pressurized liquid extraction method was developed for lipid extraction and applied it to dinoflagellate biomass. The study examined four different extraction methods, including Soxhlet, Folch methods, PLE with *n*-hexane, and PLE with ethanol. The results showed that the highest efficiency of lipid extraction can be obtained from PLE with 90% ethanol concentration. It showed that pressurized liquid extraction with payment is a green and efficient technology to lipid extraction [25]. The micro-solid-phase microextraction (SPME) probes had a good concentration effect on lipids [26]. The SPME method was used for the study of untargeted lipidomics in cell line cultures [27]. Seventy-seven lipids were detected using SPME technology, of which 63 were upregulated. This method has a good separation effect and can obtain more known species and subclasses. Moreover, compared with Bligh and Dyer method, SPME has a higher ionization efficiency for low abundance lipids. In addition, SPME methods are used in conjunction with GC/MS for lipid detection in food science in order to evaluate food quality [28, 29].

The extraction of lipids still faces many difficulties and challenges. At present, the sample preparation process still needs manual labor, which is very time-consuming and will greatly affect the reproducibility of analysis results. Therefore, it is necessary to develop an automated sample extraction solvent system for high-throughput analysis.

17.2.2 Data Collection and Analysis of Lipidomics

According to the different purposes of analysis, lipidomics based on mass spectrometry can be divided into several subfields, namely, non-targeted lipidomics, targeted lipidomics, and pseudotargeted lipidomics. These methods are driven by mass spectrometry innovations. Mass spectrometry and related techniques have been widely used in scientific research because of their high efficiency, accuracy, and sensitivity, which are helpful to the biochemical research of lipids.

17.2.2.1 Untargeted Lipidomics

Untargeted lipidomics, also known as global lipidomics, aims to unbiased qualitative identification and quantitative analysis of all lipids in the substrate and is suitable for testing as many lipids in as many samples as possible without any known information. A large amount of lipid data can be obtained by untargeted lipidomics. After further software processing, these data can be converted into lipid fingerprints, providing abundant lipid species and lipid metabolism information. In addition, further intergroup comparison between the blank control group and the sample group could be used to assess the metabolic differences of lipids under different physiological conditions. Some scholars have studied lipids in the serum of patients with coronary artery disease using non-targeted lipidomics methods and found that there are notable associations between 34 metabolite signals serum HDL cholesterol, while 34 metabolite signals correspond to a set of sphingomyelins [30]. The results showed that sphingomyelins were directly correlated with serum lipoprotein and total cholesterol levels in patients with coronary heart disease, which supports the basic role of sphingomyelin in regulating blood lipid levels and emphasizes the importance of identifying new targets in the metabolic pathways of sphingomyelin in anti-atherosclerosis therapy.

17.2.2.2 Targeted Lipidomics

In contrast, targeted lipidomics focuses on the characterization of specific lipids. Untargeted lipidomics is used to obtain an overview of lipid molecules in the system, whereas targeted lipidomics may be better fit to solving complex biological problems. Multiple reaction monitoring (MRM), characterized by high sensitivity, high selectivity, and wide linear range, is the most commonly used platform for targeted analysis [31]. Some scholars used high-performance liquid chromatography (HPLC)–MRM high coverage targeted lipidomics method to measure serum lipid spectrum of individuals with normal glucose regulation (NGR) and evaluated the changes of lipid synergistic regulation before T2DM [32]. The experimental results show that six kinds of lipids, including the TGs, lysophosphatidylinositols, phosphatidylcholines, polyunsaturated fatty acid–plasmalogen phosphatidylethanolamines, and cholesteryl esters, could be predicted lipid biomarkers of T2DM. New lipid predictors were determined in T2DM patients with NGR, which increased the predictors of diabetes incidence. Targeted lipidomics analysis has good sensitivity and specificity. Combined with MRM mode, targeted lipidomics can obtain high-quality data information, especially lipid analysis with low abundance, which has been widely used in the analysis and identification of key pathways and target components of lipid metabolism. Some scholars adopted targeted lipidomics strategy and quantitatively analyzed lysophosphatidylcholine (LPC) through LC–MS/MS parameter prediction and multivariate statistical analysis and successfully determined 60 LPCs simultaneously in plasma. Multiple LPCs have been found to be biomarkers for many types of cancer and can be used to distinguish among different types of cancer [33]. Compared with untargeted analysis, targeted analysis showed better repeatability and higher lipid coverage.

17.2.2.3 Pseudotargeted Lipidomics

Combined analysis of targeted and untargeted strategies to explore the relationship between lipid fluctuations and physiological and pathological changes can clarify the important lipid regulatory mechanisms that affect the health of diseases. The emerging pseudotargeted lipid metabolomics strategies combine the advantages of non-targeted and targeted lipidomics strategies [34], that is, based on information from untargeted methods, targeting technologies to achieve high coverage lipidomics data collection, which can be used for large-scale lipidomics analysis. Some scholars used pseudotargeted lipidomics methods to randomly select blood lipid extracts from 30 diabetic patients and healthy controls, respectively, for UHPLC/QQQ MRM MS analysis [35]. TGs, DGs, PEs, and LPEs were significantly increased and HexCers, PE–OS, PC–OS, and PC–OS were significantly decreased in diabetic patients compared with healthy controls, suggesting abnormal lipid metabolism during the onset and progression of diabetes. Compared with targeted analysis, the quasi-targeted method can include both qualitative and non-qualitative metabolites by collecting secondary mass spectrometry metabolism characteristics of mixed biological samples, thus achieving extensive coverage. Compared with untargeted analysis, MRM mode has high data stability and wide linear range, can obtain high-quality data information, can focus on more metabolites with low abundance, and shows better repeatability and higher lipid coverage. This method combines the advantages of both target and non-target analysis strategies and introduces the algorithm to automatically determine the material and mass spectrum parameters for target analysis. Pseudotargeted lipidomics can detect more potential markers and permits more accurate quantification.

17.3 Mass Spectrometry Technique in Lipidomics

The mass spectrometer consists of vacuum system, ion source, mass analyzer, and detector. In the process of lipid analysis, lipid molecules are usually ionized first, and then the ionized products are introduced into mass spectrometry for identification and analysis according to the mass charge ratio. With its excellent qualitative and quantitative abilities, mass spectrometry has become the core driving force of lipidomics [36]. The development of lipidomics is largely due to the rapid development of mass spectrometry and other technologies. With the help of mass spectrometry, lipid analysis can be performed effectively according to lipid properties.

17.3.1 Ionization Technology

Various ionization techniques can be used for lipid analysis, but the most widely used ionization techniques are soft ionization techniques. Electrospray ionization (ESI) and matrix-assisted laser desorption ionization (MALDI) have become the mainstream ionization methods for lipid tissue analysis [37]. On this basis, shotgun lipidomics, mass spectrometry, and other novel lipid analytical techniques were developed and widely utilized.

ESI/MS has gradually developed into a core technique for lipidomics analysis [38–41], on which many new techniques have been developed. Recent attention has turned to high-throughput lipidomics analysis techniques for large sample sizes. A method for the analysis of lipids in biological samples was developed by combining ultra-efficient supercritical fluid chromatography with ESI–MS [38]. It successfully separated 30 different types of lipids in six minutes, which exhibits high resolution and precision, and it is suitable for lipid analysis of large-scale biological samples.

The advantages of nano-ESI mass spectrometry include less sample consumption, simple sample processing, and short analysis time [42]. Some researchers quantitatively analyzed jasmonate in various plant tissues and body fluids by nano-ESI–MS/MS technology [43], and this method could absolutely quantify multiple jasmonates and showed good selectivity and sensitivity. Other scholars used nano-ESI–MS/MS technology to explore the cellular components in human amniotic membrane that has been found to be a potential source of pulmonary surfactant in fetal lungs and to promote the development of respiratory function in fetal lungs [44]. Nano-ESI has a good performance in lipid quantification [45] and has the potential to be applied in large-scale lipid quantification.

Desorption electrospray ionization (DESI) is an ionization method combining ESI and desorption ionization (DI). DESI is often used in the analysis and evaluation of glycerophospholipids and sphingomyelins. Compared with other ionization methods, DESI has the advantages of simple sample, easy ionization of lipids, simple operation, etc. It is also suitable for the direct analysis of ionizable lipid molecules and lipid mass spectrometry imaging analysis [46–48].

Mass spectrometry of atmospheric pressure chemical ionization (APCI) and atmospheric-pressure photoionization (APPI) is also commonly used to ionize neutral molecules and has good resolution [49–52]. In order to investigate the analysis of different optimization techniques, the positive phase liquid chromatography was used to analyze the different sources lipid classes of lipids in the sample [53]. The chromatographic results show that both methods are suitable for detecting different lipids, and APPI has better sensitivity to lower concentrations of lipids. Moreover, compared with APCI, the sensitivity of APPI has been significantly improved.

MALDI–MS has many advantages in lipid analysis, for example, fast analysis speed, easy operation, high sensitivity, and the ability to analyze almost all types of lipids, and then it is greatly effective in lipid analysis. MALDI–MS has been used in phospholipid analysis of cell membranes to identify potential biomarkers [53, 54]. The identification and analysis of cellular phospholipids in cancer showed that it had a good development prospect in screening specific biomarkers [54], and the analysis of phospholipids may reflect the state of organisms and contribute to the development of disease treatment strategies. MALDI–MS has been effectively applied in mass spectrometry imaging analysis to determine lipid distribution [55–59]. The researchers investigated lipid distribution in the kidney using MALDI imaging mass spectrometry (IMS). It showed that the changes of lipid distribution were correlated with renal toxicity and renal protection, and the physiological status of the kidney could be evaluated by lipidomics method.

17.3.2 Mass Analyzer

The mass analyzer is used to separate ionized ions according to different mass-to-charge ratios. In lipid analysis, the commonly used mass analyzers mainly include quadrupole (Q), time of flight (TOF), and ion trap. Q–MS has excellent quantitative capability and is commonly used in GC–MS [60]. The quadrupole mass analyzer can complete a full scan from low to high mass-to-charge ratio in a very short time, providing a large number of ion fragments for lipid compound analysis. Triple-quadrupole mass spectrometry (QqQ–MS), which uses three quadrupoles in series, is very common and can qualitatively and quantitatively analyze various lipids by selective reflection monitoring/multiple reflection monitoring (SRM/MRM), neutral loss scanning (NLS), precursor ion scanning (PIS), and product ion analysis.

TOF–MS has the advantages of wide analytical mass range, fast analytical speed, simple instrument structure, etc. However, easy dispersion of initial kinetic energy and initial position of ions is the core factor affecting its resolution. However, mass spectrometers using only TOF mass analyzer are difficult to use MS/MS for analysis. Hybrid mass spectrometers such as quadrupole-time-of-flight (Q-TOF) tandem mass spectrometry, linear ion triple TOF, and triple TOF are needed [7, 61, 62]. Among them, Q-TOF mass spectrometer is the most commonly used, which can meet most of the requirements of lipidomics analysis. However, its shortcoming is that the dynamic linear range of detector is narrow, and quantitative analysis can only be conducted within a small dynamic range. The development of hybrid mass spectrometers has greatly improved the product ion analysis technique [2, 63]. Ultra-performance liquid chromatography (UPLC)–Q–Exactive Orbitrap/MS method was established for three kinds of honey bee pollen plant by lipidomics analysis, including phosphatidylcholine, phosphatidyl glycerol, phosphatidyl ethanolamine, phosphatidylserine, hemolysis phosphatidyl choline, ceramide, etc. [64].

17.3.3 Tandem Mass Spectrometry

There are four tandem mass spectrometry models: product ion analysis, NLS, PIS, and SRM. In the product ion analysis mode, a two-stage mass analyzer is used to select and detect specific precursor ions and obtain the structure information of precursor ions through the obtained product ion information. The scanning mode can be used to analyze the lipidome, which is helpful to determine the structure characteristics of each lipid molecule and to discover new lipid molecules. In the neutral-loss scanning mode, there is a constant quality difference between two mass analyzers, which are suitable for detecting lipids with the same neutral loss debris. In the PIS mode, the second-level mass analyzer detects specific ions after CID, and the first-level mass analyzer detects precursor ions capable of producing such fragment ions, which is suitable for detecting lipids capable of dissociating the same product ions. In the selected reaction monitoring mode, the mass analyzer detects specific precursor and product ions, respectively. When a two-stage mass analyzer is set up to detect multiple ions, this mode is called multi-reflection detection, which has been widely used.

17.3.4 Ion Mobility–Mass Spectrometry

Ion mobility–mass spectrometry (IM–MS) is a technique based on the separation of ions in the drift tube from the colliding gas with different cross sections and shows great potential in the field of lipidomics [65]. Different types of ions have different migration time in the migration tube, and the different types of lipids can be distinguished according to different migration time [66, 67]. In addition, the size and character of ions are different, and the collision cross sections (CCS) are different, which can be used to judge the mass and structure of ions [68]. However, the number of CCS values obtained only from experimental results is limited, so a comprehensive lipid CCS value database needs to be established [69]. The researchers used machine learning, high-precision IM–MS, and other methods to establish a comprehensive

CCS database to facilitate the identification of lipids and their metabolites and to promote the application of IM–MS [69–71]. IM–MS enhanced the separation of complex lipids, the separation degree of isomers, and the ability to identify lipids [72]. This method is highly reproducible and can significantly improve the accuracy of lipid analysis by combining with advantages such as the ratio of mass to the nucleus or retention time.

17.3.5 Mass Spectrometry Imaging

Mass spectrometry is a molecular imaging technique using mass spectrometry to analyze the partial distribution characteristics of hundreds of molecules simultaneously on the same tissue slice. IMS technology is used for lipid analysis, which can provide spatial distribution information of lipid molecules in biological samples for visual analysis and dynamic data acquisition [73, 74]. It does not require labeling or complex sample pretreatment, but its detection sensitivity is limited [75]. IMS method is very important for the understanding and diagnosis of diseases, the determination of treatment methods, and the evaluation of curative effect. Cisplatin antitumor therapy may produce nephrotoxicity. Some scholars used mass spectrometry imaging method to identify and analyze lipid distribution in the kidney [76, 77]. It showed that the lipid distribution in the kidney of rats changed significantly and was significantly correlated with the quality of cisplatin. Most of the lipid-imaging samples were taken from the brain of rats or mice [78–85]. It has also been conducted to detect the lipid distribution level in the brain tissue and to deepen the pathological understanding of Alzheimer's disease (AD) [83, 84].

17.4 Application of Mass Spectrometry-Based Lipidomics for Herbal Medicine

Mass spectrometry-based lipidomics can be used to discover the perturbed pathways and lipid biomarkers as potential targets to reveal the therapeutic effects of herbal medicine. Bufalin (BFL) and cinobufagin (CBF) are the main active constituents of *Chansu*, a widely used traditional Chinese medicine (TCM). The lipidomics and mass spectrometry imaging was used to reveal the underlying mechanisms of their cotreatment with BFL and CBF [85]. The cotreatment with BFL and CBF modulated the sphingolipid metabolism and glycerophospholipid metabolism. It showed that the cotreatment with BFL and CBF confers superior anticancer efficacy compared with monotreatment. The high-throughput lipidomics using UHPLC–Q–TOF/MS combined with multivariate statistical analysis reveal the molecular mechanisms related to diabetic retinopathy (DR) development and discover the effect of Keluoxin (KLX) capsule against DR in db/db mice [86]. A total of 30 lipid biomarkers were identified, and 7 metabolic pathways such as arachidonic acid metabolism and steroid hormone biosynthesis were discovered. The preventive effect of KLX intervention can regulate 22 biomarkers such as prostaglandin D2, cortisol, LysoPA(16 : 0/0 : 0), linolenic acid, etc. The UPLC–quadrupole/Orbitrap–mass spectrometry (MS)-based lipidomics analysis was to evaluate the mechanisms of oleanolic acid (OA) regulating lipid metabolism in spontaneously hypertensive rats (SHRs) [87]. Partial least squares-discriminant analysis performed on the liver lipidome revealed a total of 14 endogenous metabolites that were significantly changed. It indicated that treatment with OA resulted in significant improvements in blood pressure and associated abnormalities in the lipid metabolites.

Serum lipidomics was conducted to reveal the action mechanism of rhodiola crenulata extract (RCE) on AD by Fourier transform ion cyclotron resonance mass spectrometry [88]. The differences of lipid metabolism profiles were evaluated by multivariate statistical analysis, the potential lipid biomarkers were identified, and the possible mechanism of RCE on AD was elucidated. A total of 28 lipids with significant differences were screened out and 19 lipids were regulated by RCE treatment. It indicated that the protective effects of RCE on AD might be closely related to the regulation of linoleic acid metabolism, sphingolipid metabolism, and α-linoleic acid metabolism. The pathological development of AD is closely related to lipid metabolism disorders. To reveal the influence of Kai–Xin–San (KXS) on lipid metabolism in APP/PSI transgenic mice and potential therapeutic targets for treating AD, brain tissue samples were collected and analyzed by high-throughput lipidomics based on UPLC–Q/TOF–MS [89]. Compared with the control wild-type mouse group, nine potential lipid biomarkers were found in the AD model group, of which seven were up-regulated and two were down-regulated. Orally administrated KXS can reverse the changes in these potential biomarkers. Compared with the model group, a total of six differential metabolites showed a recovery trend and may be potential targets for KXS to treat AD.

Dohongsamul-tang (DST) is a traditional herbal formula and widely used in the treatment of chronic liver diseases patients in Korea and China. UPLC/Q–TOF MS-based lipidomics were applied to further elucidate the effect of DST on hepatic lipid metabolism [90]. The score plots of partial least squares-discriminant analysis showed that DST changed the lipid metabolic pattern of high-fat and high-cholesterol diet mice. Twenty-two lipid metabolites were selected as biomarkers regulated by DST, and pathway analysis showed that glycerophospholipid metabolism and sphingolipid metabolism were associated with the effect of DST on non-alcoholic fatty liver diseases (NAFLDs). It suggested that DST might have an effect on NAFLD by regulating the metabolism of lipids such as phospholipids and sphingolipids and demonstrated that lipidomic profiling is useful to investigate the therapeutic effects of herbal decoctions. Danggui Buxue decoction (DBD) is often used to treat anemia in China. However, its underlying therapeutic mechanism is unclear. Ultra-high-performance liquid chromatography coupled with Q-Exactive mass spectrometry-based metabolomics and lipidomics was further utilized to screen out differential spleen metabolites associated with DBD treatment [91]. A total of 26 differential metabolites including 8 polar metabolites and 18 lipids were firstly obtained to relate with anemia mice. Seven polar metabolites and 10 lipids among them were reversed by DBD, and the regulation of pyrimidine metabolism and glycerophospholipid metabolism were mainly associated to the anti-anemia effect of DBD.

17.5 Conclusions

The excellent performance of mass spectrometry in structure identification and detection sensitivity makes it a core analysis method in the field of lipidomics. The lipid analysis by mass spectrometry is helpful to elucidate the physiological and biochemical significance. However, future large-scale lipid analysis requires higher lipid coverage, higher sensitivity, and robustness of analytical methods. According to the different analysis purposes, comparing and optimizing different analysis methods and combining multiple analysis methods to give full play to the analysis advantages will become an inevitable requirement for the development of lipidomics. With the continuous development of lipidomics technology, the integrated research methods of lipidomics and other omics are expected to provide a complete metabolic map that will greatly facilitate the study of diseases and better reveal the therapeutic effects of herbal medicine.

Acknowledgments

This work was supported by grants from the Key Program of Natural Science Foundation of State (Grant No. 81973745, 81302905), Natural Science Foundation of Heilongjiang Province (YQ2019H030), and Heilongjiang Touyan Innovation Team Program.

References

1 Muller, M.P., Jiang, T., Sun, C. et al. (2019). Characterization of lipid-protein interactions and lipid-mediated modulation of membrane protein function through molecular simulation. *Chemical Reviews* 119: 6086–6161, https://doi.org/10.1021/acs.chemrev.8b00608.

2 Zhang, H., Zhang, A., Zhou, X. et al. (2018). High-throughput lipidomics reveal mirabilite regulating lipid metabolism as anticancer therapeutics. *RSC Advances* 8 (62): 35600–35610.

3 Zhang, C., Wang, K., Yang, L. et al. (2018). Lipid metabolism in inflammation-related diseases. *Analyst* 143: 4526–4536, https://doi.org/10.1039/c8an01046c.

4 Liu, C., Zong, W., Zhang, A. et al. (2018). Lipidomic characterisation discovery for coronary heart disease diagnosis based on high-throughput ultra-performance liquid chromatography and mass spectrometry. *RSC Advances* 8: 647–654.

5 Liang, Q., Liu, H., Jiang, Y. et al. (2016). Discovering lipid phenotypic changes of sepsis-induced lung injury using high-throughput lipidomic analysis. *RSC Advances* 6: 38233–38237.

6 Cao, W.J., Wang, T.T., Gao, Y.F. et al. (2019). Serum lipid metabolic derangement is associated with disease progression during chronic HBV infection. *Clinical Laboratory* 65, https://doi.org/10.7754/Clin.Lab.2019.190525.

7 Liang, Q., Zhu, Y., Liu, H. et al. (2017). High-throughput lipidomics enables discovery of the mode of action of huaxian capsule impacting the metabolism of sepsis. *RSC Advances* 7 (71): 44990–44996.

8 Gao, H.L., Zhang, A.H., Yu, J.B. et al. (2018). High-throughput lipidomics characterize key lipid molecules as potential therapeutic targets of Kaixinsan protects against Alzheimer's disease in APP/PS1 transgenic mice. *Journal of Chromatography B: Analytical Technologies in the Biomedical and Life Sciences* 1092: 286–295.

9 Yan F, Wen Z, Wang R, et al. (2017). Identification of the lipid biomarkers from plasma in idiopathic pulmonary fibrosis by lipidomics. *BMC Pulmonary Medicine* 17: 174, https://doi.org/10.1186/s12890-017-0513-4.

10 Sethi, S. and Brietzke, E. (2017). Recent advances in lipidomics: analytical and clinical perspectives. *Prostaglandins & Other Lipid Mediators* 128–129: 8–16, https://doi.org/10.1016/j.prostaglandins.2016.12.002.

11 Zhang, A., Sun, H., Yan, G. et al. (2016). Mass spectrometry-based metabolomics: applications to biomarker and metabolic pathway research. *Biomedical Chromatography* 30: 7–12, https://doi.org/10.1002/bmc.3453.

12 Aldana, J., Romero-Otero, A., and Cala, M.P. (2020). Exploring the lipidome: current lipid extraction techniques for mass spectrometry analysis. *Metabolites* 10, https://doi.org/10.3390/metabo10060231.

13 Satomi, Y., Hirayama, M., and Kobayashi, H. (2017). One-step lipid extraction for plasma lipidomics analysis by liquid chromatography mass spectrometry. *Journal of Chromatography B: Analytical Technologies in the Biomedical and Life Sciences* 1063: 93–100, https://doi.org/10.1016/j.jchromb.2017.08.020.

14 Iriondo, A. et al. (2019). Isopropanol extraction for cerebrospinal fluid lipidomic profiling analysis. *Talanta* 195: 619–627, https://doi.org/10.1016/j.talanta.2018.11.101.

15 Calderón, C., Sanwald, C., Schlotterbeck, J. et al. (2019). Comparison of simple monophasic versus classical biphasic extraction protocols for comprehensive UHPLC–MS/MS lipidomic analysis of Hela cells. *Analytica Chimica Acta* 1048: 66–74, https://doi.org/10.1016/j.aca.2018.10.035.

16 Wong, M.W.K., Braidy, N., Pickford, R. et al. (2019). Comparison of single phase and biphasic extraction protocols for lipidomic studies using human plasma. *Frontiers in Neurology* 10: 879, https://doi.org/10.3389/fneur.2019.00879.

17 Sostare, J., Di Guida, R., Kirwan, J. et al. (2018). Comparison of modified Matyash method to conventional solvent systems for polar metabolite and lipid extractions. *Analytica Chimica Acta* 1037: 301–315, https://doi.org/10.1016/j.aca.2018.03.019.

18 Ulmer, C.Z., Jones, C.M., Yost, R.A. et al. (2018). Optimization of Folch, Bligh–Dyer, and Matyash sample-to-extraction solvent ratios for human plasma-based lipidomics studies. *Analytica Chimica Acta* 1037: 351–357, https://doi.org/10.1016/j.aca.2018.08.004.

19 Claassen, C., Kuballa, J., and Rohn, S. (2019). Polar lipids in starch-rich commodities to be analyzed with LC–MS-based metabolomics-optimization of ionization parameters and high-throughput extraction protocols. *Metabolites* 9, https://doi.org/10.3390/metabo9080167.

20 Breil, C., Abert Vian, M., Zemb, T. et al. (2017). "Bligh and Dyer" and Folch methods for solid–liquid–liquid extraction of lipids from microorganisms. Comprehension of solvatation mechanisms and towards substitution with alternative solvents. *International Journal of Molecular Sciences* 18, https://doi.org/10.3390/ijms18040708.

21 Zghaibi, N., Omar, R., Kamal, S.M.M. et al. (2019). Microwave-assisted brine extraction for enhancement of the quantity and quality of lipid production from microalgae Nannochloropsis sp. *Molecules* 24, https://doi.org/10.3390/molecules24193581.

22 Zghaibi, N., Omar, R., Kamal, S.M.M. et al. (2020). Kinetics study of microwave-assisted brine extraction of lipid from the microalgae Nannochloropsis sp. *Molecules* 25, https://doi.org/10.3390/molecules25040784.

23 Tsiaka, T., Lantzouraki, D.Z., Siapi, E. et al. (2018). Macular carotenoids in lipid food matrices: DOE-based high energy extraction of egg yolk xanthophylls and quantification through a validated APCI(+) LC–MS/MS method. *Journal of Chromatography B: Analytical Technologies in the Biomedical and Life Sciences* 1096: 160–171, https://doi.org/10.1016/j.jchromb.2018.08.010.

24 Jeevan Kumar, S.P. and Banerjee, R. (2019). Enhanced lipid extraction from oleaginous yeast biomass using ultrasound assisted extraction: a greener and scalable process. *Ultrasonics Sonochemistry* 52: 25–32, https://doi.org/10.1016/j.ultsonch.2018.08.003.

25 He, Y., Huang, Z., Zhong, C. et al. (2019). Pressurized liquid extraction with ethanol as a green and efficient technology to lipid extraction of Isochrysis biomass. *Bioresource Technology* 293: 122049, https://doi.org/10.1016/j.biortech.2019.122049.

26 Yang, Y., Deng, J., Liu, Y. et al. (2019). A microscale solid-phase microextraction probe for the in situ analysis of perfluoroalkyl substances and lipids in biological tissues using mass spectrometry. *Analyst* 144: 5637–5645, https://doi.org/10.1039/c9an01195a.

27 Birjandi, A.P., Bojko, B., Ning, Z. et al. (2017). High throughput solid phase microextraction: a new alternative for analysis of cellular lipidome? *Journal of Chromatography B: Analytical Technologies in the Biomedical and Life Sciences* 1043: 12–19, https://doi.org/10.1016/j.jchromb.2016.09.034.

28 Martin-Rubio, A.S., Sopelana, P., and Guillén, M.D. (2019). The key role of ovalbumin in lipid bioaccessibility and oxidation product profile during the in vitro digestion of slightly oxidized soybean oil. *Food and Function* 10: 4440–4451, https://doi.org/10.1039/c9fo00598f.

29 Clarke, H.J., Mannion, D.T., O'Sullivan, M.G. et al. (2019). Development of a headspace solid-phase microextraction gas chromatography mass spectrometry method for the quantification of volatiles associated with lipid oxidation in whole milk powder using response surface methodology. *Food Chemistry* 292: 75–80, https://doi.org/10.1016/j.foodchem.2019.04.027.

30 Zalloua, P., Kadar, H., Hariri, E. et al. (2019). Untargeted mass spectrometry lipidomics identifies correlation between serum sphingomyelins and plasma cholesterol. *Lipids in Health and Disease* 18: 38, https://doi.org/10.1186/s12944-018-0948-5.

31 Lee, H.C. and Yokomizo, T. (2018). Applications of mass spectrometry-based targeted and non-targeted lipidomics. *Biochemical and Biophysical Research Communications* 504: 576–581, https://doi.org/10.1016/j.bbrc.2018.03.081.

32 Lu, J., Lam, S.M., Wan, Q. et al. (2019). High-coverage targeted lipidomics reveals novel serum lipid predictors and lipid pathway dysregulation antecedent to type 2 diabetes onset in normoglycemic Chinese adults. *Diabetes Care* 42: 2117–2126, https://doi.org/10.2337/dc19-0100.

33 Zhang, Q., Xu, H., Liu, R. et al. (2019). A novel strategy for targeted lipidomics based on LC-tandem-MS parameters prediction, quantification, and multiple statistical data mining: evaluation of lysophosphatidylcholines as potential cancer biomarkers. *Analytical Chemistry* 91: 3389–3396, https://doi.org/10.1021/acs.analchem.8b04715.

34 Lv, W., Wang, L., Xuan, Q. et al. (2020). Pseudotargeted method based on parallel column two-dimensional liquid chromatography–mass spectrometry for broad coverage of metabolome and lipidome. *Analytical Chemistry* 92: 6043–6050, https://doi.org/10.1021/acs.analchem.0c00372.

35 Xuan, Q., Hu, C., Yu, D. et al. (2018). Development of a high coverage pseudotargeted lipidomics method based on ultra-high performance liquid chromatography–mass spectrometry. *Analytical Chemistry* 90: 7608–7616, https://doi.org/10.1021/acs.analchem.8b01331.

36 Liang, Q., Zhang, T., Jiang, Y. et al. (2016). Untargeted lipidomics study of coronary artery disease by FUPLC–Q–TOF–MS. *Analytical Methods* 8: 1229–1334.

37 Aerts, J., Artola, M., van Eijk, M. et al. (2019). Glycosphingolipids and infection. Potential new therapeutic avenues. *Frontiers in Cell and Development Biology* 7: 324, https://doi.org/10.3389/fcell.2019.00324.

38 Lísa, M. and Holčapek, M. (2018). UHPSFC/ESI–MS analysis of lipids. *Methods in Molecular Biology* 1730: 73–82, https://doi.org/10.1007/978-1-4939-7592-1_5.

39 Zhang, A., Qiu, S., Sun, H. et al. (2016). Scoparone affects lipid metabolism in primary hepatocytes using lipidomics. *Scientific Reports* 6: 28031.

40 Liang, Q., Wang, C., Li, B. et al. (2015). Lipidomics analysis based on liquid chromatography mass spectrometry for hepatocellular carcinoma and intrahepatic cholangiocarcinoma. *RSC Advances* 5: 63711–63718.

41 Wozny, K., Lehmann, W.D., Wozny, M. et al. (2019). A method for the quantitative determination of glycerophospholipid regioisomers by UPLC–ESI–MS/MS. *Analytical and Bioanalytical Chemistry* 411: 915–924, https://doi.org/10.1007/s00216-018-1517-5.

42 Deng, J., Li, W., Yang, Q. et al. (2018). Biocompatible surface-coated probe for in vivo, in situ, and microscale lipidomics of small biological organisms and cells using mass spectrometry. *Analytical Chemistry* 90: 6936–6944, https://doi.org/10.1021/acs.analchem.8b01218.

43 Herrfurth, C. and Feussner, I. (2085). Quantitative Jasmonate profiling using a high-throughput UPLC–NanoESI–MS/MS method. *Methods in Molecular Biology* 2020: 169–187, https://doi.org/10.1007/978-1-0716-0142-6_13.

44 Lemke, A., Castillo-Sánchez, J.C., Prodinger, F. et al. (2017). Human amniotic membrane as newly identified source of amniotic fluid pulmonary surfactant. *Scientific Reports* 7: 6406, https://doi.org/10.1038/s41598-017-06402-w.

45 Hofmann, T. and Schmidt, C. (2019). Instrument response of phosphatidylglycerol lipids with varying fatty acyl chain length in nano-ESI shotgun experiments. *Chemistry and Physics of Lipids* 223: 104782, https://doi.org/10.1016/j.chemphyslip.2019.05.007.

46 Liu, C., Qi, K., Yao, L. et al. (2019). Imaging of polar and nonpolar species using compact desorption electrospray ionization/postphotoionization mass spectrometry. *Analytical Chemistry* 91: 6616–6623, https://doi.org/10.1021/acs.analchem.9b00520.

47 Nguyen, S.N., Kyle, J.E., Dautel, S.E. et al. (2019). Lipid coverage in nanospray desorption electrospray ionization mass spectrometry imaging of mouse lung tissues. *Analytical Chemistry* 91: 11629–11635, https://doi.org/10.1021/acs.analchem.9b02045.

48 Unsihuay, D., Qiu, J., Swaroop, S. et al. (2020). Imaging of triglycerides in tissues using nanospray desorption electrospray ionization (nano-DESI) mass spectrometry. *International Journal of Mass Spectrometry* 448, https://doi.org/10.1016/j.ijms.2019.116269.

49 Gachumi, G. and El-Aneed, A. (2017). Mass spectrometric approaches for the analysis of phytosterols in biological samples. *Journal of Agricultural and Food Chemistry* 65: 10141–10156, https://doi.org/10.1021/acs.jafc.7b03785.

50 Navas-Madroñal, M., Rodriguez, C., Kassan, M. et al. (2019). Enhanced endoplasmic reticulum and mitochondrial stress in abdominal aortic aneurysm. *Clinical Science* 133: 1421–1438, https://doi.org/10.1042/cs20190399.

51 Tessema, E.N., Gebre-Mariam, T., Frolov, A. et al. (2018). Development and validation of LC/APCI–MS method for the quantification of oat ceramides in skin permeation studies. *Analytical and Bioanalytical Chemistry* 410: 4775–4785, https://doi.org/10.1007/s00216-018-1162-z.

52 Vrkoslav, V., Rumlová, B., Strmeň, T. et al. (2018). Applicability of low-flow atmospheric pressure chemical ionization and photoionization mass spectrometry with a microfabricated nebulizer for neutral lipids. *Rapid Communications in Mass Spectrometry* 32: 639–648, https://doi.org/10.1002/rcm.8086.

53 Abreu, S., Solgadi, A., and Chaminade, P. (2017). Optimization of normal phase chromatographic conditions for lipid analysis and comparison of associated detection techniques. *Journal of Chromatography A* 1514: 54–71, https://doi.org/10.1016/j.chroma.2017.07.063.

54 Bandu, R., Mok, H.J., and Kim, K.P. (2018). Phospholipids as cancer biomarkers: mass spectrometry-based analysis. *Mass Spectrometry Reviews* 37: 107–138, https://doi.org/10.1002/mas.21510.

55 Ucal, Y., Durer, Z.A., Atak, H. et al. (2017). Clinical applications of MALDI imaging technologies in cancer and neurodegenerative diseases. *Biochimica et Biophysica Acta Proteins and Proteomics*: 795–816, https://doi.org/10.1016/j.bbapap.2017.01.005.

56 Fernández, R., Garate, J., Abad, B. et al. (2018). Mapping lipid distribution in rat sciatic nerve using imaging mass spectrometry. *Methods in Molecular Biology* 1791: 51–65, https://doi.org/10.1007/978-1-4939-7862-5_5.

57 Holzlechner, M., Bonta, M., Lohninger, H. et al. (2018). Multisensor imaging-from sample preparation to integrated multimodal interpretation of

LA–ICPMS and MALDI MS imaging data. *Analytical Chemistry* 90: 8831–8837, https://doi.org/10.1021/acs.analchem.8b00816.

58 Rocha, B., Cillero-Pastor, B., Blanco, F.J., and Ruiz-Romero, C. (2017). MALDI mass spectrometry imaging in rheumatic diseases. *Biochimica et Biophysica Acta Proteins and Proteomics* 1865: 784–794, https://doi.org/10.1016/j.bbapap.2016.10.004.

59 Svirkova, A., Turyanskaya, A., Perneczky, L. et al. (2018). Multimodal imaging of undecalcified tissue sections by MALDI MS and μXRF. *Analyst* 143: 2587–2595, https://doi.org/10.1039/c8an00313k.

60 Bueno, M., Resconi, V.C., Campo, M.M. et al. (2019). Development of a robust HS-SPME–GC–MS method for the analysis of solid food samples. Analysis of volatile compounds in fresh raw beef of differing lipid oxidation degrees. *Food Chemistry* 281: 49–56, https://doi.org/10.1016/j.foodchem.2018.12.082.

61 Li, H., Song, Y., Zhang, H. et al. (2020). Comparative lipid profile of four edible shellfishes by UPLC-triple TOF–MS/MS. *Food Chemistry* 310: 125947, https://doi.org/10.1016/j.foodchem.2019.125947.

62 Zhang, Y.Y., Qin, L., Liu, Y.X. et al. (2018). Evaluation of lipid profile in different tissues of Japanese abalone Haliotis discus hannai Ino with UPLC–ESI–Q–TOF–MS-based lipidomic study. *Food Chemistry* 265: 49–56, https://doi.org/10.1016/j.foodchem.2018.05.077.

63 Napolitano, A., Cerulli, A., Pizza, C., and Piacente, S. (2018). Multi-class polar lipid profiling in fresh and roasted hazelnut (Corylus avellana cultivar "Tonda di Giffoni") by LC–ESI/LTQOrbitrap/MS/MS^n. *Food Chemistry* 269: 125–135, https://doi.org/10.1016/j.foodchem.2018.06.121.

64 Li, Q., Liang, X., Zhao, L. et al. (2017). UPLC–Q–exactive orbitrap/MS-based lipidomics approach to characterize lipid extracts from bee pollen and their *in vitro* anti-inflammatory properties. *Journal of Agricultural and Food Chemistry* 65: 6848–6860, https://doi.org/10.1021/acs.jafc.7b02285.

65 Zhou, Z., Shen, X., Chen, X. et al. (2019). LipidIMMS analyzer: integrating multi-dimensional information to support lipid identification in ion mobility–mass spectrometry based lipidomics. *Bioinformatics* 35: 698–700, https://doi.org/10.1093/bioinformatics/bty661.

66 Blaženović, I., Shen, T., Mehta, S.S., et al. (2018). Increasing compound identification rates in untargeted lipidomics research with liquid chromatography drift time-ion mobility mass spectrometry. *Analytical Chemistry* 90: 10758–10764, https://doi.org/10.1021/acs.analchem.8b01527.

67 Paglia, G. and Astarita, G. (2017). Metabolomics and lipidomics using traveling-wave ion mobility mass spectrometry. *Nature Protocols* 12: 797–813, https://doi.org/10.1038/nprot.2017.013.

68 George, A.D., Gay, M.C.L., Wlodek, M.E. et al. (2020). Untargeted lipidomics using liquid chromatography–ion mobility–mass spectrometry reveals novel triacylglycerides in human milk. *Scientific Reports* 10: 9255, https://doi.org/10.1038/s41598-020-66235-y.

69 Zhou, Z., Tu, J., and Zhu, Z.J. (2018). Advancing the large-scale CCS database for metabolomics and lipidomics at the machine-learning era. *Current Opinion in Chemical Biology* 42: 34–41, https://doi.org/10.1016/j.cbpa.2017.10.033.

70 Leaptrot, K.L., May, J.C., Dodds, J.N., and McLean, J.A. (2019). Ion mobility conformational lipid atlas for high confidence lipidomics. *Nature Communications* 10: 985, https://doi.org/10.1038/s41467-019-08897-5.

71 Zhou, Z., Tu, J., Xiong, X. et al. (2017). LipidCCS: prediction of collision cross-section values for lipids with high precision to support ion mobility–mass spectrometry-based lipidomics. *Analytical Chemistry* 89: 9559–9566, https://doi.org/10.1021/acs.analchem.7b02625.

72 Chen, X., Yin, Y., Zhou, Z. et al. (2020). Development of a combined strategy for accurate lipid structural identification and quantification in ion-mobility mass spectrometry based untargeted lipidomics. *Analytica Chimica Acta* 1136: 115–124.

73 Barré, F.P.Y., Claes, B.S.R., Dewez, F. et al. (2018). Specific lipid and metabolic profiles of R-CHOP-resistant diffuse large B-cell lymphoma elucidated by matrix-assisted laser desorption ionization mass spectrometry imaging and *in vivo* imaging. *Analytical Chemistry* 90: 14198–14206, https://doi.org/10.1021/acs.analchem.8b02910.

74 Karlsson, O. and Hanrieder, J. (2017). Imaging mass spectrometry in drug development and toxicology. *Archives of Toxicology* 91: 2283–2294, https://doi.org/10.1007/s00204-016-1905-6.

75 Luberto, C., Haley, J.D., and Del Poeta, M. (2019). Imaging with mass spectrometry, the next frontier in sphingolipid research? A discussion on where we stand and the possibilities ahead. *Chemistry and Physics of Lipids* 219: 1–14, https://doi.org/10.1016/j.chemphyslip.2019.01.001.

76 Moreno-Gordaliza, E., Esteban-Fernández, D., Lázaro, A. et al. (2017). MALDI-LTQ-Orbitrap mass spectrometry imaging for lipidomic analysis in kidney under cisplatin chemotherapy. *Talanta* 164: 16–26, https://doi.org/10.1016/j.talanta.2016.11.026.

77 Moreno-Gordaliza, E., Esteban-Fernández, D., Lázaro, A. et al. (2018). Lipid imaging for visualizing cilastatin amelioration of cisplatin-induced nephrotoxicity. *Journal of Lipid Research* 59: 1561–1574, https://doi.org/10.1194/jlr.M080465.

78 Genangeli, M., Heijens, A.M.M., Rustichelli, A. et al. (2019). MALDI-mass spectrometry imaging to investigate lipid and bile acid modifications caused by lentil extract used as a potential hypocholesterolemic treatment. *Journal of the American Society for Mass Spectrometry* 30: 2041–2050, https://doi.org/10.1007/s13361-019-02265-9.

79 Wäldchen, F., Mohr, F., Wagner, A.H., and Heiles, S. (2020). Multifunctional reactive MALDI matrix enabling high-lateral resolution dual polarity MS imaging and lipid C=C position-resolved MS(2) imaging. *Analytical Chemistry* 92: 14130–14138, https://doi.org/10.1021/acs.analchem.0c03150.

80 Prentice, B.M., McMillen, J.C., and Caprioli, R.M. (2019). Multiple TOF/TOF events in a single laser shot for multiplexed lipid identifications in MALDI imaging mass spectrometry. *International Journal of Mass Spectrometry* 437: 30–37, https://doi.org/10.1016/j.ijms.2018.06.006.

81 Mallah, K., Quanico, J., Trede, D. et al. (2018). Lipid changes associated with traumatic brain injury revealed by 3D MALDI-MSI. *Analytical Chemistry* 90: 10568–10576, https://doi.org/10.1021/acs.analchem.8b02682.

82 Wang, J., Wang, C., and Han, X. (2018). Enhanced coverage of lipid analysis and imaging by matrix-assisted laser desorption/ionization mass spectrometry via a strategy with an optimized mixture of matrices. *Analytica Chimica Acta* 1000: 155–162, https://doi.org/10.1016/j.aca.2017.09.046.

83 O'Rourke, M.B., Smith, C.C., De La Monte, S.M. et al. (2019). Higher mass accuracy MALDI–TOF/TOF lipid imaging of human brain tissue in Alzheimer's disease. *Current Protocols in Molecular Biology* 126: e86, https://doi.org/10.1002/cpmb.86.

84 Kaya, I., Brinet, D., Michno, W. et al. (2017). Novel trimodal MALDI imaging mass spectrometry (IMS3) at 10 μm reveals spatial lipid and peptide correlates implicated in Aβ plaque pathology in Alzheimer's disease. *ACS Chemical Neuroscience* 8: 2778–2790, https://doi.org/10.1021/acschemneuro.7b00314.

85 Zhang, J., Hong, Y., Xie, P. et al. (2021). Spatial lipidomics reveals anticancer mechanisms of bufalin in combination with cinobufagin in tumor-bearing mice. *Frontiers in Pharmacology* 11: 593815. https://doi.org/10.3389/fphar.2020.593815.

86 Ge, N., Kong, L., Zhang, A.-h. et al. (2021). Identification of key lipid metabolites during metabolic dysregulation in the diabetic retinopathy disease mouse model and efficacy of Keluoxin capsule using an UHPLC–MS-based non-targeted lipidomics approach. *RSC Advances* 11: 5491.

87 Zhang, S., Liu, Y., Wang, X. et al. (2020). Antihypertensive activity of oleanolic acid is mediated via downregulation of secretory phospholipase A2 and fatty acid synthase in spontaneously hypertensive rats. *International Journal of Molecular Medicine* 46 (6): 2019–2034. https://doi.org/10.3892/ijmm.2020.4744.

88 Sun, W., Liu, C., Zhou, X. et al. (2020). Serum lipidomics study reveals protective effects of rhodiola crenulata extract on Alzheimer's disease rats. *Journal of Chromatography B: Analytical Technologies in the Biomedical and Life Sciences* 1158: 122346. https://doi.org/10.1016/j.jchromb.2020.122346.

89 Zhang, A.H., Ma, Z.M., Kong, L. et al. (2020). High-throughput lipidomics analysis to discover lipid biomarkers and profiles as potential targets for evaluating efficacy of Kai–Xin–San against APP/PS1 transgenic mice based on UPLC–Q/TOF–MS. *Biomedical Chromatography* 34 (2): e4724. https://doi.org/10.1002/bmc.4724.

90 Park, S.H., Lee, J.E., Lee, S.M. et al. (2020). An unbiased lipidomics approach identifies key lipid molecules as potential therapeutic targets of Dohongsamultang against non-alcoholic fatty liver diseases in a mouse model of obesity. *Journal of Ethnopharmacology* 260: 112999. https://doi.org/10.1016/j.jep.2020.112999.

91 Liu, Y., Li, X., Li, A. et al. (2020). UHPLC Q-Exactive MS-based spleen metabolomics and lipidomics to explore the effect mechanisms of Danggui Buxue decoction in anemia mice. *Journal of Pharmaceutical and Biomedical Analysis* 185: 113234. https://doi.org/10.1016/j.jpba.2020.113234.

Index

Mass Spectrometry-Based Metabolomics in Clinical and Herbal Medicines: Strategies, Technologies, and Applications, First Edition. Edited by Aihua Zhang and Wanying Wang.

n